Contents

BARRON'S

AP*

ENVIRONMENTAL SCIENCE

5TH EDITION

Gary S. Thorpe, M.S.
AP Environmental Science Teacher (ret.)
Beverly Hills High School
Beverly Hills, CA

Acknowledgments

I would like to thank my wife, Patti, and my two daughters, Kris and Erin, for their patience and understanding while I was writing this book. This book is dedicated in tribute to the late Dr. Jerry Bobrow of Bobrow Test Preparation Services, who was not only my publishing agent but a dear friend for many years. A special thanks also goes to the many professional and dedicated high school APES teachers across the United States who contributed their finest work in making this book possible.

All inquiries should be addressed to:
Barron's Educational Series, Inc.
250 Wireless Boulevard
Hauppauge, New York 11788
www.barronseduc.com

ISBN: 978-1-4380-0132-6 (book only)
ISBN: 978-1-4380-7260-9 (book and CD-ROM package)
ISSN: 2157-3824 (book only)
ISSN: 2157-3832 (book and CD-ROM package)

Printed in the United States of America

9 8 7 6

**10%
POST-CONSUMER
WASTE**
Paper contains a minimum
of 10% post-consumer
waste (PCW). Paper used
in this book was derived
from certified, sustainable
forestlands.

As you review the content in this book and work toward earning that **5** on your AP ENVIRONMENTAL SCIENCE exam, here are five things that you **MUST** do:

1 **Practice writing your own answers to the Free-Response Questions (FRQs).** After you review each FRQ, close this book and write your own response to the question. Then, compare your answer to the one in the book. The more you practice writing *your own* essays, the higher your score will be on the actual APES exam.

2 **Be sure to review the Case Studies presented in this book.** You will find several questions on the exam that focus on specific events that have occurred throughout history and involve environmental science principles and issues. When possible, reference these Case Studies in your responses to the FRQs.

3 **Thoroughly review Chapters 9 (Pollution) and 10 (Impacts on the Environment and Human Health).** About 25–30% of the multiple-choice questions on the APES exam will test your knowledge of the topics in these chapters. In particular,

- Be familiar with the differences between industrial and photochemical smog.
- Know the causes and effects of acid deposition.
- Know the causes and effects of global warming.
- Familiarize yourself with remediation and reduction strategies for environmental pollution.

4 **Become comfortable with doing the types of math problems found in this book, especially those involving energy calculations.** These problems are found in Chapter 8 and in both practice exams. Historically, question 1 of the FRQ section of the APES exam requires you to complete mathematical calculations.

5 **Review the relevant acts, laws, and treaties detailed in each chapter.** You will find several questions on the APES exam that test your knowledge on them; you can also reference them in your answers to the FRQs when appropriate.

What's New in the 5th Edition?

This 5th edition includes many new and important features: (1) a diagnostic test that consists of 100 brand new multiple-choice questions with all answers explained. At the end of the diagnostic test, a predictive rubric developed by the College Board will allow you to predict your final score on the APES exam; (2) all chapters have been fully updated to reflect recent changes in environmental laws; (3) case studies have been updated to reflect events that have recently occurred; e.g., 2011 earthquake and tsunami that hit Japan and the severe tornadoes that occurred in the United States in 2011–2012; (4) new charts have been re-formatted for students to make reviewing for the APES exam easier; (5) "Quick Review Checklists," which list the most important topics covered in each chapter, have been added. You can use the checklists to mark your progress in reviewing each topic; (6) a new "Vocabulary Review" for each chapter provides you with a handy list of the most important vocabulary words and definitions that you can review just before the exam, and finally (7) the 5th edition is available as a book/CD set that contains two *additional* APES exams with fully explained answers. Combine this 5th edition book and CD with Barron's AP Environmental Science Flash Cards, and you'll have everything you need to fully prepare yourself for the APES exam. No other APES review on the market today comes close to all that is included in this book.

Introductory Material

The introductory material contains information on the AP exam itself. It includes commonly asked questions, strategies for taking the exam, and techniques for writing outstanding essays.

Specific Topics

The majority of this book is divided into seven units. Each contains specific chapters on material covered on the APES exam. Each chapter provides a condensed review of key information, updated case studies, relevant and updated environmental laws, a topic-specific vocabulary list including definitions, multiple-choice questions, and free-response essay questions similar to the questions found on actual APES exams. Many of the free-response questions and explanations have been contributed by award-winning APES teachers from across the United States.

Practice Exams

The last part of this book contains two complete practice APES exams. All questions are thoroughly explained and the free-response questions provide a rubric for scoring your answers along with tips on how to score even higher.

If you have purchased a book that contains the CD-ROM, the software contains two additional practice tests and answers so you can practice taking the test under timed conditions and receive immediate scoring.

Format of the APES Exam

The APES exam is three hours long and is divided equally in time between a multiple-choice section and a free-response section. The multiple-choice section, which makes up 60% of the final score and lasts 90 minutes, consists of 100 multiple-choice questions and is designed to cover your knowledge and understanding of environmental science. Thought-provoking problems and questions based on fundamental ideas are included along with questions based on the recall of basic facts and major concepts. The number of multiple-choice questions taken from each major topic area is reflected in the percentages shown in "Topics Covered on the Exam" later in this section. For example, in Earth Systems and Resources, you will see "(10–15%)." Expect to find 10 to 15 multiple-choice questions from this area on the actual APES exam. Therefore, spend about 10–15% of your time reviewing this area.

The free-response section, which is also 90 minutes long, makes up 40% of your final score. It contains one data set question, one document-based question, and two synthesis and evaluation questions. The differences between these types of free-response questions will be explained later in this introductory section. You must organize your answers to demonstrate reasoning and analytical skills, as well as the ability to write clearly and concisely.

Test-Taking Strategies

The Educational Testing Service will send you your AP Environmental Science score in July. Depending upon your choice, the scores may also be sent to colleges and universities. The scores are reported on the following scale:

- 5—Extremely well qualified
- 4—Well qualified
- 3—Qualified
- 2—Possibly qualified
- 1—No recommendation

Most colleges and universities accept a score of 4 or 5 for credit and placement, while some may accept a score of 3. Scores of 1 or 2 are not accepted by colleges and universities for either credit or placement.

The rule of thumb in determining how well you will probably do on the AP Environmental Science Exam is to look at the number of multiple-choice questions you answer correctly on the practice exams in this book. If you consistently get between 50 and 60% correct, you should be able to score a minimum of a 3. If you score consistently between 65 and 75% of the multiple-choice questions correct, you should be able to achieve a 4, and if you score 80 to 100% correctly, you should be looking at a 5. This assumes, of course, that you adequately answer the questions in the free-response section.

The Multiple-Choice Questions

- Because there is no scoring penalty for guessing, it is to your advantage to answer as many multiple-choice questions as possible.
- When you reach a question that you are not quite sure of and need more time, place a "+" next to the number.
- When you reach a question that you cannot answer, place a "−" next to the number.
- After you have answered all of the questions you know for sure, go back to only your "+" questions. Remember, as there is no scoring penalty for guessing, eliminate any obvious incorrect answers and then guess.
- If time remains, scan through the "−" questions. Again, there is no scoring penalty for guessing, so eliminate any obvious incorrect answers and then guess.

TYPES OF MULTIPLE-CHOICE QUESTIONS

The AP Environmental Science Exam relies on a variety of multiple-choice questions including identification and analysis. Those kinds of questions can be further identified as generalizations, comparing and contrasting concepts and events, sequencing a series of related ideas or events, cause-and-effect relationships,

TIP

Always read the entire question. Underline key words in the question such as:

- *all of the following EXCEPT*
- *which of the following*
- *increases*
- *decreases*
- *are commonly used*
- *is responsible*
- *most accurately compares*
- *best describes*
- *is correct*
- *results*
- *reflects*
- *is most likely*
- *least likely*
- *which best DEFINES*

definitions, solutions to a problem, hypothetical situations, chronological problems, multiple correct answers, and negative questions. More than 75% of the multiple-choice questions fit into these categories.

In addition to identification and analysis questions, there are stimulus-based questions that rely on your interpretation and understanding of maps, graphs, charts, tables, pictures, flowcharts, photographs or sketches, cartoons, short narrative passages, surveys and poll data, quotations that come from primary source documents, and passages from environmental legislation and statutes, as well as questions that relate directly to laboratory and field investigations.

Identification and Analysis Questions

Question Type	Example
Definitional	Any factor that influences a natural process under study is a(n) (A) independent variable (B) dependent variable (C) control (D) placebo (E) experimental value
Cause-and-effect relationships	_____ contributes to the formation of _____ and thereby compounds the problem of _____ . (A) ozone, carbon dioxide, acid rain (B) carbon dioxide, carbon monoxide, ozone depletion (C) sulfur dioxide, acid deposition, global warming (D) nitrous oxide, ozone, industrial smog (E) nitric oxide, ozone, photochemical smog

Sequencing a series of related ideas or events	Which of the answers below correctly describes the order in which environmental legislation would pass through Congress?
	
	   I. Reports the bill out of the appropriate committee
	  II. Debates the bill on the floor of the respective houses
	 III. Rejects or accepts amendments to the bill
	 IV. Resolves any differences in a conference committee
	
	(A) I, II, III, IV
	(B) I, III, IV, II
	(C) II, IV, I, III
	(D) III, I, II, IV
	(E) IV, III, II, I

Question Type	Example

Generalization

What is generally considered to be the most significant factor in terms of being a causative agent for cancer?

(A) Smoking
(B) Diet
(C) Stress
(D) Heredity
(E) Pollution

Solution to a problem

A field biologist had been keeping density counts of the number of coastal side-blotched lizards (*Uta stansburiana hesperis* Richardson) that had recently been introduced into a large section of California desert. The yearly counts per acre were: 1998: 4; 1999: 8; 2000: 16. Assuming that the carrying capacity had not been reached, which of the following would be true?

(A) In 2001, there should be 20 lizards due to arithmetic growth.
(B) In 2001, there should be 20 lizards due to exponential growth.
(C) In 2001, there should be 32 lizards due to exponential growth.
(D) In 2001, there should be around 16 lizards due to population stability.
(E) It is impossible to predict how many lizards there would be in 2001.

Hypothetical situation

Converting to a solar-hydrogen energy source could theoretically be achieved by

(A) attracting private investors.
(B) passing legislation that would fund "seed money" for entrepreneurs.
(C) passing legislation that would discontinue government subsidies of fossil fuels.
(D) educating the public as to the environmental benefits of solar-hydrogen fuel sources.
(E) all of the above.

Question Type	Example

Chronological problem

The following events are related to major case studies of environmental pollution. Place the events in chronological order.

   I. Bhopal, India
  II. *Exxon Valdez*
 III. Donora, Pennsylvania
 IV. Three Mile Island
  V. Chernobyl, Ukraine

(A) I, II, III, IV, V
(B) V, IV, III, II, I
(C) I, III, V, II, IV
(D) I, V, II, IV, III
(E) III, IV, V, I, II

Comparing and contrasting concepts and events

Which of the following acts or treaties accurately compares the political, environmental, and economic goals of the participating nations to the goal of reducing global greenhouse gas emissions?

(A) Clean Air Act of 1955
(B) National Environmental Policy Act (NEPA) 1969
(C) Comprehensive Environmental Response, Compensation, and Liability (Superfund) Act (CERCLA) 1980
(D) Pollution Prevention Act of 1990
(E) Kyoto Protocol of 1997

Multiple correct answers

Nitrogen is assimilated in plants in what form?

(A) NO_2^-
(B) NH_3
(C) NH_4^+
(D) NO_3^-
(E) Choices B, C, and D

Negative questions

An effective method to decrease the amount of pesticide use would include all of the following EXCEPT

(A) using monoculture techniques
(B) rotating crops
(C) using pheromones
(D) using polyculture techniques
(E) using insect-resistant crops

Stimulus-Based Multiple-Choice Questions

Question Type	Example

Short narrative passage

"Human beings are adaptable. They have survived and conquered when other species have become extinct. Granted, that we as humans have a long way to go in order to live in harmony with nature, nevertheless, the future looks bright for man to conquer all environmental problems. There is no limit to what problems humans can solve. If we all work together, we can solve any problem. The future is bright."

The quote above would be most closely associated with

(A) Malthusian principles.
(B) neo-Luddites.
(C) cornucopian fallacy.
(D) nihilists.
(E) utilitarianism.

Short quotation

As the troposphere warms, the stratosphere begins to cool and causes the formation of ice-clouds above the South Pole. The ice-clouds tend to accelerate the decrease of ozone above the South Pole by providing a surface for chemical reactions to occur. This is an example of

(A) serendipity.
(B) a positive feedback loop.
(C) a negative feedback loop.
(D) synergy.
(E) chaos.

Environmental case law

Which act established the first clear water purity standards?

(A) Water Resources Planning Act
(B) Water Resources Development Act
(C) Safe Drinking Water Act
(D) Surface Water Treatment Rule
(E) Water Quality Act

Sketch or diagram

The following diagram indicates

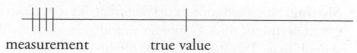

measurement   true value

(A) a measurement that is accurate and precise.
(B) a measurement that is accurate but not precise.
(C) a measurement that is precise but not accurate.
(D) a measurement that is neither accurate nor precise.
(E) a conclusion cannot be made regarding this distribution.

The Elimination Strategy

Take advantage of being able to mark in your test booklet. As you go through the "+" questions, eliminate choices from consideration by marking them out in your question booklet. Mark with question marks any choices you wish to consider as possible answers. See the following example:

A.

B. ?

C.

D.

E. ?

This technique will help you avoid reconsidering those choices that you have already eliminated and will thus save you time. It will also help you narrow down your possible answers. Remember, there is no penalty for guessing, so use this strategy to eliminate obvious incorrect choices!

Question Type	Example

Graph or table interpretation

Two varieties of the same species of voles (meadow mice), albino and red-backed, were used in an experiment. Both varieties were subjected to the predation of a hawk, under controlled laboratory conditions. During the experiment, the floor of the test room was covered on alternate days with white ground cover that matched the albino voles and red-brown cover that matched the red-backed voles. The results of fifty trials are shown in the following tabulation.

NUMBER OF VOLES CAPTURED

Variety	White Cover	Red-Brown Cover	Total
Albino	35	57	92
Red-backed	60	40	100
Total	95	97	192

Which result would have been most likely if only red-brown floor covering had been used?

(A) Ninety-seven red-backed voles would have survived.
(B) Ninety-five albino voles would have survived.
(C) The survival rate of the albino voles would have decreased markedly.
(D) A greater number of red-backed voles would not have survived.
(E) There would have been no change in the results.

A FINAL WORD ABOUT THE MULTIPLE-CHOICE QUESTIONS

Throughout the APES course, your teacher will be giving you a variety of sample multiple-choice questions with five possible answers. Other methods include:

- **Building:** Build up your own database of questions by using questions from this book and accompanying CD.
- **Using:** Use *Barron's AP Environmental Science Flash Cards*, available from your local bookstore.
- **Sharing:** Share and collect questions that other students come up with (the Internet is wonderful for this). Create reciprocal agreements with APES classes from other schools (e.g., "We'll give you 300 of our questions if you give us 300 of yours").
- **Creating:** Create a class database where everyone contributes ten questions per chapter in all question formats. Or make your own class APES web site—post questions, answers with explanations, class projects, photos, and so on.
- **Obtaining:** Obtain past questions from the Educational Testing Service—the people who create the APES exam. Yes, past questions are available!

- **Searching:** Search the Internet for web sites or social networking pages from other APES classes (I counted over sixty when I recently searched). Many of these web sites have practice questions posted.
- **Visiting:** Visit the web sites of publishers of environmental science textbooks. Often you will find many practice questions available for free on their sites.
- **Debriefing:** Right after you finish the AP exam, come immediately back to your APES teacher and let him or her know the type of questions that you found easy and the ones that you found more difficult. This will help your teacher to prepare for next year. After a few years of student feedback, your teacher will have a pretty good idea of what to expect.

The Free-Response Questions

There are three types of free-response questions: (1) data analysis, (2) document based, and (3) synthesis and evaluation.

The data analysis (or data set) question presents data in tabular or graphical form and measures your ability to interpret and analyze data. The document-based question (DBQ) presents material in the form of real-life documents (newspaper clippings, product advertisements, etc.). The synthesis and evaluation questions are in-depth essay questions, often with multiple parts.

In scoring the free-response essays, points are awarded only for arguments that are supported by scientific facts and principles. Each question is scored on a scale from 1 to 10, with 10 being the highest.

All four essay questions must be answered within 90 minutes. If you spend an equal amount of time on each essay, this averages to about 23 minutes per essay. Lists, outlines, and unlabeled drawings are not acceptable and will not be awarded credit.

THESIS STATEMENT

The thesis statement is a sentence or two toward the beginning of your response that tells the AP graders what your answer is going to be about. A thesis statement:

- tells the grader <u>how you will interpret</u> the significance of the subject matter under discussion;
- tells the grader <u>what to expect</u> from your response;
- and <u>presents your argument</u> to the grader. The rest of your response provides the evidence to support your thesis statement.

Examples of APES thesis statements:

1. Many of the problems faced by less-developed countries, such as poverty, hunger, and environmental destruction, are the consequences of excessive population growth.

2. Family structure and the status of women in society are the prime determinants of fertility and population growth.

3. Governmental run "food aid" programs often funnel tax dollars to large, worldwide corporate agribusinesses while increasing the influence of foreign food aid organizations; often at the expense of small, local food growers.

TIP

The best way to earn the highest score possible on the AP Exam is to look at other students' essays from past APES exams and see how their essays were graded by the College Board readers. Your teacher has access to these past exams at *www.apcentral. collegeboard.com.* This is the best kept secret there is for getting a 5.

The free-response questions often include phrases that should guide how you write your essays. The following table lists many of these key terms and what they actually mean.

Key Phrases

Key Term	Meaning	Example
Analyze the effect . . .	Evaluate the impact	Analyze the effects of major global wind patterns in determining the distribution of biomes.
Assess the accuracy . . .	Determine the truth	Assess the accuracy of the following statement, "A high birth rate corresponds to a high standard of living."
Compare the strengths and weaknesses . . .	Show differences	Compare the strengths and weaknesses of three government intervention programs to combat pollution.
Critically evaluate . . .	Give examples that agree and/or disagree	Critically evaluate evidence that both supports and refutes the claim that greenhouse gases are responsible for the hole in the ozone layer.
Define and evaluate . . .	Give a definition and analyze the point of view	Define ecofeminism and evaluate the contention that male-dominated societies are responsible for environmental imbalance and social oppression.
Discuss . . .	Give examples that illustrate	Discuss the rapid increase of African "killer bees" in the United States as an example of the process of natural selection.
Evaluate . . .	Determine the validity	Evaluate the claim that increasing the standard of living and improving the role of women will result in a lower birthrate.
Explain . . .	Offer the meaning, cause, effect, and influence	Explain three different theories of moral responsibility to the environment. Include in your discussion the major proponents of each theory and their specific ideas(s) or "school."
To what extent . . .	Explain the relationship and role	To what extent does moisture determine a soil profile? Provide specific soil types, overview of soil horizons, and geographic distribution.

Topics Covered on the Exam*

I. Earth Systems and Resources (10–15%)

A. **Earth Science Concepts**—Geologic time scale, plate tectonics, earthquakes, volcanism, seasons, solar intensity and latitude.

B. **The Atmosphere**—Composition, structure, weather, climate, atmospheric circulation and the Coriolis effect, atmosphere-ocean interactions, and ENSO.

C. **Global Water Resources and Use**—Freshwater, saltwater, ocean circulation, agriculture, industrial and domestic use, surface and groundwater issues, global problems, and conservation.

D. **Soil and Soil Dynamics**—Rock cycle, formation, composition, physical and chemical properties, main soil types, erosion and other soil problems, and soil conservation.

II. The Living World (10–15%)

A. **Ecosystem Structure**—Biological populations and communities, ecological niches, interactions among species, keystone species, species diversity, and edge effects, and major terrestrial and aquatic biomes.

B. **Energy Flow**—Photosynthesis and cellular respiration, food webs, and trophic levels, and ecological pyramids.

C. **Ecosystem Diversity**—Biodiversity, natural selection, evolution, and ecosystem services.

D. **Natural Ecosystem Change**—Climate shifts, species movement, and ecological succession.

E. **Natural Biogeochemical Cycles**—Carbon, nitrogen, phosphorus, sulfur, water, and conservation of matter.

III. Population (10–15%)

A. **Population Biology Concepts**—Population ecology, carrying capacity, reproductive strategies, and survivorship.

B. **Human Population**

1. Human population dynamics—Historical population sizes, distribution, fertility rates, growth rates and doubling times, demographic transition, and age-structure diagrams.

2. Population size—Strategies for sustainability, case studies, and national policies.

3. Impacts of population growth—Hunger, disease, economic effects, resource use, and habitat destruction.

---

*Advanced Placement Program Description—Environmental Science, 2010, AP College Board.

IV. Land and Water Use (10–15%)

A. Agriculture

1. Feeding a growing population—Human nutritional requirements, types of agriculture, green revolution, genetic engineering and crop production, deforestation, irrigation, and sustainable agriculture.
2. Controlling pests—Types of pesticides, costs and benefits of pesticide use, integrated pest management, and relevant laws.

B. Forestry—Tree plantations, old-growth forests, forest fires, forest management, and national forests.

C. Rangelands—Overgrazing, deforestation, desertification, rangeland management, and federal rangelands.

D. Other Land Use

1. Urban land development—Planned development, suburban sprawl, and urbanization.
2. Transportation infrastructure—Federal highway system, canals and channels, roadless areas, and ecosystem impacts.
3. Public and federal lands—Management, wilderness areas, national parks, wildlife refuges, forests, and wetlands.
4. Land conservation options—Preservation, remediation, mitigation, and restoration.
5. Sustainable land use strategies.

E. Mining—Mineral formation, extraction, global reserves, and relevant laws and treaties.

F. Fishing—Fishing techniques, overfishing, aquaculture, and relevant laws and treaties.

G. Global Economics—Globalization, World Bank, *Tragedy of the Commons*, and relevant laws and treaties.

V. Energy Resources and Consumption (10–15%)

A. Energy Concepts—Energy forms, power, units, conversions, and laws of thermodynamics.

B. Energy Consumption

1. History—Industrial Revolution, exponential growth, and energy crisis.
2. Present global energy use.
3. Future energy needs.

C. Fossil Fuel Resources and Use—Formation of coal, oil, and natural gas; extraction/purification methods; world reserves and global demand; synfuels; and environmental advantages/disadvantages of sources.

D. Nuclear Energy—Nuclear fission process, nuclear fuel, electricity production, nuclear reactor types, environmental advantages/disadvantages, safety issues, radiation and human health, radioactive wastes, and nuclear fusion.

E. Hydroelectric Power—Dams, flood control, salmon, silting, and other impacts.

F. **Energy Conservation**—Energy efficiency, CAFE standards, hybrid electric vehicles and mass transit.

G. **Renewable Energy**—Solar energy, solar electricity, hydrogen fuel cells, biomass, wind energy, small-scale hydroelectric, ocean waves and tidal energy, geothermal, and environmental advantages/disadvantages.

VI. Pollution (25–30%)

A. Pollution Types

1. Air pollution—Sources (primary and secondary) of major air pollutants, measurement units, smog, acid deposition—causes and effects, heat islands and temperature inversions, indoor air pollution, remediation and reduction strategies, Clean Air Act, and other relevant laws.
2. Noise pollution—Sources, effects, and control measures.
3. Water pollution—Types, sources, causes and effects, cultural eutrophication, groundwater pollution, maintaining water quality, water purification, sewage treatment/septic systems, Clean Water Act, and other relevant laws.
4. Solid waste—Types, disposal, and reduction.

B. Impacts on the Environment and Human Health

1. Hazards to human health—Environmental risk analysis, acute and chronic effects, dose-response relationships, air pollutants, smoking, and other risks.
2. Hazardous chemicals in the environment—Types of hazardous waste, treatment/disposal of hazardous waste, cleanup of contaminated sites, biomagnification, and relevant laws.

C. Economic Impacts—Cost-benefit analysis, externalities, marginal costs, and sustainability.

VII. Global Change (10–15%)

A. **Stratospheric ozone**—Formation of stratospheric ozone, ultraviolet radiation, causes of ozone depletion, effects of ozone depletion, strategies for reducing ozone depletion, and relevant laws and treaties.

B. **Global Warming**—Greenhouse gases and the greenhouse effect, impacts and consequences of global warming, reducing climate change, and relevant laws and treaties.

C. **Loss of Biodiversity**

1. Habitat loss—Overuse, pollution, introduced species, and endangered and extinct species.
2. Maintenance through conservation.
3. Relevant laws and treaties.

Questions Commonly Asked About the APES Exam

Q: What's the best way to prepare for the APES exam?

A: Don't wait until the last few weeks in April or May to start "cramming," as that seldom works. Instead, highly successful students use *Barron's AP Environmental Science* starting from the very first day of their high school AP Environmental Science class. They use this book *along with* their textbook and class lectures, covering each topic concurrently with their course schedule. This steady, slower-paced approach, which allows you to cover more information in greater depth, has proven highly successful.

Q: How are the multiple-choice questions graded?

A: Section I, worth 60% of the total score, is 90 minutes long and consists of 100 multiple-choice questions. The total score for Section I is the number of correct answers. If you leave a question unanswered, it does not count at all. You will need to answer around 50% of the multiple-choice questions correctly to obtain a 3 on the exam. The multiple-choice questions are based on recall of basic facts and major concepts of environmental science.

Q: How are the essay or free-response questions graded?

A: Your essays are read several times by both college and high school faculty who have taught environmental science for many years. These readers use a set of written guidelines or standards known as a rubric to come to a consensus on the overall score for each essay that you write.

Q: Will I find questions about labs?

A: Although the College Board does not specify the number or type of labs that must be completed in an APES course, there are several questions on the exam that will draw upon your lab experiences and your ability to analyze experimental data and make valid conclusions.

Q: What do the scores mean?

A: The APES exam is graded on a 5-point scale:

5: Passing. ~9% earn this score.
4: Passing. ~25% earn this score.
3: Passing. ~15% earn this score.
2: Not passing. ~25% earn this score.
1: Not passing. ~26% earn this score.

In summary, about half pass and half do not.

Q: What materials do I take with me to the exam?

A: Bring your admission ticket, an official photo including signature I.D., your social security number, several sharpened #2 pencils with non-smudging erasers, black or blue erasable pens for the free-response questions, and a watch. Do not bring food or drink, colored pencils, highlighters, rulers, or cell phones. Calculators are NOT allowed.

Q: Should I guess?

A: Yes. There is no penalty for wrong answers in the multiple-choice section, so you should answer all multiple-choice questions. Even if you have no idea of the correct answer, you should try to eliminate any obvious incorrect choices, and then guess!

Q: Can I cancel my scores?

A: Yes. Your request must be received in writing by June 15. You may also request that one or more of your AP grades NOT be sent to colleges.

Q: Can I write on the test booklet?

A: Yes. You will definitely want to make notes or brief outlines on the test booklet to help organize your free-response answers. Several examples of this technique are presented in this book.

Q: How do I get more information?

A: Log on to *http://apcentral.collegeboard.com* or *http://collegeboard.com/apstudents*

Answer Sheet

DIAGNOSTIC TEST

1 Ⓐ Ⓑ Ⓒ Ⓓ Ⓔ	26 Ⓐ Ⓑ Ⓒ Ⓓ Ⓔ	51 Ⓐ Ⓑ Ⓒ Ⓓ Ⓔ	76 Ⓐ Ⓑ Ⓒ Ⓓ Ⓔ
2 Ⓐ Ⓑ Ⓒ Ⓓ Ⓔ	27 Ⓐ Ⓑ Ⓒ Ⓓ Ⓔ	52 Ⓐ Ⓑ Ⓒ Ⓓ Ⓔ	77 Ⓐ Ⓑ Ⓒ Ⓓ Ⓔ
3 Ⓐ Ⓑ Ⓒ Ⓓ Ⓔ	28 Ⓐ Ⓑ Ⓒ Ⓓ Ⓔ	53 Ⓐ Ⓑ Ⓒ Ⓓ Ⓔ	78 Ⓐ Ⓑ Ⓒ Ⓓ Ⓔ
4 Ⓐ Ⓑ Ⓒ Ⓓ Ⓔ	29 Ⓐ Ⓑ Ⓒ Ⓓ Ⓔ	54 Ⓐ Ⓑ Ⓒ Ⓓ Ⓔ	79 Ⓐ Ⓑ Ⓒ Ⓓ Ⓔ
5 Ⓐ Ⓑ Ⓒ Ⓓ Ⓔ	30 Ⓐ Ⓑ Ⓒ Ⓓ Ⓔ	55 Ⓐ Ⓑ Ⓒ Ⓓ Ⓔ	80 Ⓐ Ⓑ Ⓒ Ⓓ Ⓔ
6 Ⓐ Ⓑ Ⓒ Ⓓ Ⓔ	31 Ⓐ Ⓑ Ⓒ Ⓓ Ⓔ	56 Ⓐ Ⓑ Ⓒ Ⓓ Ⓔ	81 Ⓐ Ⓑ Ⓒ Ⓓ Ⓔ
7 Ⓐ Ⓑ Ⓒ Ⓓ Ⓔ	32 Ⓐ Ⓑ Ⓒ Ⓓ Ⓔ	57 Ⓐ Ⓑ Ⓒ Ⓓ Ⓔ	82 Ⓐ Ⓑ Ⓒ Ⓓ Ⓔ
8 Ⓐ Ⓑ Ⓒ Ⓓ Ⓔ	33 Ⓐ Ⓑ Ⓒ Ⓓ Ⓔ	58 Ⓐ Ⓑ Ⓒ Ⓓ Ⓔ	83 Ⓐ Ⓑ Ⓒ Ⓓ Ⓔ
9 Ⓐ Ⓑ Ⓒ Ⓓ Ⓔ	34 Ⓐ Ⓑ Ⓒ Ⓓ Ⓔ	59 Ⓐ Ⓑ Ⓒ Ⓓ Ⓔ	84 Ⓐ Ⓑ Ⓒ Ⓓ Ⓔ
10 Ⓐ Ⓑ Ⓒ Ⓓ Ⓔ	35 Ⓐ Ⓑ Ⓒ Ⓓ Ⓔ	60 Ⓐ Ⓑ Ⓒ Ⓓ Ⓔ	85 Ⓐ Ⓑ Ⓒ Ⓓ Ⓔ
11 Ⓐ Ⓑ Ⓒ Ⓓ Ⓔ	36 Ⓐ Ⓑ Ⓒ Ⓓ Ⓔ	61 Ⓐ Ⓑ Ⓒ Ⓓ Ⓔ	86 Ⓐ Ⓑ Ⓒ Ⓓ Ⓔ
12 Ⓐ Ⓑ Ⓒ Ⓓ Ⓔ	37 Ⓐ Ⓑ Ⓒ Ⓓ Ⓔ	62 Ⓐ Ⓑ Ⓒ Ⓓ Ⓔ	87 Ⓐ Ⓑ Ⓒ Ⓓ Ⓔ
13 Ⓐ Ⓑ Ⓒ Ⓓ Ⓔ	38 Ⓐ Ⓑ Ⓒ Ⓓ Ⓔ	63 Ⓐ Ⓑ Ⓒ Ⓓ Ⓔ	88 Ⓐ Ⓑ Ⓒ Ⓓ Ⓔ
14 Ⓐ Ⓑ Ⓒ Ⓓ Ⓔ	39 Ⓐ Ⓑ Ⓒ Ⓓ Ⓔ	64 Ⓐ Ⓑ Ⓒ Ⓓ Ⓔ	89 Ⓐ Ⓑ Ⓒ Ⓓ Ⓔ
15 Ⓐ Ⓑ Ⓒ Ⓓ Ⓔ	40 Ⓐ Ⓑ Ⓒ Ⓓ Ⓔ	65 Ⓐ Ⓑ Ⓒ Ⓓ Ⓔ	90 Ⓐ Ⓑ Ⓒ Ⓓ Ⓔ
16 Ⓐ Ⓑ Ⓒ Ⓓ Ⓔ	41 Ⓐ Ⓑ Ⓒ Ⓓ Ⓔ	66 Ⓐ Ⓑ Ⓒ Ⓓ Ⓔ	91 Ⓐ Ⓑ Ⓒ Ⓓ Ⓔ
17 Ⓐ Ⓑ Ⓒ Ⓓ Ⓔ	42 Ⓐ Ⓑ Ⓒ Ⓓ Ⓔ	67 Ⓐ Ⓑ Ⓒ Ⓓ Ⓔ	92 Ⓐ Ⓑ Ⓒ Ⓓ Ⓔ
18 Ⓐ Ⓑ Ⓒ Ⓓ Ⓔ	43 Ⓐ Ⓑ Ⓒ Ⓓ Ⓔ	68 Ⓐ Ⓑ Ⓒ Ⓓ Ⓔ	93 Ⓐ Ⓑ Ⓒ Ⓓ Ⓔ
19 Ⓐ Ⓑ Ⓒ Ⓓ Ⓔ	44 Ⓐ Ⓑ Ⓒ Ⓓ Ⓔ	69 Ⓐ Ⓑ Ⓒ Ⓓ Ⓔ	94 Ⓐ Ⓑ Ⓒ Ⓓ Ⓔ
20 Ⓐ Ⓑ Ⓒ Ⓓ Ⓔ	45 Ⓐ Ⓑ Ⓒ Ⓓ Ⓔ	70 Ⓐ Ⓑ Ⓒ Ⓓ Ⓔ	95 Ⓐ Ⓑ Ⓒ Ⓓ Ⓔ
21 Ⓐ Ⓑ Ⓒ Ⓓ Ⓔ	46 Ⓐ Ⓑ Ⓒ Ⓓ Ⓔ	71 Ⓐ Ⓑ Ⓒ Ⓓ Ⓔ	96 Ⓐ Ⓑ Ⓒ Ⓓ Ⓔ
22 Ⓐ Ⓑ Ⓒ Ⓓ Ⓔ	47 Ⓐ Ⓑ Ⓒ Ⓓ Ⓔ	72 Ⓐ Ⓑ Ⓒ Ⓓ Ⓔ	97 Ⓐ Ⓑ Ⓒ Ⓓ Ⓔ
23 Ⓐ Ⓑ Ⓒ Ⓓ Ⓔ	48 Ⓐ Ⓑ Ⓒ Ⓓ Ⓔ	73 Ⓐ Ⓑ Ⓒ Ⓓ Ⓔ	98 Ⓐ Ⓑ Ⓒ Ⓓ Ⓔ
24 Ⓐ Ⓑ Ⓒ Ⓓ Ⓔ	49 Ⓐ Ⓑ Ⓒ Ⓓ Ⓔ	74 Ⓐ Ⓑ Ⓒ Ⓓ Ⓔ	99 Ⓐ Ⓑ Ⓒ Ⓓ Ⓔ
25 Ⓐ Ⓑ Ⓒ Ⓓ Ⓔ	50 Ⓐ Ⓑ Ⓒ Ⓓ Ⓔ	75 Ⓐ Ⓑ Ⓒ Ⓓ Ⓔ	100 Ⓐ Ⓑ Ⓒ Ⓓ Ⓔ

Diagnostic Test

The following diagnostic test will allow you to pinpoint your current strengths and weaknesses as you work toward the AP Environmental Science exam.

If you're just getting underway with your APES course, you may realize that you've already seen some of the topics found on the APES exam in other courses (AP Biology, AP Chemistry, etc.), but for many of you this will be your first exposure to this information.

Questions found on this diagnostic test reflect the actual percentage of questions that you will find for each topic; i.e., 10–15% of the questions on the actual APES exam cover earth science concepts (geology, tectonics, earthquakes, soil dynamics, etc.)—likewise, 10–15% of the questions on this diagnostic also cover these same topics). Furthermore, the questions on this diagnostic test are grouped into categories to make it easy for you and your teacher to identify areas in which you might need extra study. As on the actual APES exam, there is no penalty for guessing on this diagnostic test; however, for this practice diagnostic test *ONLY*, it is suggested that if you are not familiar with the question or answer choices, that you leave it blank. This will give you a clearer picture of where you currently are. Some students take this same diagnostic test again just before the actual APES exam, to see how far they have progressed—teachers refer to this as a "post" test.

An answer key with full explanations can be found following this diagnostic test. Also included in the scoring section is a predictive scale of what your actual APES score might be based on your performance on this diagnostic test. Free-Response Questions (FRQs) are not included in this diagnostic test as that would involve subjective grading.

For this Diagnostic Test *only*, if you do not know the answer for a particular question, do *not* guess; this will allow you to determine your strengths and weaknesses. If you know the answer or are pretty sure of the answer, then answer the question.

1. Approximately how old is the Earth?

   (A) 4,500,000 years
   (B) 45,000,000 years
   (C) 450,000,000 years
   (D) 4,500,000,000 years
   (E) 45,000,000,000 years

2. An earthquake of Richter magnitude 2 releases _____ times less energy than an earthquake of Richter magnitude 4?

   (A) 0.01
   (B) 0.5
   (C) two
   (D) 31.7×2 or 63.4
   (E) 100

3. The interface where tectonic plates slide or grind past each other is known as a

   (A) transform plate boundary
   (B) convergent plate boundary
   (C) divergent plate boundary
   (D) subduction zone
   (E) trench

4. Climate on Earth is primarily determined by

   (A) the distance between the Earth and the Sun during a particular season
   (B) the tilt of the Earth's axis as it rotates around the Sun
   (C) the amount of cloud cover over a particular region of the Earth
   (D) ocean temperatures
   (E) Earth's longitude

5. Which of the following gases found in the troposphere are greenhouse gases and are also gases produced by domestic livestock?

    I. CO
    II. CO_2
    III. CH_4
    IV. O_2
    V. H_2O

    (A) I and V
    (B) I and III
    (C) II and III
    (D) II, III, and V
    (E) II, III, IV, and V

6. Which area listed below would be found in the regions of the Earth influenced by Ferrel cells?

    (A) Ecuador—near the equator
    (B) Antarctica
    (C) Greenland
    (D) Mid-western states of the United States (e.g., Nebraska, Kansas, etc.)
    (E) Southern Africa

7. Which of the following events about El Niño are TRUE?

    I. Just prior to an El Niño, the sea surface warms in the eastern equatorial Pacific Ocean
    II. Just prior to an El Niño, trade winds weaken
    III. During an El Niño, the number of hurricanes in the Atlantic Ocean increases
    IV. During an El Niño, upwelling increases along the western coast of South America resulting in large schools of fish

    (A) I only
    (B) II only
    (C) I and II
    (D) III
    (E) I, III, and IV

8. Water in areas of upwelling are generally

    (A) warm and lacking in nutrients
    (B) warm and high in nutrients
    (C) cold and lacking in nutrients
    (D) cold and high in nutrients
    (E) very hot due to thermal vents and contain high amounts of sulfur

9. Which of the following choices below lists the amount of freshwater available on Earth in the correct descending order?
    (A) Groundwater > soil moisture > atmospheric moisture > lakes > rivers and streams > glaciers and ice caps
    (B) Glaciers and ice caps > atmospheric moisture > groundwater > rivers and streams > soil moisture > lakes
    (C) Glaciers and ice caps > groundwater > lakes > soil moisture > atmospheric moisture > rivers and streams
    (D) Lakes > groundwater > atmospheric moisture > glaciers and ice caps > soil moisture > rivers and streams
    (E) Atmospheric moisture > rivers and streams > ground water > soil moisture > lakes > glaciers and ice caps

10. Freshwater used in developing countries is primarily used for

    (A) agriculture
    (B) industrial purposes
    (C) drinking water
    (D) domestic needs
    (E) all of the above are used in approximate equal proportion

11. Currently, the use of freshwater by humans is increasing at about _____rate of worldwide population growth.

    (A) one-half the
    (B) the same
    (C) twice the
    (D) four times the
    (E) eight times the

For Questions 12–13 choose from the following

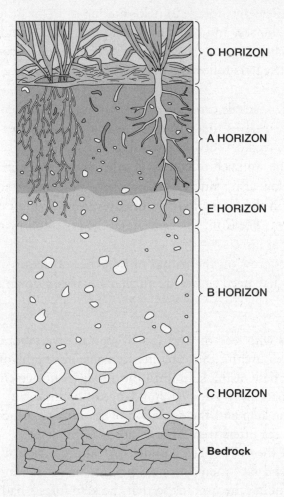

(A) A
(B) B
(C) C
(D) O
(E) E

12. Which soil layer consists of clay, iron oxides, and other components that came from the zone of leaching?

13. Which soil layer consists of about equal amounts of organic material and minerals?

14. Tropical rain forests have a rich biodiversity primarily because

   (A) warm, moist climates make it easier for living forms to survive
   (B) they are generally very large allowing for many life-forms to coexist
   (C) evolution occurs faster in warmer and more humid environments
   (D) historically, there has been less human settlement and disturbance in tropical rain forests
   (E) many diverse niche opportunities and habitats occur in tropical rain forests

15. Which of the following is NOT an example of commensalism?

   (A) Birds known as cattle egrets forage in fields among cattle or other livestock. As cattle, horses, and other livestock graze on the field, they cause movements that stir up various insects. As the insects are stirred up, the cattle egrets following the livestock catch and feed upon them.

   (B) In India, lone golden jackals expelled from their pack have been known to attach themselves to a particular tiger, trailing it at a safe distance in order to feed on the big cat's kills.

   (C) Bacteria living in the stomach of cows and other ruminant animals.

   (D) Birds will often follow army ant raids on a forest floor. As the army ant colony travels on the forest floor, it stirs up various flying insect species. As the insects flee from the army ants, the birds following the ants catch the fleeing insects.

   (E) Orchids and mosses grow on the trunks or branches of trees. They get the light they need as well as nutrients that run down along the tree.

16. "Sea otters are mammals who feed on a variety of marine invertebrates but who feed especially on sea urchins. Sea urchins are voracious herbivores that tend to feed on the base of the kelp until the whole plant detaches from the bottom and floats away. An overabundance of urchins can lead to overgrazing of the kelp and the depletion of sea urchins. Once abundant in California, sea otters were hunted down for their pelts to near extinction. After they were placed under federal protection to increase their numbers and save them from extinction, sea otters were able to begin naturally controlling the urchins so that the kelp forests had a chance to recover."

   In this example, sea otters are considered to be a(n)

   (A) indicator species
   (B) keystone species
   (C) ecotone species
   (D) niche specialist
   (E) niche generalist

17. Which of the following represents intraspecific interaction(s)?

   (A) mutualism
   (B) commensalism
   (C) parasitism
   (D) territoriality
   (E) predation

18. The marine zone that is essentially open ocean, and whose flora include surface seaweeds and whose fauna includes many species of fish and some mammals, such as whales and dolphins, is known as the

    (A) intertidal zone
    (B) pelagic zone
    (C) benthic zone
    (D) abyssal zone
    (E) hadal zone

19. Photosynthesis produces sugars from

    (A) carbon dioxide, water, and oxygen
    (B) oxygen, water, and energy
    (C) glucose, oxygen, and energy
    (D) carbon dioxide, water, and energy
    (E) oxygen, carbon dioxide, and energy

Use the following diagram for Questions 20 and 21.

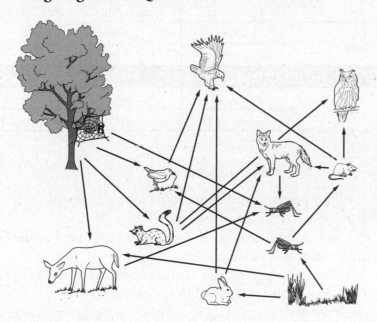

20. In the food web above, which of the following statements would be TRUE for the owl?

    I. The owl is an omnivore
    II. The owl is a primary consumer
    III. The owl is a secondary consumer
    IV. The owl is a tertiary consumer
    V. The owl is a primary producer

    (A) II only
    (B) III only
    (C) III and IV only
    (D) I, IV, and V only
    (E) II, III, and IV only

21. Energy enters this system as sunlight. If a leaf from the plant contains 10 calories of energy and the leaf is totally consumed by the grasshopper, which is then totally consumed by the songbird, which is then totally consumed by the owl, about how much of the energy from the leaf is eventually available as energy for the owl to use?

    (A) 0.0001 calories
    (B) 0.001 calories
    (C) 10 calories
    (D) 100 calories
    (E) 1,000 calories

22. An APES class performed an experiment of releasing and recapturing light and dark moths in the city and the country. The results are presented below:

Comparison of Light and Dark Moth Recaptures in the City and Country		Number of Light Moths	Number of Dark Moths
Location		Number of Light Moths	Number of Dark Moths
Country	Released	300	300
Country	Recaptured	200	100
City	Released	400	400
City	Recaptured	200	300

Which of the following statements is TRUE?

    (A) A higher percentage of light moths were recaptured in the country than were recaptured in the city.
    (B) A lower percentage of dark moths were recaptured in the city than were recaptured in the country.
    (C) A lower percentage of light moths were recaptured in the country than dark moths recaptured in the country.
    (D) A higher percentage of light moths were recaptured in the city than dark moths recaptured in the city.
    (E) All statements are true.

23. In New York City, where the quality of drinking water had fallen below standards required by the U.S. Environmental Protection Agency (EPA), authorities opted to restore the polluted Catskill Watershed. Once the input of sewage and pesticides to the watershed area was reduced, natural abiotic processes such as soil adsorption and filtration of chemicals, together with biotic recycling via root systems and soil microorganisms, water quality improved to levels that met government standards. The cost of this investment in natural capital was estimated between $1–1.5 billion, which contrasted dramatically with the estimated $6–8 billion cost of constructing a water filtration plant plus the $300 million annual running costs.

    The Catskill Watershed provides _____ to New York City.

    (A) natural resource services
    (B) market services
    (C) social services
    (D) economic services
    (E) ecosystem services

24. In response to climate shifts due to global warming, terrestrial animals can adapt, migrate, or die. Which of the following factors below contribute to surviving the effects of global warming?

    I.   Long generational times
    II. Being widely dispersed
    III. Having narrow climatic tolerances
    IV. Being an *r*-strategist rather than a *K*-strategist
    V.   Being confined to one geographic location

    (A) I, II, and IV
    (B) II and IV
    (C) I, IV, and V
    (D) I, II, and V
    (E) All of the above

25. Which of the following would be subject to primary succession?

    (A) rock exposed by a retreating glacier
    (B) an abandoned farm
    (C) a forest that had been clear-cut
    (D) newly flooded agricultural land used to create a reservoir
    (E) a forest that has been burned

26. Which of the following would be TRUE regarding succession?

    I. Growth of lichen on rocks after a fire destroyed a forest is an example of primary succession
    II. Growth of grasses on a newly formed sand dune is an example of secondary succession
    III. Growth of algae on cooled lava rock is an example of secondary succession
    IV. *K*-selected organisms are typical of primary succession
    V. *r*-selected organisms are typical of secondary succession

    (A) III
    (B) I and III
    (C) II, IV, and V
    (D) I, III, IV, and V
    (E) None are true

27. The first compound that ammonia is converted into in the nitrogen cycle is

    (A) nitrate
    (B) urea
    (C) protein
    (D) nitrite
    (E) amino acid

28. Which of the following below are examples of density-independent factors that control population growth?

    I. Droughts and floods
    II. Fires
    III. Use of pesticides
    IV. Release of a pollutant
    V. Overhunting and fishing

    (A) I
    (B) I and II
    (C) III, IV, and V
    (D) IV
    (E) All choices are correct examples

29. The graph below is of a paramecium population growing in culture over a period of 30 hours. If food is a limiting factor for this population, then increasing the amount of food available in the water at the beginning of the experiment will . . .

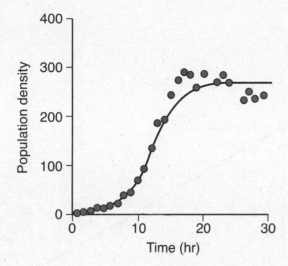

(A) increase the population density between 0–15 hours
(B) decrease the population density after the carrying capacity had been reached
(C) have no effect on the population density between 0 and 15 hours
(D) decrease the time it takes for the population to reach its carrying capacity
(E) All of the above are true

30. Consider the following organisms

I. Elephant
II. Weeds
III. Mice
IV. Acorn tree
V. Mosquitos

Which of the above would be considered *K*-strategists?

(A) I
(B) II, III, and V
(C) I and IV
(D) II and V
(E) V

31. In wild populations, individuals most often show a _____ pattern of dispersion.

    (A) clumped
    (B) uniform
    (C) random
    (D) scattered
    (E) chaotic

32. Examine the following diagram below.

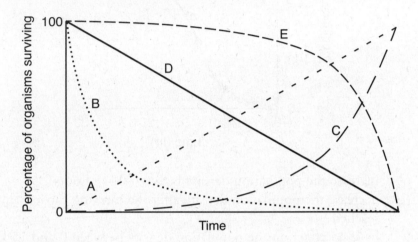

Which of the choices above would be a characteristic survivorship curve for a songbird?

    (A) A
    (B) B
    (C) C
    (D) D
    (E) E

33. Examine the diagram below.

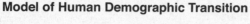

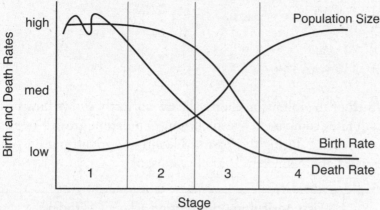

Which of the following statements are TRUE of Stage 2?

I.   This stage would be typical for countries today such as Afghanistan and much of sub-Saharan Africa.

II. Decline in death rates is primarily due to improvements in the food supply brought about by higher yields in agricultural practices and better transportation, which prevent death due to starvation and lack of water. Agricultural improvements included crop rotation, selective breeding, and seed drill technology.

III. Improvements in water supply, sewage treatment, food handling, and general personal hygiene following from growing scientific knowledge of the causes of disease and the improved education and social status of mothers are primarily responsible for decreasing the death rate.

IV. Growth in population is primarily due to an increase in death rates, not fertility rates.

V.   The age structure of the population becomes increasingly older as more and more people live longer.

(A) I and IV
(B) I, II, and III
(C) I, II, IV, and V
(D) III, IV, and V
(E) All statements are true

34. A country with an annual population growth rate of 5% would double its population in approximately how many years?

   (A) 5 years
   (B) 7 years
   (C) 14 years
   (D) 70 years
   (E) 350 years

35. Which of the following choices below correctly shows how birthrates and death rates compare between current population growth trends and trends that occurred during the First or Neolithic Agricultural Revolution that occurred approximately 10,000 years ago?

	First Agricultural Revolution	**Today**
(A)	Birthrate: High	Birthrate: Low
	Death Rate: Low	Death Rate: High
(B)	Birthrate: Low	Birthrate: High
	Death Rate: High	Death Rate: Low
(C)	Birthrate: High	Birthrate: High
	Death Rate: Low	Death Rate: Low
(D)	Birthrate: High	Birthrate: Low
	Death Rate: High	Death Rate: Low
(E)	Birthrate: High	Birthrate: High
	Death Rate: Low	Death Rate: High

36. Examine the age distribution diagrams below.

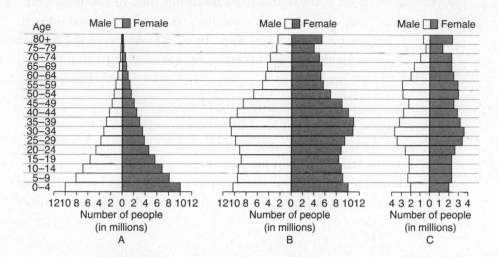

Which of the following statements would be TRUE?

I. A is experiencing rapid growth, B is experiencing slow growth, and C is experiencing a decline in growth

II. A is experiencing slow growth, B is experiencing rapid growth, and C is experiencing a decline in growth

III. A is experiencing a decline in growth, B is experiencing a rapid growth, and C is experiencing a slow growth

IV. A would be typical for Nigeria, B would be typical for Germany, and C would be typical for the United States

V. A would be typical for the United States, B would be typical for Nigeria, and C would be typical for Germany

(A) I

(B) I and III

(C) I and V

(D) II and IV

(E) III and IV

37. The most populous countries in the world are India, China, and

(A) Russia

(B) United States

(C) Mexico

(D) Indonesia

(E) Pakistan

38. The world's current population of approximately 7 billion people with a growth rate of about 1.3% represents a doubling time of about 54 years. At this rate, according to current projections, the global population will likely reach between 7.5 and 10.5 billion by 2050. Which of the following would NOT be a viable strategy for ensuring adequate nutrition for a population of that size?

    (A) Where feasible, switch to drip irrigation
    (B) Increase the efficiency of food distribution networks
    (C) Produce more meat products while decreasing the amount of land devoted to grain and crop production
    (D) Develop new hybrid crops that are more drought and pest resistant
    (E) Expand the use of non-toxic and environmentally-friendly methods of pest control

39. Worldwide human population growth within the last 100 years is primarily due to

    (A) increase in the birthrate
    (B) decrease in the death rate
    (C) a lower infant mortality rate
    (D) immigration
    (E) a desire to have more children

40. Which of the following was NOT a major component of the First Green Revolution?

    (A) Genetic engineering of food crops
    (B) Widespread use of synthetic nitrogen-based fertilizers
    (C) Widespread use of pesticides
    (D) Development of major modern irrigation projects
    (E) Development of new crop varieties through selective breeding technology

41. Which type of insecticide developed during the 19th century affects the nervous system of an insect by disrupting the enzyme that regulates acetylcholine, a neurotransmitter? They usually are not persistent in the environment. Common examples include parathion and malathion.

    (A) Carbamates
    (B) Pyrethroids
    (C) Organophosphates
    (D) DDT
    (E) Organochlorines

42. Which of the following is NOT a principle of integrated pest management (IPM)?

   (A) Establish acceptable pest thresholds
   (B) Selecting crop varieties best for local growing conditions
   (C) Visual inspection and using insect and spore traps to monitor pest levels while keeping accurate records
   (D) Using broad, non-specific insecticides to reduce or eliminate all pest populations before they have a chance to get out of control
   (E) Using mechanical methods to reduce pest populations, which include handpicking, erecting insect barriers, using traps, vacuuming, and tillage to disrupt breeding.

43. The type of timber-harvesting method shown below is known as

   (A) seed tree harvesting
   (B) group selection
   (C) clear-cutting
   (D) single tree selection
   (E) shelterwood cutting

44. Which of the choices below are effective method(s) of remediating the effects of desertification?

   I. Reforestation
   II. Fixating the soil through the use of shelter belts, woodlots, and windbreaks or sand fences
   III. Planting legumes
   IV. Enabling native sprouting tree growth through selective pruning of shrub growth
   V. Planting market gardens to keep soil from blowing away

   (A) I and II
   (B) I, III, IV
   (C) II, IV, V
   (D) I, III, IV, V
   (E) I, II, III, IV

45. Which of the following is a characteristic of urban sprawl?

    (A) Multi-use zoning
    (B) High-density zoning
    (C) Car-independent zoning
    (D) Job sprawl
    (E) High unemployment

46. During which U.S. presidency did the first federal highway system develop?

    (A) Abraham Lincoln (1860–1865)
    (B) Franklin Roosevelt (1933–1945)
    (C) Harry Truman (1945–1953)
    (D) Dwight Eisenhower (1953–1961)
    (E) Ronald Reagan (1981–1989)

47. Which of the following lists land use in order from the smallest collective area on Earth to the largest?

    (A) Agricultural < forests < cities
    (B) Forests < cities < agricultural
    (C) Cities < forests < agricultural
    (D) Forests < agricultural < cities
    (E) Agricultural < cities < forests

48. Which agency listed below is responsible for the management of public lands?

    (A) Bureau of Land Management
    (B) Environmental Protection Agency
    (C) National Park Service
    (D) Fish and Wildlife Service
    (E) Department of Agriculture

49. Rainwater runoff from US Route 7 in South Burlington, Vermont polluted local streams that fed into Lake Champlain, a source of drinking water. The city of South Burlington and the Vermont Agency of Natural Resources constructed a storm water treatment system along Bartlett Brook to control the pollution. The project also included the cleanup and widening of a stream channel and creation of a new wetland in the area from an area that had once been occupied by an industrial factory. This is an example of environmental

    (A) preservation
    (B) remediation
    (C) mitigation
    (D) restoration
    (E) all of the above

50. The type of mining that produces 85% of minerals (excluding petroleum and natural gas) in the United States, including 98% of metallic ores, is known as

    (A) surface mining
    (B) subsurface mining
    (C) *in-situ* leaching
    (D) long wall mining
    (E) room and pillar mining

51. The type of fishing illustrated below is

    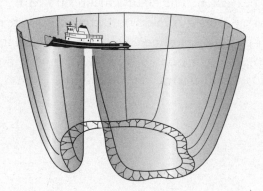

    (A) gill netting
    (B) long lining
    (C) reef netting
    (D) trolling
    (E) purse seining

52. The "Tragedy of the Commons" illustrates how

    (A) individual rationality results in group irrationality
    (B) the pursuit of short-term individual gain often leads to long-term collective costs
    (C) the cost of a product also includes the cost of utilization of common property
    (D) "Commons" are real resources that are owned collectively by people occupying the area
    (E) common people who work for a living are often victims of environmental exploitation

53. A store uses one-hundred 200-watt lightbulbs for 10 hours per day. How many kilowatt-hours of electrical energy are used in one year if the store is open 50 weeks per year, 6 days per week?

    (A) 50,000
    (B) 60,000
    (C) 6,000,000
    (D) 30,000,000
    (E) 60,000,000

54. The gas mileage of one model of the 2003 Hummer 2 was determined to be 8 miles per gallon. The 2012 Mitsubishi i-MiEV is an all-electric vehicle and when using a conversion factor to convert the fuel economy into miles per gallon of gasoline equivalent (MPGe), it is rated at 112 miles per gallon of gasoline. How much more efficient in fuel consumption is the all-electric vehicle over the Hummer 2?

    (A) 13%
    (B) 14%
    (C) 130%
    (D) 140%
    (E) 1300%

55. In an internal combustion automobile engine, only about 20% of the high-quality chemical energy available in the gasoline is converted to mechanical energy used to propel the car; the remaining 80% is degraded to low-quality heat that is released into the environment. In addition, about 50% of the mechanical energy produced is also degraded to low-quality heat energy through friction, so that 90% of the energy in gasoline is wasted and not used to move the car. This best illustrates which principle below?

    (A) First Law of Thermodynamics
    (B) Second Law of Thermodynamics
    (C) Third Law of Thermodynamics
    (D) Enthalpy
    (E) None of the above

56. Per capita, the United States uses about _____ amount of oil per day than China.

    (A) half the
    (B) the same
    (C) twice the
    (D) ten times the
    (E) one hundred times the

57. In the United States, on average, _____ energy is used to heat homes than is used to cool them.
    (A) 25% less
    (B) 50% less
    (C) 50% more
    (D) 100% more
    (E) 250% more

58. Which sector listed below currently uses the largest amount of energy in the world?

    (A) Commercial (i.e., stores, restaurants, offices, etc.)
    (B) Industrial (i.e., manufacturing plants)
    (C) Residential (i.e., homes, apartments, etc.)
    (D) Transportation (i.e., cars, planes, etc.)
    (E) All of the above in roughly equal proportions

59. The estimated natural gas reserves in the United States are estimated at 2,170 trillion cubic feet (Tcf). The rate of consumption of natural gas in the United States is approximately 62 billion cubic feet per day. Approximately how long will the United States natural gas reserves last at the current rate of consumption?

    (A) 10 years
    (B) 25 years
    (C) 50 years
    (D) 100 years
    (E) 200 years

60. Which of the following has the largest potential negative environmental impact on global warming?

    (A) Coal
    (B) Oil
    (C) Natural gas
    (D) Solar
    (E) Nuclear

61. Which country listed below obtains the greatest percentage of its electricity from nuclear power?

    (A) United States
    (B) Russia
    (C) China
    (D) Japan
    (E) France

62. Examine the diagram of a nuclear power plant below

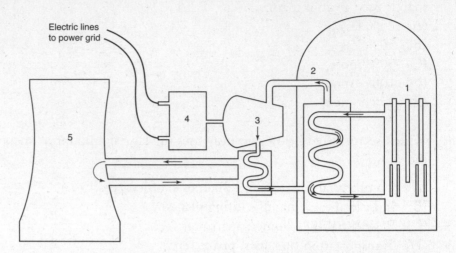

What is the name and function of the part labeled "5"?

(A) Heat exchanger—transfers heat from the fission reaction to water, which then turns to steam capable of powering steam turbines

(B) Generator—converts mechanical energy from the steam turbine into electrical energy

(C) Turbine—steam under very high pressure and temperature strikes blades or vanes mounted on a wheel enabling the large wheel to turn

(D) Moderator—the container holding a material, such as heavy water or graphite, used for slowing down neutrons in the cores of nuclear reactors so that they have a higher probability of inducing nuclear fission

(E) Cooling tower—cools or recycles water and releases unwanted heat to the environment

63. Which of the following current methods to produce 1 kWh of electricity is the MOST expensive?

(A) Nuclear
(B) Coal
(C) Natural gas
(D) Wind
(E) Hydroelectric

64. Which of the following is the LEAST likely environmental consequence of building hydroelectric dams?

(A) Dams change water temperatures downstream
(B) Dams increase the risk of predation
(C) Dams increase silting downstream
(D) Dams decrease the oxygen content of water downstream
(E) Dams decrease the flow of water downstream

65. Which government agency administers and regulates CAFE standards?

    (A) Environmental Protection Agency (EPA)
    (B) National Highway Traffic Safety Administration (NHTSA)
    (C) National Science Foundation (NSF)
    (D) National Transportation Safety Board (NTSB)
    (E) Office of Weights and Measures (OWM)

66. A 42 watt CFL (compact fluorescent light) bulb produces 2,600 lumens of light, lasts approximately 10,000 hours, and costs $13.00. A 150-watt incandescent lightbulb produces the same amount of light, lasts only 1,000 hours and costs $2.00. If electricity runs $0.10 per kWh, how much will the CFL save over the lifetime of the bulb compared to using the incandescent bulbs?

    (A) $7.00
    (B) $13.00
    (C) $42.00
    (D) $108.00
    (E) $115.00

67. Which of the following RENEWABLE energy sources listed below is used the LEAST in the United States?

    (A) Solar
    (B) Biomass
    (C) Wind
    (D) Hydroelectric
    (E) Geothermal

68. Which of the following would be classified as a secondary air pollutant?

    (A) Carbon monoxide
    (B) Acid rain
    (C) Carbon dioxide
    (D) Sulfur dioxide
    (E) Lead

69. Which of the following would NOT be classified as a criteria pollutant?

    (A) Carbon monoxide
    (B) Sulfur dioxide
    (C) Particulates
    (D) Ozone
    (E) Carbon dioxide

70. 1,500 parts per billion would be the same concentration as _____ parts per million.

    (A) 0.15
    (B) 1.5
    (C) 15
    (D) 150
    (E) 1,500

71. Which one of the following criteria pollutants listed below contributes to the formation of both acid deposition AND photochemical smog?

    (A) CO
    (B) NO_2
    (C) O_3
    (D) Particulates
    (E) SO_2

72. During summer months, Los Angeles, California often experiences a temperature inversion. During this event

    (A) Cold air that is heavier lies above a mass of warmer air that wants to rise, trapping particulates close to the ground.
    (B) Stable warm air lies above a mass of colder air, trapping air particulates close to the ground.
    (C) There is no colder air mass to provide a difference in temperature; therefore, the air stalls and traps particulates close to the ground.
    (D) Warmer air is trapped between two cold fronts, trapping the particulates within the warm air mass.
    (E) Warm air rises while colder air sinks. Particulates within the cold air are then brought closer to the ground and their concentration increases, producing smog.

73. Given the following pollutants

    I.   CFCs
    II. VOCs
    III. Radon
    IV. PANs
    V.   Asbestos

    Which are NOT naturally occurring indoor air pollutants?

    (A) I and II
    (B) I and IV
    (C) I, II, and IV
    (D) III, V
    (E) I, III, IV, V

74. Which of the following were the criteria used for acceptable exposure to specific air pollutants as addressed in the National Ambient Air Quality Standards set forth by the Clean Air Act?

    I.   The amount of time exposed to the air pollutant
    II.   The concentration of the air pollutant
    III. The toxicity of the air pollutant
    IV.   The LD_{50} of the air pollutant
    V.   The number of people that would be at risk if exposed to the air pollutant

    (A) I and II
    (B) II and III
    (C) I, II, and III
    (D) I, III, and IV
    (E) III, IV, and V

75. Which of the following statement(s) regarding noise pollution is/are TRUE?

    I.   Noise pollution is a significant occupational hazard and is regulated and enforced by the Environmental Protection Agency.
    II.   Sound measured at 40 decibels is twice as loud as sound measured at 20 decibels.
    III. Noise pollution is not a serious threat in the oceans as organisms that live underwater do not hear.
    IV.   The Office of Noise Abatement and Control of the Environmental Protection Agency (EPA) is currently charged with regulating and overseeing noise-abatement activities.

    (A) I and II
    (B) I and III
    (C) II, III, and IV
    (D) I, III, and IV
    (E) All statements are false

76. Which of the following forms of water pollution affects the largest number of people worldwide?

    (A) Brackish water resulting from flooding
    (B) Water that is not properly processed through sewage treatment plants
    (C) Water containing pollutants from industrial factories
    (D) Waterborne pathogens and diseases
    (E) Groundwater contamination resulting from illegal dumping of toxic materials

77. An APES class was doing a field study of effluent water entering a river through a large drainage pipe from a large building. Which of the following would be a logical conclusion based upon their observations?

    (A) The students observed green algae covering a small area of the river a mile away from the factory and concluded that the factory was dumping hot water into the river as algae grow profusely in warm water.

    (B) The students detected measurable amounts of coliform bacteria in the river near the drainage pipe and concluded that the building could be a meatpacking plant.

    (C) The students detected high concentrations of metal ions in the river bottom upstream of the factory and found nothing but anaerobic bacteria, fungi, and sludge worms living in the mud. They concluded that the building was probably a factory producing inorganic fertilizer and releasing some of the fertilizer into the water.

    (D) The students observed a large number of insect larvae in the water 2 miles from the drainage pipe and concluded that the water was polluted by a factory that was releasing large amounts of food wastes into the water.

    (E) The students measured the turbidity of the water exiting the drainage pipe and found it to be extremely high and concluded that the building was probably a factory that produced pharmaceuticals.

78. Groundwater contamination is addressed in all of the following EXCEPT the

    (A) Clean Water Act
    (B) Safe Drinking Water Act
    (C) Resource Conservation and Recovery Act
    (D) Superfund
    (E) Groundwater contamination is addressed in all of the above

Questions 79 and 80 refer to the following graph.

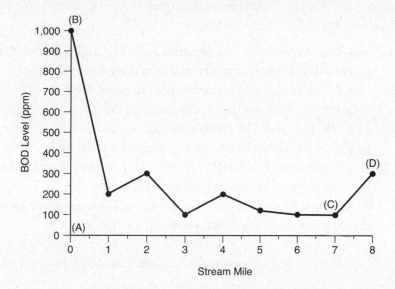

An APES class measured BOD along an 8-mile stretch of a stream. Their results are presented above.

79. BOD indicates

(A) the level of pollution in a sample of water
(B) the rate at which oxygen is being produced by aquatic plants
(C) the rate at which oxygen is being consumed by microorganisms living in the water
(D) the potential amount of oxygen that the water sample could contain given the pH, temperature, and volume of the sample
(E) the concentration of oxygen in a water sample that a living organism requires to survive

80. At which point on the graph would you expect to find the highest concentration of anaerobic bacteria?

(A) A
(B) B
(C) C
(D) D
(E) Anaerobic bacteria would not be found in a stream

81. Sewage treatment plants have three general processes that water undergoes. Which choice below describes the correct sequence of these three processes?

    (A) Separation of solids—disinfection with UV light, chlorine, or ozone—breakdown of organic material by bacteria
    (B) Separation of solids—breakdown of organic material by bacteria—disinfection with UV light, chlorine, or ozone
    (C) Disinfection with UV light, chlorine, or ozone—separation of solids—breakdown of organic material by bacteria
    (D) Disinfection with UV light, chlorine, or ozone—breakdown of organic material by bacteria—separation of solids
    (E) Breakdown of organic material by bacteria—separation of solids—disinfection with UV light, chlorine, or ozone

82. Most solid municipal wastes in the United States are disposed of by

    (A) taking the wastes far from land and dumping it into the sea
    (B) burning it at large municipal incinerators
    (C) burying it in giant pits
    (D) taking it to sanitary landfills
    (E) burning the wastes at the site (home, school, factory, etc.)

83. Three "environmentally friendly" methods of dealing with waste disposal are to reuse, reduce, and recycle. Which of the following choices below places these methods in order of increasing amounts of energy required for that particular process?

    (A) Reuse < reduce < recycle
    (B) Reuse < recycle < reduce
    (C) Recycle < reduce < reuse
    (D) Recycle < reuse < reduce
    (E) Reduce < reuse < recycle

84. A group of workers at a nuclear power plant received an acute exposure of radiation. This means

    (A) the workers received a high dosage of radiation over a long time period
    (B) the workers received a high dosage of radiation in a very short time period
    (C) the workers received a low dosage of radiation over a long time period
    (D) the workers received a low dosage of radiation over a short time period
    (E) the workers received a lethal dosage of radiation

85. A laboratory was testing the effectiveness of a new insecticide. The laboratory determined that the LD_{50} dosage level for rats was 150 milligrams per kilogram of body mass. Based on this information, which of the following statements would be most accurate?

    (A) Fifty out of 100 rats receiving 300 milligrams of the new insecticide per kilogram of body mass would die.

    (B) Fifty out of 100 rats receiving 150 milligrams of the new insecticide per kilogram of body mass would die.

    (C) Any amount more than 150 milligrams of the new insecticide per kilogram of body mass would be lethal to humans.

    (D) Out of the 100 rats tested, 50 of them would be perfectly normal with no ill-effects and 50 of them would eventually die from the new insecticide.

    (E) Fifty percent of any population of organism would die if given more than 150 milligrams of the new insecticide.

Questions 86–87 refer to the following dose-response curve.

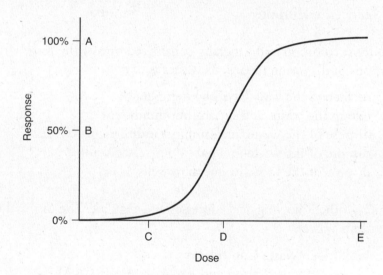

86. Which point represents the LD_{50}?

    (A) A
    (B) B
    (C) C
    (D) D
    (E) E

87. Which point represents the threshold?

    (A) A
    (B) B
    (C) C
    (D) D
    (E) E

88. Of the indoor pollutants that are listed below, which one has the highest total health care costs, affects the most people, and causes the most deaths?

    (A) asbestos
    (B) exposure to radon
    (C) exposure to smog
    (D) cigarette smoke
    (E) ozone

89. DDT, a synthetic agricultural pesticide, was very effective in reducing incidents of malaria but caused shell thinning in certain species of birds particularly birds of prey such as the bald eagle, brown pelican, peregrine falcon, and osprey. DDT is classified as what type of hazardous waste?

    (A) neurotoxin
    (B) carcinogen
    (C) teratogen
    (D) mutagen
    (E) endocrine disrupter

90. The most common and generally considered one of the most effective methods of disposing hazardous wastes is

    (A) incinerate the waste at a high temperature
    (B) entrap the waste in a surface impoundment
    (C) dispose of the waste in a sanitary landfill
    (D) dispose of the waste in a hazardous waste landfill
    (E) dispose of the waste in injection wells

91. Which of the following acts addresses and regulates the disposal of hazardous wastes?

    (A) Hazardous Waste Control Act
    (B) Resource Conservation and Recovery Act (RCRA)
    (C) Toxic Substance Control Act
    (D) Comprehensive Response, Compensation, and Liability Act (CERCLA) – Superfund
    (E) Pollution Prevention Act (PPA)

92. If the price of a barrel of oil from the Middle East were to suddenly double, it would

    (A) make it economically feasible to explore for new sources of oil within the United States; thereby increasing the supply, which would decrease the price for oil
    (B) make it economically unfeasible to explore for new sources of oil within the United States; thereby decreasing the supply, which would cause a decrease in demand for oil
    (C) make it economically feasible to use alternative sources of energy within the United States, thereby increasing the demand for oil and consequently increasing the supply of oil
    (D) make it economically unfeasible to use alternative sources of energy within the United States, thereby increasing the demand for oil, which would result in a decrease in the supply of gasoline
    (E) have no effect on exploring for new sources of oil within the United States, therefore the supply of oil in the United States would decrease and the price would then also decrease

93. Currently, the gas contributing most to stratospheric ozone breakdown is

    (A) Chlorofluorocarbons (CFCs) used in air conditioners
    (B) Carbon tetrachloride (CCl_4) used as an industrial solvent and in fire extinguishers
    (C) Methyl chloroform (CH_3CCl_3) used as a solvent for adhesives, for metal degreasing, and in textile processing
    (D) Methyl bromide (CH_3Br) used as a fumigant
    (E) Nitrous oxide (N_2O) released from livestock manure, sewage treatment, and industrial processes

94.	An increase in greenhouse gases increases tropospheric temperatures. If ice melts due to global warming, it affects the Earth's albedo. Which of the following statements would be true regarding the effect of melting ice and its effect on the Earth's albedo?

(A)	Melting ice would increase the Earth's albedo resulting in less heat absorbed, which in turn would result in more ice melting in what would be known as a positive feedback loop.

(B)	Melting ice would decrease the Earth's albedo resulting in more heat being absorbed, resulting in more ice melting in what would be known as a positive feedback loop.

(C)	Melting ice would increase the Earth's albedo resulting in less heat absorbed, which in turn would result in more ice melting in what would be known as a negative feedback loop.

(D)	Melting ice would increase the Earth's albedo resulting in more heat absorbed, which in turn would result in less ice melting in what would be known as a negative feedback loop.

(E)	Melting ice would decrease the Earth's albedo resulting in less heat being absorbed, resulting in less ice melting in what would be known as a negative feedback loop.

95.	Which of the following was instrumental in significantly reducing the production of chlorofluorocarbons?

(A)	Clean Air Act
(B)	Montreal Protocol
(C)	Kyoto Protocol
(D)	National Environmental Policy Act
(E)	Toxic Substances Control Act

96.	The majority of the natural greenhouse effect is due to which gas listed below?

(A)	Carbon dioxide (CO_2)
(B)	Chlorofluorocarbons (CFCs)
(C)	Water vapor (H_2O)
(D)	Nitrous oxide (N_2O)
(E)	Methane (CH_4)

97.	Which of the following factors are primary reasons why sea level is rising?

I.	Thermal expansion of the oceans
II.	More frequent rain and flooding
III.	Melting of sea ice
IV.	Melting of land-based ice

(A)	I and III
(B)	II and IV
(C)	III and IV
(D)	I and IV
(E)	I, II, III, and IV

98. Which countries listed below have recently either stated they would not enter a second round of carbon cuts under a new Kyoto Protocol due to go into effect in 2012 or never signed the original Kyoto Protocol of 1997?

    I.   Russia
    II. Canada
    III. United States
    IV. Japan
    V.   France

    (A) I, III
    (B) II, IV, and V
    (C) III, IV, and V
    (D) I, III, and IV
    (E) I, II, III, IV, and V

99. The greatest threat to a species survival is generally considered

    (A) lack of food
    (B) pollution
    (C) poaching
    (D) invasive species
    (E) loss of habitat

100. Which international agreement has been helpful in protecting endangered animals and plants by listing those species and products whose international trade is controlled?

    (A) Endangered Species Act
    (B) CITES
    (C) Fish and Wildlife Act
    (D) National Environmental Policy Act
    (E) International Treaty on Endangered and Threatened Species

DIAGNOSTIC TEST ANSWERS AND EXPLANATIONS

Directions: On the line to the left of each number, place a "C" if you got the question Correct, place an "X" if you answered the question incorrectly, and leave the line blank if you did not answer the question.

I. Earth Systems and Resources (10–15%)

_____ 1. **(D)** The age of the Earth is approximately 4.54 billion years. This age is based on evidence from radiometric age dating of meteorite material and is consistent with the ages of the oldest-known terrestrial and lunar samples. *Earth Science Concepts: Geologic Time Scale*

_____ 2. **(D)** The Richter scale is logarithmic, meaning that whole-number jumps indicate a tenfold increase (or decrease). In this case, the increase is in wave amplitude. That is, the wave amplitude in a level 6 earthquake is 10 times greater than in a level 5 earthquake, and the amplitude increases 100 times between a level 7 earthquake and a level 9 earthquake. However, the amount of energy released increases (or decreases) 31.7 times between whole number Richter values. *Earth Science Concepts: Earthquakes*

_____ 3. **(A)** Transform plate boundaries are locations where two plates slide past one another. The fracture zone that forms a transform plate boundary is known as a transform fault. Most transform faults are found in the ocean basin. Transform faults are locations of recurring earthquake activity and faulting. The earthquakes are usually shallow because they occur within and between plates that are not involved in subduction. Volcanic activity is normally not present because the typical magma sources of an upwelling convection current or a melting subducting plate are not present. *Earth Science Concepts: Plate Tectonics*

_____ 4. **(B)** The tilt of the Earth is responsible for the yearly cycle of seasonal-weather changes. Two factors change during the course of a year to give seasonal variations in temperatures: (1) the angle at which sunlight enters the atmosphere and strikes the ground and (2) the number of daylight hours. If sunlight enters the atmosphere at a direct angle, as it does in the Northern Hemisphere during the summer, it will go through less of the atmosphere. Sunlight entering the atmosphere at the Southern Hemisphere at this time has to pass through more of the atmosphere. Over time, the angle of Earth's tilt varies between 22.2 and 24.5 degrees. A larger tilt means that the summer hemisphere will receive more solar radiation, while the winter hemisphere will receive less. The Sun appears lower in the sky during winter, when the hemisphere affected is tilted away from it. This decreases the number of hours for the sun to heat the ground and lower temperatures result. *The Atmosphere: Climate*

_____ 5. **(D)** Products of cellular respiration are CO_2 and H_2O ($C_6H_{12}O_6 + 6O_2 \rightarrow 6H_2O + 6CO_2$). In addition to water and carbon dioxide being produced, livestock also produce methane (CH_4) through a process known as enteric fermentation. CO_2, H_2O, and CH_4 are all greenhouse gases. *The Atmosphere: Composition*

_____ 6. **(E)** The overall movement of surface air in the Ferrel cell occurs between the 30th to the 60th north and south latitudes. Ecuador and the majority of the United States including mid-western states occur in the Hadley cell, which lies between 30° north and 30° south latitudes while Antarctica and Greenland occurs in areas influenced by the polar cells. *The Atmosphere: Atmospheric Circulation*

_____ 7. **(C)** El Niño episodes are defined as sustained warming of the central and eastern tropical Pacific Ocean and results in a decrease in the strength of the Pacific trade winds. A deeper thermocline (often observed during El Niño years) limits the amount of nutrients brought to shallower depths by upwelling processes, greatly impacting the year's fish crop. *The Atmosphere: ENSO*

_____ 8. **(D)** Upwelling is a phenomenon that involves wind-driven motion of dense, cooler, and usually nutrient-rich water toward the surface, replacing the warmer, usually nutrient-depleted surface water. The increased availability in upwelling regions results in high levels of primary productivity and thus fishery production. Approximately 25% of the total global marine fish catches come from five upwellings that occupy only 5% of the total ocean area. Upwellings that are driven by coastal currents or diverging open ocean have the greatest impact on nutrient-enriched waters and global fishery yields. *The Atmosphere: ENSO*

_____ 9. **(C)** See diagram below. *Global Water Resources and Use: Freshwater/Saltwater*

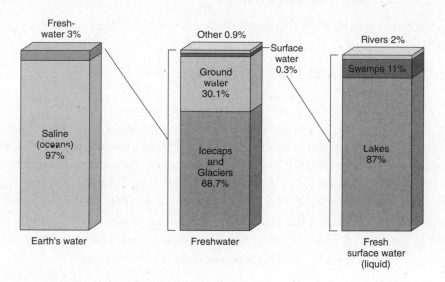

_____ 10. **(A)** The agricultural sector is the most dominant user of water, accounting for up to 70% of the world's freshwater usage. Agricultural water use is especially heavy in the developing world, with some countries in Asia, Africa, and South America using more than 79% of their total freshwater supply for agricultural purposes. In fact, 85–90% of all the freshwater that is used in Africa and Asia is for agriculture. Farmers in developing countries use up to twice as much water than their counterparts in industrialized countries, and often obtain crop yields that are three times less. *Global Water Resources and Use: Agricultural*

_____ 11. **(C)** Currently, 7 billion people use nearly 30% of the world's total accessible renewal supply of water. By 2025, with an estimated 8 billion people, that value is estimated to reach 70%. *Global Water Resources and Use: Global Problems*

_____ 12. **(B)** The B horizon is commonly referred to as "subsoil," and consists of mineral layers that may contain concentrations of clay or minerals such as iron or aluminum oxides or organic material that got there by leaching. Accordingly, this layer is also known as the "zone of accumulation." Plant roots penetrate through this layer, but it has very little humus. It is usually brownish or red because of the clay and iron oxides washed down from the A horizon. *Soil and Soil Dynamics: Main Soil Types*

_____ 13. **(A)** The A horizon is the top layer of the soil horizons and is often referred to as "topsoil." This layer has a layer of dark decomposed organic materials, which is called "humus." The A horizon is the zone in which most biological activity occurs. Soil organisms such as earthworms, arthropods, nematodes, fungi, and many species of bacteria are concentrated here, often in close association with plant roots. *Soil and Soil Dynamics: Main Soil Types*

RECAP FOR EARTH SYSTEMS AND RESOURCES

_____ Total number correct for this section

_____ Percent correct for this section
(number correct for this section / 13) × 100%

_____ Number wrong for this section (questions you thought you knew).
Do NOT count answers left blank.

_____ Number left blank for this section (questions you did not know)

II. The Living World (10–15%)

_____ 14. **(E)** As a general rule, diversity and ecosystem productivity increase with the amount of solar energy available to the system. The stable tropical rain forest environment promotes diversity by allowing plants and animals to interact all year round without needing to develop protection against severe weather conditions. In addition, because the sun shines all year long providing plants with the energy to manufacture food through photosynthesis, there is no seasonal food shortage in the ecosystem. Over the course of millions of years, with abundant food, tropical rain forest species have adapted to take full advantages of all the available niches. *Ecosystem Structure: Biological Populations and Communities*

_____ 15. **(C)** Commensalism is a class of relationship between two organisms where one organism benefits but the other is unaffected (there is no harm or benefit). Cows eat grass and hay almost exclusively. Cows, and other ruminant animals, have a special type of stomach called a rumen, which is home to billions of microbes that consume grass and hay. These bacteria, fungi, and protists provide nutrients that the cow can digest. Without these microbes, the cow would die. The ungulates benefit from the cellulase produced by the bacteria, which facilitates digestion; the bacteria benefit from having a stable supply of nutrients in the host environment. *Ecosystem Structure: Interactions Among Species*

_____ 16. **(B)** A keystone species is a species that has a disproportionately large effect on its environment relative to its abundance. Such species play a critical role in maintaining the structure of an ecological community, affecting many other organisms in an ecosystem, and helping to determine the types and numbers of various other species in the community. Similarly, an ecosystem may experience a dramatic shift if a keystone species is removed, even though that species was a small part of the ecosystem by measures of biomass or productivity. *Ecosystem Structure: Keystone Species*

_____ 17. (**D**) Territoriality is behavior by a single animal, mating pair, or group of animals of the same species of an area occupied and often vigorously defended against intruders, especially those of the same species. All of the other choices are interspecific interactions. *Ecosystem Structure: Interactions Among Species*

_____ 18. (**B**) See diagram below *Ecosystem Structure: Aquatic Biomes*

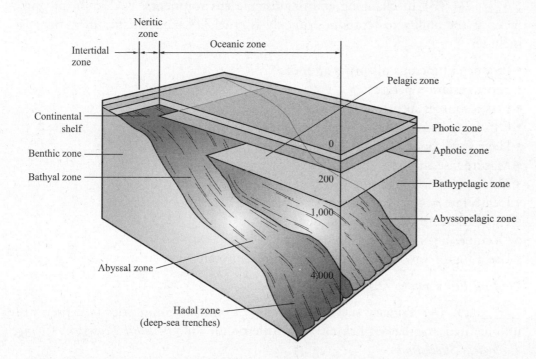

_____ 19. (**D**) Photosynthesis (plants): $6CO_2 + 6H_2O$ + energy (sunlight) $\rightarrow$ $C_6H_{12}O_6 + 6O_2$

Cellular Respiration (animals): $C_6H_{12}O_6 + 6O_2 \rightarrow 6CO_2 + 6H_2O$ + energy *The Living World: Photosynthesis and Cellular Respiration*

_____ 20. (**C**) Producers are plants. Primary consumers eat plants. Secondary consumers eat primary consumers. Tertiary consumers eat secondary consumers. Two possible examples are provided below. *Energy Flow: Food Webs*

Secondary Consumer: plant (producer) > rabbit (primary consumer) > owl (secondary consumer)

Tertiary Consumer: plant (producer) > grasshopper (primary consumer) > songbird (secondary consumer) > owl (tertiary consumer)

_____ 21. (**B**) A general rule of thumb is that approximately 10% of the energy entering one level of a trophic level passes onto the next higher level. Leaf (10 calories) > grasshopper (1 calories) > songbird (0.1 calorie) > owl (0.001 calorie). This loss of energy is consistent with the Second Law of Thermodynamics, which states that no energy conversion is 100% efficient. The loss occurs as energy is trapped in indigestible parts and energy (heat) lost in metabolic processes. *Energy Flow: Trophic Levels*

_____ 22. **(A)** To find the percentage of moths recaptured, divide the number of moths recaptured by the number of moths released. *Ecosystem Diversity: Natural Selection*

_____ 23. **(E)** Humankind benefits from a multitude of resources and processes that are supplied by natural ecosystems. Collectively, these benefits are known as ecosystem services and include products like clean drinking water and processes such as the decomposition of wastes. *Ecosystem Diversity: Ecosystem Services*

_____ 24. **(B)** In changing and/or unstable environments, *r*-selection predominates as the ability to reproduce quickly is crucial. Characteristics of *r*-strategists include:

• Produce numerous offspring at once
• Short gestation period
• Less resources spent per offspring
• Offspring hatch or born capable of surviving on their own
• Have small bodies
• Mature fast and have short life span
• Able to disperse offspring widely
• Death rate generally not correlated with density of population
• Population size fluctuates and not stable
• Occupies a generalist role in ecology
• Use a high reproductive rate to high mortality rate

Natural Ecosystem Change: Climate Shifts

_____ 25. **(A)** Primary succession occurs in biotic communities in a previously uninhabited and barren habitat with little or no soil. *Natural Ecosystem Change: Ecological Succession*

_____ 26. **(E)** Primary succession occurs in biotic communities in a previously uninhabited and barren habitat with little or no soil. Secondary succession is a process started by an event (e.g., forest fire, harvesting, hurricane) that reduces an already established ecosystem (e.g., a forest or a wheat field) to a smaller population of species, and as such, secondary succession occurs on preexisting soil whereas primary succession usually occurs in a place lacking soil. Secondary succession is usually faster than primary succession as soil is already present, so there is no need for pioneer species and seeds, roots and underground vegetative organs of plants may still survive in the soil. In unstable or unpredictable environments, *r*-selection predominates as the ability to reproduce quickly is crucial. There is little advantage in adaptations that permit successful competition with other organisms because the environment is likely to change again. Traits that are thought to be characteristic of *r*-selection include: high fecundity, small body size, early maturity onset, short generation time, and the ability to disperse offspring widely. Examples include: bacteria and diatoms, insects, and weeds. In stable or predictable environments, *K*-selection predominates as the ability to compete successfully for limited resources is crucial. Traits that are thought to be characteristic of *K*-selection include: large body size, long life expectancy, and the production of fewer offspring, which require extensive parental care until they mature. Organisms with *K*-selected traits include large organisms such as elephants, trees, humans, and whales. *Natural Ecosystem Change: Ecological Succession*

_____27. **(D)** The conversion of ammonium (NH_4^+) to nitrate (NO_3^-) is performed primarily by soil-living bacteria and other nitrifying bacteria. In the primary stage of nitrification, the oxidation of ammonium (NH_4^+) is performed by bacteria such as the *Nitrosomonas* species, which converts ammonia (NH_3) to nitrites (NO_2^-). Other bacterial species, such as *Nitrobacter*, are responsible for the oxidation of the nitrites (NO_2^-) into nitrates (NO_3^-). It is important for the nitrites to be converted to nitrates because accumulated nitrites are toxic to plant life. *Natural Biogeochemical Cycles: Nitrogen Cycle*

RECAP FOR THE LIVING WORLD

_____ Total number correct for this section

_____ Percent correct for this section
(number correct for this section / 14) $\times$ 100%

_____ Number wrong for this section (questions you thought you knew).
Do NOT count answers left blank.

_____ Number left blank for this section (questions you did not know)

III. Population (10–15%)

_____ 28. **(E)** A density-independent factor is one where the effect of the factor on the size of the population is independent of and does NOT depend upon the original density or size of the population; e.g., a severe storm and flood coming through an area can just as easily wipe out a large population as a small one. *Population Biology Concepts: Population Ecology*

_____ 29. **(C)** Having extra food in the water between 0 and 15 hours will not affect the population density during this time period. Assuming ideal conditions with more than enough food available during this time period, the paramecium are already reproducing at their maximum capacity. However, if there is extra food available at all times during the 30 hours of the experiment, then the time period of exponential growth is extended, increasing the population density until a point is reached that another limiting factor such as oxygen content in the water begins to affect the density of paramecium that can survive; at which point a new carrying capacity is established. *Population Biology Concepts: Carrying Capacity*

_____ 30. **(C)** In stable or predictable environments, *K*-selection predominates as the ability to compete successfully for limited resources is crucial and populations of *K*-selected organisms typically are very constant and close to the maximum that the environment can bear (unlike *r*-selected populations, where population sizes can change much more rapidly). Traits that are thought to be characteristic of *K*-selection include: large body size, long life expectancy, and the production of fewer offspring, which require extensive parental care until they mature. Examples of organisms with *K*-selected traits include large organisms such as elephants, trees, humans, and whales. *Population Biology Concepts: Reproductive Strategies*

_____ 31. **(A)** In clumped distribution, the distance between neighboring individuals is minimized and is found in environments that are characterized by patchy resources. Clumped distribution is the most common type of dispersion found in nature because animals need certain resources to survive, and when these resources become scarce during certain parts of the year, animals tend to "clump" together around these crucial resources. Individuals might be clustered together in an area due to social factors such as herds or family groups. Organisms that usually serve as prey form clumped distributions in areas where they can hide and detect predators easily. Other causes of clumped distributions are the inability of offspring to independently move from their habitat. Clumped distribution in species also acts as a mechanism against predation as well as an efficient mechanism to trap or corner prey. *Population Biology Concepts: Survivorship*

_____ 32. (**D**) Choice D is known as a Type II survivorship curve and is typical of species in which a fairly constant mortality rate is experienced regardless of age. Type I survivorship curves (E) are characterized by high survival in early and middle life, followed by a rapid decline in survivorship in later life. Humans are one of the species that show this pattern of survivorship. In Type III curves (B), the greatest mortality is experienced early on in life, with relatively low rates of death for those surviving this bottleneck. This type of curve is characteristic of species that produce a large number of offspring. One example of a species that follows this type of survivorship curve is the frog. *Population Biology Concepts: Survivorship*

_____ _33. (**B**) Stage 2 sees a rise in population caused by a decline in the death rate while the birth rate remains high, or perhaps even rises slightly. The decline in the death rate is due initially to two factors: (1) improvements in food supply brought about by higher yields as agricultural practices were improved in the Agricultural Revolution of the 18th century and (2) significant improvements in public health that reduced mortality, particularly in childhood. As a consequence, there is an increasingly rapid rise in population growth (a "population explosion") as the gap between deaths and births grows wider; not due to an increase in fertility (or birthrates) but to a decline in deaths, which results in the increasing survival of children. Hence, the age structure of the population becomes increasingly youthful. This trend is intensified as these increasing numbers of children enter into reproduction while maintaining the high fertility rate of their parents. *Human Population: Demographic Transition*

_____ 34. (**C**) The Rule-of-70 provides a simple way to calculate the approximate number of years it takes for the level of a variable growing at a constant rate to double. This rule states that the approximate number of years n for a variable growing at the constant growth rate of R percent, to double is

$$n = \frac{70}{r} = \frac{70}{5} = 14$$

Human Population: Doubling Times

_____ 35. (**D**) Prior to the First or Neolithic Agricultural Revolution, humans were hunters-gatherers. During the First Agricultural Revolution, humans began to domesticate animals and raise crops, which allowed a steady food supply. With a steady food supply, the population increased. However, without the current advancements in medical care, sanitation, food preservation, etc., both birth and death rates were high. Today's birth and death rates in many areas of the world are significantly lower than 10,000 years ago due to these advancements along with women joining the workforce, family planning, etc. *Human Population: Historical Population Sizes*

_____ 36. (**A**) Four general types of age-structure diagrams have been identified by the fertility and mortality rates of a country.

(1) Stable pyramid—A population pyramid showing an unchanging pattern of fertility and mortality.

(2) Stationary pyramid—A population pyramid typical of countries with low fertility and low mortality, very similar to a constrictive pyramid.

(3) Expansive pyramid—A population pyramid showing a broad base, indicating a high proportion of children, a rapid rate of population growth, and a low proportion of older people. This wide base indicates a large number of children. A steady upward narrowing shows that more people die at each higher age band. This type of pyramid indicates a population in which there is a high birthrate, a high death rate, and a short life expectancy. This is the typical pattern for less economically developed countries, due to little access to and incentive to use birth control, negative environmental factors (for example, lack of clean water), and poor access to health care. This age structure diagram would be typical for a country such as Nigeria.

(4) Constrictive pyramid—A population pyramid showing lower numbers or percentages of younger people. The country will have a greying population, which means that people are generally older, as the country has long life expectancy, a low death rate, but also a low birthrate. This pyramid is often a typical pattern for a very developed country, a high over-all education, and easy access and incentive to use birth control, good health care, and few or no negative environmental factors. This age structure diagram would be typical for a country such as Germany and to a lesser extent the United States. *Human Population: Age-Structure Diagram*

_____ 37. (**B**) *Human Population: Population Size*

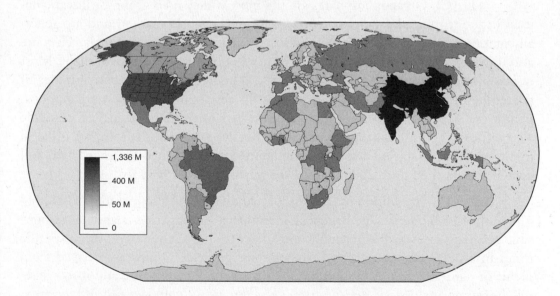

_____ 38. (**C**) It takes up to 16 pounds of grain to produce just 1 pound of meat and raising animals for food (including land used for grazing and land used to grow feed crops) uses 30% of the Earth's landmass. Raising animals for food is grossly inefficient, because while animals eat large quantities of grain, soybeans, oats, and corn, they only produce comparatively small amounts of meat, dairy products, or eggs in return. That is why more than 70% of the grain and cereals that are grown in the United States are fed to farmed animals. In addition, it takes more than 11 times as much fossil fuel to produce 1 calorie from animal protein as it does to produce 1 calorie from plant protein. *Human Population: Impacts of Population Growth*

_____ 39. (**B**) The combination of decreasing death rate due to the march of progress in sanitation and medicine, coupled later with the decrease in birthrate due to changes in the economies, has led to a significant change in the population growth in the developed world. *Human Population: Human Population Dynamics*

RECAP FOR POPULATION

_____ Total number correct for this section

_____ Percent correct for this section
(number correct for this section / 12) × 100%

_____ Number wrong for this section (questions you thought you knew).
Do NOT count answers left blank.

_____ Number left blank for this section (questions you did not know)

IV. Land and Water Use (10–15%)

_____ 40. (**A**) Genetically modified crops are crops that have been genetically modified in a lab setting. The majority of processed foods in the United States contain food products that have been genetically modified. The DNA of the organism is spliced and swapped out for "better" genes, which are theoretically designed to create stronger, more resistant, and more productive plants. *Agriculture: Feeding a Growing Population: Green Revolution*

_____ 41. (**C**) Organophosphates are the most widely used group of insecticides used in the world today (approximately 40% of the world market) and are generally among the most acutely toxic of all pesticides to vertebrate animals. They are also unstable and therefore break down relatively quickly in the environment. Most organophosphate pesticides are insecticides, although there are also a number of related herbicide and fungicide compounds. In developing countries, organophosphates are widely used because they are less expensive than the newer pesticide alternatives. Their use is banned or restricted in 23 countries, their import is illegal in a total of 50 countries, and they were banned in the United States in 2000 and have not been used since 2003. *Controlling Pests: Types of Pesticides*

_____ 42. (**D**) In addition to (A), (B), (C), and (E), IPM also includes (1) biological controls, which promote beneficial insects that eat target pests and (2) responsible pesticide use, which is generally only used as required and often only at specific times in a pest's life cycle. Many of the newer pesticide groups are derived from plants or naturally occurring substances (e.g., nicotine, pyrethrum, and insect juvenile hormone analogues). *Agriculture: Controlling Pests: Integrated Pest Management*

_____ 43. (**E**) In a shelterwood cut, mature trees are removed in two or three harvests over a period of 10 to 15 years. This method allows regeneration of medium to low shade-tolerant species because a "shelter" is left to protect them. *Forestry: Forest Management*

_____ 44. (**E**) Market gardening is growing of vegetables and flowers on suburban land of high value for the supply of nearby cities. Heavy fertilizing, regular watering, and the planting of successive crops are employed to obtain continuous returns from the acreage. *Rangelands: Desertification*

_____ 45. (**D**) Job sprawl is defined as low-density, geographically spread-out patterns of employment, where the majority of jobs in a given metropolitan area are located outside of the main city's central business district, and increasingly in the suburban periphery. It is often the result of urban disinvestment, the geographic freedom of employment location allowed by predominantly car-dependent commuting patterns of many American suburbs, and many companies' desire to locate in low-density areas that are often more affordable and offer potential for expansion. *Other Land Use: Urban Land Development: Suburban Sprawl*

_____ 46. (**D**) In 1919 when Dwight Eisenhower was a young officer in the U.S. Army, it took two months for a military convoy to go from Washington, D.C. to San Francisco. Later, when Eisenhower was a general during WWII, he had first-hand knowledge of the German autobahn. Through his first-hand knowledge and urging, the expanded Federal-Aid Highway Act of 1956 authorized a budget of $25 billion to build an extended federal highway system network of 41,000 miles (66,000 km) that included: a minimum of two lanes in each direction; lanes that were 12 feet (3.7 m) in width; a 10-foot (3-m) right paved shoulder and design speeds of 50–70 mph (80–113 km per hour). Today the network stretches 47,000 miles (75,000 km). *Other Land Use: Transportation Infrastructure: Federal Highway System*

_____ 47. (**C**) Cities occupy approximately 5% of the Earth's land surface; forests occupy approximately one-third of the Earth's surface, and land used for agricultural purposes occupies approximately half of all the Earth's land surface. *Other Land Use: Public and Federal Lands: Management*

_____ 48. (**A**) The Bureau of Land Management (BLM) is an agency within the U.S. Department of the Interior, which administers America's public lands, totaling approximately 253 million acres (1,020,000 sq. km), or one-eighth of the landmass of the United States. The BLM also manages 700 million acres (2,800,000 sq. km) of subsurface mineral rights underlying federal, state, and private lands. Most public lands are located in western states, including Alaska. The BLM's stated mission is to sustain the health, diversity, and productivity of the public lands for the use and enjoyment of present and future generations. *Other Land Use: Public and Federal Lands: Management*

_____ 49. (**C**) Environmental mitigation involves steps taken to avoid or minimize negative environmental impacts. Mitigation can include: avoiding the impact by not taking a certain action; minimizing impacts by limiting the degree or magnitude of the action; rectifying the impact by repairing or restoring the affected environment; reducing the impact by protective steps required with the action; and/or compensating for the impact by replacing or providing substitute resources. The example is not preservation as nothing was preserved. Environmental remediation deals with the removal of pollution or contaminants from environmental media such as soil, groundwater, sediment, or surface water. Restoration involves bringing back to a former or original state. *Other Land Use: Land Conservation Options: Preservation, Remediation, Mitigation, Restoration*

_____ 50. (**A**) Surface mining is done by removing (stripping) surface vegetation, dirt, and if necessary, layers of bedrock in order to reach buried ore deposits. Techniques of surface mining include open-pit mining, which consists of recovery of materials from an open pit in the ground, quarrying or gathering building materials from an open pit mine; strip mining, which consists of stripping surface layers off to reveal ore/seams underneath; and mountaintop removal, commonly associated with coal mining, which involves taking the top of a mountain off to reach ore deposits at depth. *Other Land Use: Mining: Extraction*

_____ 51. (**E**) Purse seining establishes a large wall of netting to encircle schools of fish. Fishermen pull the bottom of the netting closed—like a drawstring purse—to herd fish into the center. This method is used to catch schooling fish, such as sardines, or species that gather to spawn, such as squid. *Other Land Use: Fishing: Fishing Techniques*

_____ 52. (**B**) "The Tragedy of the Commons," written by ecologist Garrett Hardin in 1968, illustrates the argument that free access and unrestricted demand for a finite resource ultimately reduces the resource through overexploitation. This occurs because the benefits of exploitation accrue to individuals or groups, each of whom is motivated to maximize use of the resource to the point in which they become reliant on it, while the costs of the exploitation are borne by all those to whom the resource is available (which may be a wider class of individuals than those who are exploiting it). This, in turn, causes demand for the resource to increase, which causes the problem to increase to the point that the resource is depleted. *Global Economics: Tragedy of the Commons*

RECAP FOR LAND AND WATER USE

_____ Total number correct for this section

_____ Percent correct for this section
(number correct for this section / 13) × 100%

_____ Number wrong for this section (questions you thought you knew).
Do NOT count answers left blank.

_____ Number left blank for this section (questions you did not know)

V. Energy/Resources and Consumption (10–15%)

_____ 53. (**B**) *Energy Concepts: Conversions*

$$\frac{100 \text{ lightbulbs}}{1} \times \frac{200 \text{ watts}}{1 \text{ lightbulb}} \times \frac{10 \text{ hours}}{1 \text{ day}} \times \frac{6 \text{ days}}{1 \text{ week}} \times \frac{50 \text{ weeks}}{1 \text{ year}} \times \frac{1 \text{ kilowatt}}{1,000 \text{ watts}} = \frac{60,000 \text{ kilowatt} \cdot \text{hrs}}{\text{year}}$$

_____ 54. (**E**) The percent change is calculated as $\% = \frac{V_f - V_i}{V_i} \times 100\%$ where V_f is the final value and V_i is the initial V_i value. Therefore, $\% = \frac{112 - 8}{8} \times 100\% = 1300\%$. *Energy Concepts: Conversions*

_____ 55. (**B**) The Second Law of Thermodynamics states that energy varies in its quality or ability to do useful work. For useful work to occur energy must move or flow from a level of high-quality (more concentrated) energy to a level of lower-quality (less concentrated) energy. The chemical potential energy concentrated in a tank of gasoline and the concentrated heat energy at a high temperature are forms of high-quality energy. Because the energy in gasoline is concentrated, it has the ability to perform useful work in moving or changing matter. In contrast, less concentrated heat energy at a low temperature has little remaining ability to perform useful work. *Energy Concepts: Laws of Thermodynamics*

_____ 56. (**D**) According to the United States Energy Information Administration, the United States consumes about 19 million barrels of oil per day; oil that is used to produce gasoline, manufacturing products, heating and power, etc. Since each barrel of oil equates to 42 gallons (160 L), this represents about 800 million gallons (3 billion L) of oil per day. With the United States population at approximately 313 million, this works out to be about 2.5 gallons (10 L) of oil per person per day. In contrast, China with a population of 1.4 billion people (almost three times that of the United States) currently consumes 8.2 million barrels of oil (345 million gallons (1.3 billion L)) per day, which represents about 0.25 gallons per person per day. However, with a rising standard of living, this figure is expected to rise, and compounded with the larger population, China is expected to become the world's primary oil consumer. Because oil is a finite resource, according to the Law of Supply and Demand, prices are expected to rise. *Energy Consumption: Present Global Energy Use*

_____ 57. (**E**) In the United States on average the amount of energy required for heating a home far outpaces that used to cool it. In fact, over the course of a year nearly one-third of a typical utility bill goes to heating. For most of the United States on average the temperature difference in the winter is about twice that of the summer to maintain an indoor temperature of 75°F (24°C). For example, during the winter, if the temperature outside is 35°F (2°C), that represents a 40°F difference. In summer, if the outdoor temperature reaches 95°F (35°C), that represents a 20°F difference in order to keep the indoor temperature at 75°F (24°C). *Energy Consumption: Present Global Energy Use*

_____ 58. (**B**) *Energy Consumption: Present Global Energy Use*

**Energy Consumption in the World
by Each Sector (Quadrillion BTJ)**

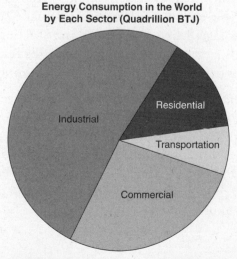

Source: U.S. Energy Information Administration

_____59. (**D**) *Fossil Fuel Resources and Use: World Reserves and Global Demand*

$$\frac{2,170 \times 10^{12} \text{ ft}^3}{62 \times 10^9 \text{ ft}^3/\text{day}} \times \frac{1 \text{ year}}{365 \text{ day}} = 35 \times 10^3 \div 365 \approx 100 \text{ years}$$

_____60. (**C**) Natural gas (mostly methane) when burned produces 45% less carbon dioxide than coal and 30% less carbon dioxide than oil. However, methane is 21 times more dangerous for greenhouse warming than carbon dioxide so incomplete combustion or any leakage of natural gas (from animals, pipelines, landfills, melting tundra, sea floor hydrates, etc.) contributes strongly to greenhouse emissions and global warming. *Fossil Fuel Resources and Use: Environmental Advantages/Disadvantages of Sources*

_____ 61. (**E**) Nuclear power provides 79% of the electricity produced in France and France's electricity price is among the lowest in Europe. Following the 2011 Fukushima nuclear accident, there have been active debates in France as to whether to upgrade the existing nuclear reactors or whether to begin phasing them out. The following represent the amount of electricity generated from nuclear power: United States (20%), Russia (18%), China (2%), and Japan (29%). *Nuclear Energy: Electricity Production*

_____ 62. (**E**) For every three units of energy produced by the reactor core of a U.S. nuclear power plant, two units are discharged to the environment as waste heat. Nuclear plants are often built on the shores of lakes, rivers, and oceans because these bodies provide the large quantities of cooling water needed to handle the waste heat discharge that results in thermal pollution. *Nuclear Energy: Electricity Production*

_____ 63. (**C**) In order (most expensive to least expensive): Natural Gas: $0.081 > Wind: $0.030 > Coal: $0.027 > Nuclear: $0.019 (*includes eventual decommissioning costs*) > Hydroelectric: $0.009. *Nuclear Energy: Electricity Production*

_____ 64. (**C**) Dams built across rivers slow down the river's flow and cause sediments in the water to deposit to the bottom of the reservoir behind the dam. As the sediments accumulate in the reservoir, the dam gradually loses its ability to store water for the purpose(s) for which it was built. The rate of reservoir sedimentation depends mainly on the size of a reservoir relative to the amount of sediment flowing into it: a small reservoir on an extremely muddy river will rapidly lose capacity whereas a large reservoir on a very clear river may take some time to lose an appreciable amount of storage. Large reservoirs in the U.S. lose storage capacity at an average rate of around 0.2% per year while most major reservoirs in China lose capacity at an annual rate over 2%. *Hydroelectric Power: Silting*

_____ 65. (**B**) First enacted by Congress in 1975, the purpose of CAFE (Corporate Average Fuel Economy) is to reduce energy consumption by increasing the fuel economy of cars and light trucks. The National Highway Traffic Safety Administration (NHTSA) administers the CAFE program and sets fuel economy standards for cars and light trucks sold in the U.S., while the Environmental Protection Agency (EPA) provides the fuel economy data and calculates the average fuel economy for each manufacturer.

_____ 66. (**E**) To begin with, you will need to buy 10 incandescent lightbulbs to last the same amount of time as the CFL: 10 × $2.00 = $20. By buying the CFL, you therefore save $20 − $13 = $7.

The total cost of energy for the incandescent bulbs would be:

$$\frac{10 \text{ incandescent lightbulbs}}{1} \times \frac{150 \text{ watts}}{1 \text{ incandescent lightbulb}} \times \frac{1 \text{ kilowatt}}{1,000 \text{ watts}} \times \frac{1,000 \text{ hours}}{1} \times \frac{\$0.10}{1 \text{ kWh}} = \$150.$$

The total cost of energy for the single CFL bulb would be:

$$\frac{10 \text{ CFL bulbs}}{1} \times \frac{42 \text{ watts}}{1 \text{ CFL bulb}} \times \frac{1 \text{ kilowatt}}{1,000 \text{ watts}} \times \frac{10,000 \text{ hours}}{1} \times \frac{\$0.10}{1 \text{ kWh}} = \$42.00$$

Therefore, you save $150.00 − $42.00 = $108.00 in energy for a total savings of $108.00 in energy + $7.00 saving in cost of bulb = $115.00 *Energy Conservation: Energy Efficiency*

_____ 67. (**A**) *Solar Energy is currently used the least in the U.S.*

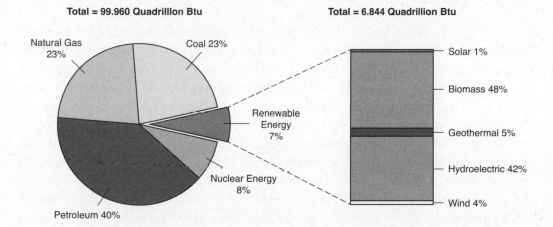

RECAP FOR ENERGY RESOURCES AND CONSUMPTION

_____ Total number correct for this section

_____ Percent correct for this section
(number correct for this section / 15) × 100%

_____ Number wrong for this section (questions you thought you knew).
Do NOT count answers left blank.

_____ Number left blank for this section (questions you did not know)

VI. Pollution (15–30%)

_____ 68. **(B)** Primary air pollutants are pollutants that are released directly into the lower atmosphere (troposphere) and are toxic. Secondary air pollutants are those pollutants that are formed by the combination or chemical reaction of primary pollutants in the atmosphere. Acid rain forms from the chemical reaction in the atmosphere. Acid rain is formed by the reaction of sulfur dioxide (SO_2) and water vapor. Another example of a secondary air pollutant would be PANs (peroxacyl nitrates), which are formed by the reaction of hydrocarbons, oxygen, and nitrogen dioxide. *Air Pollution: Sources—Primary and Secondary.*

_____ 69. **(E)** Criteria pollutants are pollutants that do the most harm to humans. In addition to those listed by the Environmental Protection Agency (EPA), nitrogen dioxide (NO_2), and lead (Pb) are also included. *Air Pollution: Major Air Pollutants*

_____ 70. **(B)** This type of problem can easily be solved by factor analysis. *Air Pollution: Measurement Units*

$$\frac{1,500 \text{ parts}}{1 \text{ billion}} \times \frac{1 \text{ billion}}{1,000,000,000 \text{ parts}} \times \frac{1,000,000 \text{ parts}}{1 \text{ million}} = 1.5 \text{ ppm}$$

_____ 71. **(B)** Nitrogen dioxide (NO_2) reacts with water vapor (H_2O) to form nitrous acid (HNO_2) and nitric acid (HNO_3), which are both components found in acid deposition (rain): $2NO_2 + H_2O \rightarrow HNO_2 + HNO_3$. Nitrogen dioxide also absorbs light energy and breaks down to form nitric oxide (NO) and atomic oxygen (O). Nitric oxide can then combine with ozone (O_3) to form nitrogen dioxide (NO_2) and oxygen (O_2): $NO + O_3 \rightarrow NO_2 + O_2$. If the concentration of nitrogen dioxide is greater than the concentration of nitric oxide, then ozone is formed and can reach dangerous levels. If the concentration of nitrogen dioxide is less than the concentration of nitric acid, then the ozone is destroyed almost instantly, keeping the concentration of ozone stable and below a harmful threshold. Sulfur dioxide is only involved in formation of acid rain, not photochemical smog. *Air Pollution: Acid Deposition and Smog.*

_____ 72. **(B)** Temperature inversions usually occur in urban areas when air pollutants are trapped close to the ground and not able to rise into the atmosphere. Under normal conditions, warmer air, which is less dense than cooler air, will rise into the cooler atmosphere above. However, during a temperature inversion, the air above a city is warm, and prevents the polluted warmer air from rising, causing smog levels to increase, which results in increased respiratory distress. *Air Pollution: Temperature Inversions*

_____ 73. **(C)** "Naturally occurring" means occurring in nature in a free state. Both radon and asbestos are found in nature and are sometimes found in indoor air samples. If radioactive radon occurs in the bedrock near a building, it can seep into a basement or crawlspace and eventually result in lung cancer. Asbestos was often used as insulation material. When asbestos fibers are inhaled, they are trapped in the lungs, which causes the lungs to produce acids. In time, this can lead to diseases such as asbestosis, lung cancer, or mesothelioma. Chlorofluorocarbons and peroxyacylnitrates (PANs) are not naturally occurring in their free state and generally not considered indoor air pollutants. VOCs (volatile organic compounds) such as formaldehyde, are indoor air pollutants and are commonly found in carpet, paneling, furniture, paint, plastics, cleaning fluids, etc., but like CFCs and PANs, they are not naturally occurring. *Air Pollution: Indoor Air Pollution*

_____ 74. (**A**) National Ambient Air Quality Standards (NAAQS) are standards established by the U.S. Environmental Protection Agency (EPA) under authority of the Clean Air Act that apply for outdoor air throughout the country. Primary standards are designed to protect human health, with an adequate margin of safety, including sensitive populations such as children, the elderly, and individuals suffering from respiratory diseases. Secondary standards are designed to protect public welfare from any known or anticipated adverse effects of a pollutant. *Air Pollution: Clean Air Act and Other Relevant Laws*

_____ 75. (**E**) Noise pollution in the workplace is regulated and enforced by OSHA (Occupational Safety and Health Administration) a branch of the U.S. Department of Labor. A difference of 20 decibels corresponds to an increase of 10 $\times$ 10 or 100 times in intensity. Reef fish, whales, dolphins, and other marine species rely on sounds to communicate. Under the Clean Air Act, the EPA established the Office of Noise Abatement and Control (ONAC) to carry out investigations and studies on noise and its effect on the public health and welfare. Through ONAC, the EPA coordinated all federal noise control activities, but in 1981 the administration concluded that noise issues were best handled at the state and local level. As a result, ONAC was closed and primary responsibility of addressing noise issues was transferred to state and local governments. However, EPA retains authority to investigate and study noise and its effect, disseminate information to the public regarding noise pollution and its adverse health effects, respond to inquiries on matters related to noise, and evaluate the effectiveness of existing regulations for protecting the public health and welfare. *Noise Pollution: Effects*

_____ 76. (**D**) The World Health Organization (WHO) has estimated that 1.1 billion people globally lack basic access to drinking water resources, while 2.4 billion people have inadequate sanitation facilities, which accounts for many water related acute and chronic diseases. Some 3.4 million people, many of them young children, die each year from water-borne diseases, such as intestinal diarrhea (cholera, typhoid fever, and dysentery), caused by microbially-contaminated water supplies that are linked to deficient or non-existent sanitation and sewage disposal facilities. Globally, water-borne diseases are the second leading cause of death in children below the age of five years, while childhood mortality rates from acute respiratory infections ranks first. *Water Pollution: Causes and Effects*

_____ 77. (**B**) In choice (A), hot water does not promote the growth of algae—phosphates and/or nitrates do. With an algal bloom being a mile away from the factory, it would be more likely that phosphates and/or nitrates were coming from another source. In choice (C), finding high concentrations of metal ions upstream of the drainage pipe would generally rule out the factory. Furthermore, metal ions would generally not be a waste product from a fertilizer plant. In choice (D), large numbers of insect larvae, also known as indicator species, probably indicates that the water this far from the factory may be very clean. Large amounts of food wastes dumped into the water at the site of the drainage pipe would supply vital nutrients to bacteria that would thrive and use up large amounts of oxygen for their metabolic requirements and deprive other organisms of oxygen that they would need (cultural eutrophication). Turbidity measures the amount of suspended material in the water and would probably not have much to do with the release of waste water from the production of pharmaceuticals. *Water Pollution: Sources*

_____ 78. (**A**) The Clean Water Act passed in 1972 was enacted to reduce point source pollution and addressed protecting surface water. It did not directly address groundwater contamination. The Safe Drinking Water Act, Resource Conservation and Recovery Act, and the Superfund Act did address groundwater contamination. *Water Pollution: Clean Water Act and Other Relevant Laws*

_____ 79. (**C**) Biochemical oxygen demand or BOD is the amount of dissolved oxygen needed by aerobic biological organisms in a body of water to break down organic material present in a given water sample at certain temperature over a specific time period. The term also refers to a chemical procedure for determining this amount. *Water Pollution: Causes and Effects*

_____ 80. (**B**) Anaerobic decomposition is a biological process in which decomposition of organic matter occurs without oxygen. The area of the graph where oxygen in the water is lowest is the area where BOD is the highest. Two processes occur during anaerobic decomposition. First, facultative acid forming bacteria use organic matter as a food source and produce volatile (organic) acids, gases such as carbon dioxide and hydrogen sulfide, stable solids, and more facultative organisms. Second, anaerobic methane formers use the volatile acids as a food source and produce methane gas, stable solids, and more anaerobic methane formers. *Water Pollution: Cultural Eutrophication*

_____ 81. (**B**) Suspended solid matter is usually removed first by screens and/or sedimentation tanks. Next, the biological material from human wastes, food wastes, soaps, etc., is removed by allowing bacteria to break down the material. Finally, the last stage uses a form of disinfection. *Water Pollution: Sewage Treatment*

_____ 82. (**D**) Sanitary landfills are one of the most popular forms of waste disposal in the United States, primarily because they are the least expensive way to dispose of waste. In a sanitary landfill, waste is spread in layers on a piece of property, usually on marginal or submarginal land. The waste is spread into layers and then compacted tightly, greatly reducing the volume of the waste. The waste is then covered by soil. Problems that are encountered in open dumping, including insects, rodents, safety hazards, and fire hazards, are usually avoided with landfilling. A landfill should not be located in areas with high groundwater tables. Leachate migration control standards must be followed in the design, construction, and operation of landfills during the use of the facility and during the post closure period. Much of the waste in a sanitary landfill will decompose through biological and chemical processes that produce solid, liquid, and gaseous products. Food wastes degrade rapidly, whereas plastics, glass, and construction wastes do not. The most common types of gas produced by the decomposition of the wastes are methane and carbon dioxide. After a landfill has reached capacity, it is closed for waste deposition and covered. In some cases it can be used as pasture, cropland, or for recreational purposes. Maintenance of the closed landfill is important to avoid soil erosion and excess runoff into desirable areas. *Solid Waste: Types*

_____ 83. (**E**) Wherever possible, waste reduction is the least expensive option. If waste is produced, every effort should be made to reuse it if practicable. Recycling is the third option in the waste management hierarchy. Although recycling does help to conserve resources and reduce wastes, there are economic and environmental costs associated with waste collection and recycling. For this reason, recycling should only be considered for waste that cannot be reduced or reused. Finally, it may be possible to recover materials or energy from waste that cannot be reduced, reused, or recycled. *Solid Waste: Reduction*

_____ 84. (**B**) An acute exposure refers to a single exposure to a harmful substance that results in severe biological harm or death. *Hazards to Human Health: Acute and Chronic Effects*

_____ 85. (**B**) The median lethal dose, LD_{50} of a toxin, radiation, or pathogen is the dose required to kill half the members of a tested population after a specified test duration. LD_{50} figures are frequently used as a general indicator of a substance's acute toxicity. *Hazards to Human Health: Dose-Response Relationships*

_____ 86. (**D**) Follow the Y axis up to 50%, then move right until you reach the curve, then read straight down. *Hazards to Human Health: Dose-Response Relationships*

_____ 87. (**C**) The first point along the graph where a response above zero is reached is referred to as a threshold dose. For most drugs, the desired effects are found at doses slightly greater than the threshold dose. At higher doses, undesired side effects appear and grow stronger as the dose increases. The more potent a particular substance is, the steeper this curve will be. *Hazards to Human Health: Dose-Response Relationships*

_____88. (**D**) Cigarette smoke causes about half a million deaths a year in the United States; this represents 1 of every 5 deaths. On average, adults who smoke cigarettes die 14 years earlier than nonsmokers. Based on current cigarette smoking patterns, an estimated 25 million Americans who are alive today will die prematurely from smoking-related illnesses, including 5 million people younger than 18 years of age. Smokers increase their risk of dying from bronchitis by nearly 10 times, from emphysema by nearly 10 times, and from lung cancer by up to 22 times. The total cost of caring for people with health problems caused by cigarette smoking—counting all sources of medical payments—is about $73 billion per year. *Hazards to Human Health: Smoking and Other Risks*

_____ 89. (**E**) Endocrine disruptors are substances that interfere with the synthesis, secretion, transport, binding, action, or elimination of natural hormones in the body that are responsible for development, behavior, fertility, and maintenance of homeostasis.

Studies have shown that endocrine disruptors can cause adverse biological effects in animals, and low-level exposures also cause similar effects in human beings. In the late 1950s, Rachel Carson, a marine biologist, turned her attention to conservation and the environmental problems caused by synthetic pesticides. Her book, *Silent Spring* (1962), brought environmental concerns to an unprecedented portion of the American public. *Silent Spring*, while met with fierce denial from chemical companies, spurred a reversal in national pesticide policy—leading to a nationwide ban on DDT and other pesticides—and led to the creation of the Environmental Protection Agency (EPA). *Hazardous Chemicals in the Environment: Types of Hazardous Waste*

_____ 90. (**D**) Following is a diagram of a hazardous waste landfill. *Hazards to Human Health: Waste Treatment/Disposal of Hazardous Waste*

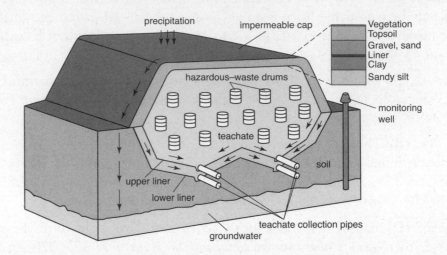

_____ 91. (**B**) Congress enacted the Resource Conservation and Recovery Act (RCRA) in 1976 to address the increasing problems the nation faced from its growing volume of municipal and industrial waste. RCRA amended the Solid Waste Disposal Act of 1965. It set national goals for:

• Protecting human health and the natural environment from the potential hazards of waste disposal.
• Energy conservation and natural resources.
• Reducing the amount of waste generated, through source reduction and recycling.
• Ensuring the management of waste in an environmentally sound manner. It is now most widely known for the regulations promulgated under RCRA that set standards for the treatment, storage, and disposal of hazardous waste in the United States. *Hazards to Human Health: Relevant Laws*

RECAP FOR POLLUTION

_____ Total number correct for this section

_____ Percent correct for this section
(number correct for this section / 24) × 100%

_____ Number wrong for this section (questions you thought you knew).
Do NOT count answers left blank.

_____ Number left blank for this section (questions you did not know)

VII. Global Change (10–15%)

_____ 92. (**A**) The marketplace forces of supply and demand determine the price of fuel. If demand grows or if a disruption in supply occurs, there will be upward pressure on prices. By the same token, if demand falls or there is an oversupply of product in the market, there will be downward pressure on prices. Recently, crude oil prices have risen dramatically, driven by rising global demand, speculation on prices, and political instability in several oil producing countries. *Economic Impacts*

_____ 93. (**E**) Currently, nitrous oxide emissions from human activities are more than twice as high as the next leading ozone-depleting gas. Nitrous oxide is emitted from natural sources and as a by-product of agricultural fertilization, livestock manure, sewage treatment, and other industrial processes. With CFCs and certain other ozone-depleting gases coming in check as a result of the 1987 Montreal Protocol, the international treaty that phased out ozone-destroying compounds, anthropogenic nitrous oxide is becoming an increasingly larger fraction of the emissions of ozone-depleting substances. Nitrous oxide is not regulated by the Montreal Protocol. Nitrous oxide is also a greenhouse gas. *Stratospheric Ozone: Causes of Ozone Depletion*

_____ 94. (**B**) Ice-albedo feedback (or snow-albedo feedback) is a positive feedback climate process whereby a change in the area of snow-covered land, ice caps, glaciers, or sea ice alters the albedo (Earth's reflectivity). Global warming tends to decrease ice cover allowing for greater exposure of darker land to increase absorption of solar energy, leading to more warming and more ice melting. *Global Warming: Impacts and Consequences of Global Warming*

_____ 95. (**B**) The Montreal Protocol on Substances That Deplete the Ozone Layer is an international treaty designed to protect the ozone layer by phasing out the production of numerous substances believed to be responsible for ozone depletion. The treaty went into effect on January 1, 1989. It is widely believed that if the international agreement is adhered to, the ozone layer may be expected to recover by 2050; however, increasing anthropogenic nitrous oxide emissions may delay that goal. *Stratospheric Ozone: Relevant Laws and Treaties*

_____ 96. (**C**) Water vapor is one of the most important elements of the climate system. As a greenhouse gas, it represents around 80% of total greenhouse gas mass in the atmosphere and 90% of greenhouse gas volume. The primary reasons why water vapor cannot be a cause of climate change are its short atmospheric residence time and a basic physical limitation on the quantity of water vapor in the atmosphere for any given temperature (its saturation vapor pressure). The addition of a large amount of water vapor to the troposphere would have little effect on global temperatures in the short term due to the thermal inertia of the climate system. *Global Warming: Greenhouse Gases and the Greenhouse Effect*

_____ 97. (**D**) Global sea level rose at an average rate of about 3.3 ± 0.4 mm per year from 1993 to 2009. Two main factors contribute to sea level rise. The first is thermal expansion: as ocean water warms, it expands. The second is from the contribution of land-based ice due to increased melting. The major store of water on land is found in glaciers and ice sheets. Melting of sea ice does not result in a rise in sea level. *Global Warming: Impacts and Consequences of Global Warming*

_____ 98. (**E**) In 1997, delegates from 194 nations met in Kyoto, Japan, and collectively promised to reduce greenhouse gas emissions by about 5% below 1990 levels by 2012, as a first step toward global cooperation on limiting carbon-driven climate change. The treaty they produced, known as the Kyoto Protocol, expires at the end of 2012. The treaty required the major industrialized nations to meet targets on emissions reduction but imposed no mandates on developing countries, including emerging economic powers and sources of global greenhouse gas emissions like China, India, Brazil, and South Africa. Major countries, including

Canada, France, Japan, and Russia, have said they will not agree to an extension of the protocol unless the unbalanced requirements of developing and developed countries are changed. That is similar to the United States' position, which is that any successor treaty must apply equally to all major economies. The United States has never been a party to the protocol. *Global Warming: Relevant Laws and Treaties*

_____ 99. (**E**) Habitat loss due to destruction, fragmentation, or degradation of habitat is the primary threat to the survival of wildlife. When an ecosystem has been dramatically changed by human activities—such as agriculture, oil and gas exploration, commercial development, or water diversion—it may no longer be able to provide the food, water, cover, and/or places for endangered species to live and raise young. *Loss of Biodiversity: Habitat Loss*

_____ 100. (**B**) CITES (the Convention on International Trade in Endangered Species of Wild Fauna and Flora) is an international agreement between governments. Its aim is to ensure that international trade in specimens of wild animals and plants does not threaten their survival. *Loss of Biodiversity: Relevant Laws and Treaties*

RECAP FOR GLOBAL CHANGE

_____ Total number correct for this section

_____ Percent correct for this section
(number correct for this section / 9) × 100%

_____ Number wrong for this section (questions you thought you knew).
Do NOT count answers left blank.

_____ Number left blank for this section (questions you did not know)

Assess Your Strengths and Weaknesses

I. Earth Systems and Resources: _____ correct

II. The Living World: _____ correct

III. Population: _____ correct

IV. Land and Water Use: _____ correct

V. Energy Resources and Consumption: _____ correct

VI. Pollution: _____ correct

VII. Global Change: _____ correct

TOTAL: _____

PREDICTED AP SCORE

Less than 50 correct:	1 or 2 (not passing)
50–60 correct:	3 on the APES exam
61–75 correct:	4 on the APES exam
76+ correct:	5 on the APES exam

UNIT I: EARTH SYSTEMS AND RESOURCES (10–15%)

Areas on Which You Will Be Tested

A. Earth Science Concepts—geologic time scale, plate tectonics, earthquakes, volcanism, seasons, solar intensity, and latitude.

B. The Atmosphere—composition, structure, weather, climate, atmospheric circulation and the Coriolis effect, atmosphere-ocean interactions, and ENSO.

C. Global Water Resources and Use—freshwater, saltwater, ocean circulation, agriculture, industrial and domestic use, surface and groundwater issues, global problems, and conservation.

D. Soil and Soil Dynamics—rock cycle, formation, composition, physical and chemical properties, main soil types, erosion and other soil problems, and soil conservation.

The Earth

GEOLOGIC TIME SCALE

Two time scales are used to measure the age of Earth. One is based on the sequence of layering of the rocks and the evolution of life. The other is the radiometric time scale, based on the natural radioactivity of chemical elements in rocks. Earth's past has been organized into various units according to events that took place in each period. Different spans of time on the time scale are usually separated by major geologic or paleontological events, such as mass extinctions. For example, the boundary between the Cretaceous period and the Paleogene period is defined by the extinction of the dinosaurs and many marine species. The largest defined unit of time is the eon. Eons are dived into eras, which are in turn divided into periods, epochs, and stages (Eon → Eras → Periods → Epochs → Stages). Key principles of the geological time scale:

1. Rock layers (strata) are laid down in succession with each strata representing a "slice" of time.

2. The principle of superposition—any given layer is probably older than those above it and younger than those below it.

Several factors complicate the geologic time scale:

1. Layers are often eroded, distorted, tilted, or even inverted after deposition.

2. Layers laid down at the same time in different areas can have entirely different appearances.

3. A layer from any given area represents only part of Earth's history.

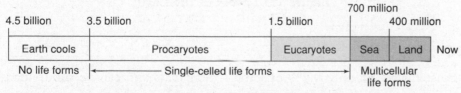

Figure 1.1 Timeline of life development

THE GEOLOGIC TIME SCALE

Era	When period began (millions of years ago)	Period	Animal life	Plant life	Major geologic events
Cenozoic		Neogene	Rise of civilizations	Increase in number of herbs and grasses	Ice Age
	2	Paleogene	Appearance of first men; dominance on land of mammals, birds, and insects	Dominance of land by flowering plants	
	65				
Mesozoic		Cretaceous		Dominance of land by conifers; first flowering plants appear	Building of the Rocky Mountains
	135	Jurassic	Age of dinosaurs		
	180	Triassic	First birds		
	225				
Paleozoic		Permian	Expansion of reptiles		Building of the Appalachian Mountains
	275	Carboniferous	Age of amphibians	Formation of great coal swamps	
	350	Devonian	Age of fishes		
	413	Silurian	Invasion of land by invertebrates	Invasion of land by primitive plants	
	430	Ordovician	Appearance of first vertebrates (fish)	Abundant marine algae	
	500	Cambrian	Abundant marine invertebrates	Appearance of primitive marine algae	
	570				
Precambrian		—	Primitive marine life		

Figure 1.2 Geologic time scale

EARTH STRUCTURE

Earth, which formed about 4.6 billion years ago, is the third planet away from the sun in the solar system and is the only planet known to support life. Earth can be divided into three sections: the biosphere, the hydrosphere, and the internal structure. The biosphere includes all forms of life (plants and animals) both on land and in the sea. The hydrosphere includes all forms of water (fresh and saltwater, snow, and ice). The internal structure of Earth is divided into the crust, mantle, and core.

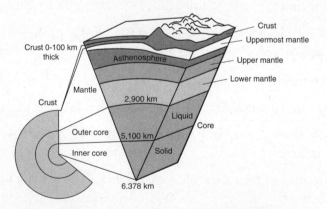

Figure 1.3 Layers of the Earth

Crust

The crust makes up only 0.5% of Earth's total mass and can be subdivided into two main parts: basalt-rich oceanic crust and granite-rich continental crust. The crust floats on top of the mantle. Oceanic crust covers about two-thirds of Earth's surface but comprises about only one-third of the crustal mass, as the continental crust is much thicker. Being relatively cold, the crust is rocky and brittle, so it can fracture in earthquakes.

CONTINENTAL CRUST

The continental crust extends from the surface of Earth down to 20–30 miles (30–50 km). The exposed parts of the continental crust are less dense than oceanic crust because oceanic crust contains minerals rich in heavier elements such as iron and magnesium. However, continental crust appears to be stratified (layered) and becomes denser with depth. It is largely composed of volcanic, sedimentary, and granite-type rocks, although the older areas are dominated by metamorphic rocks.

OCEANIC CRUST

From the surface of Earth down to about 7 miles (11 km) is the oceanic crust. The oceanic crust can be divided into ocean basins where water depth exceeds 2 miles (3 km), and the crust is layered and very uniform.

MOHO

The Mohorovicic discontinuity, usually referred to as the Moho, is the boundary between the Earth's crust and the mantle. The Moho serves to separate both oceanic crust and continental crust from underlying mantle. It lies 3 miles (5 km) below the ocean floor and 19–31 miles (30 to 50 km) beneath the continents. It was first identified in 1909, when abrupt increases in the velocity of earthquake waves (specifically P-waves) occurred in this area. During the late 1950s, Project Mohole was proposed to drill a hole through the ocean floor to reach this boundary, but the project was cancelled in 1967.

MANTLE

Most of Earth's mass is in the mantle, which is composed of iron, magnesium, aluminum, and silicon-oxygen compounds. At over 1800°F (1000°C), most of the mantle is solid. However, the upper third (known as the asthenosphere) is more plastic-like in nature.

CORE

The core is composed mostly of iron and is so hot that the outer core is molten. The inner core is under such extreme pressure that it remains solid.

PLATE TECTONICS

Plate tectonic theory arose out of two separate geological observations: continental drift and sea-floor spreading.

Latitude and Longitude

Longitude specifies east-west positions on the Earth. 0° longitude begins at the Prime Meridian in Greenwich, England. The degrees continue 180° east and 180° west until they meet and form the International Date Line in the Pacific Ocean. Latitude specifies north-south positions. 0° lati-

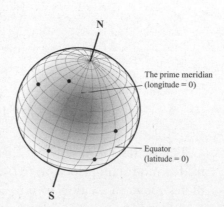

tude is the equator. 90° north latitude is the north pole while 90° south latitude is the south pole. Each degree of latitude is approximately 69 miles (111 km) apart.

The Continental Drift Theory

In 1915, Alfred Wegener proposed that all present-day continents originally formed one landmass (Pangaea). Wegener believed that this supercontinent began to break up into smaller continents around 200 million years ago. He based his theory on five factors:

1. Fossilized tropical plants were discovered beneath Greenland's icecaps.

2. Glaciated landscapes occurred in the tropics of Africa and South America.

3. Tropical regions on some continents had polar climates in the past, based on paleoclimatic data.

4. The continents fit together like pieces of a puzzle.

5. Similarities existed in rocks between the east coasts of North and South America and the west coasts of Africa and Europe.

Continental drift gained acceptance in the 1960s when the theory of plate tectonics provided a mechanism that would account for the movement of the continents.

The Seafloor Spreading Theory

During the 1960s, alternating patterns of magnetic properties were discovered in rocks found on the seafloor. Similar patterns were discovered on either side of mid-oceanic ridges found near the center of the oceanic basins. Dating of the rocks indicated that as one moved away from the ridge, the rocks became older. This suggested that new crust was being created at volcanic rift zones.

The lithosphere (crust and upper mantle, approximately 62 miles (100 km) thick) is divided into massive sections known as plates, which float and move on the viscous asthenosphere. Subduction zones are areas on the Earth where two tectonic plates meet and move toward each other, with one sliding underneath the other and moving down into the mantle.

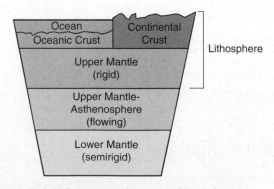

Figure 1.4 The outer layers of Earth

The plates move slowly over time and sink in areas of volcanic island chains, folded mountain belts, and trenches, and rise up from ridges and rift valleys. These plates move in relation to one another at one of three types of plate boundaries: transform, divergent, and convergent.

Figure 1.5 Earth's major plates

TRANSFORM BOUNDARIES

Transform boundaries occur where plates slide *past* each other. The friction and the stress buildup from the sliding plates frequently cause earthquakes—a common feature along transform boundaries. The San Andreas Fault, which is found near the western coast of North America, is where the Pacific and North American plates move relative to each other such that the Pacific plate is moving northwest with respect to North America. In about 50 million years, the part of California that is west of the San Andreas Fault will be a separate island near Alaska.

DIVERGENT BOUNDARIES

Divergent boundaries occur where two plates slide *apart* from each other with the space that was created being filled with molten magma from below. Examples of areas of oceanic divergent boundaries include the Mid-Atlantic Ridge and the East Pacific Rise. Examples of areas of continental divergent boundaries include the East African Great Rift Valley. Divergent boundaries can create massive fault zones in the oceanic ridge system and are areas of frequent oceanic earthquakes.

CONVERGENT BOUNDARIES

Convergent boundaries occur where two plates slide *toward* each other, commonly forming either a subduction zone (if one plate moves underneath the other) or an orogonic belt (if the two plates collide and compress). When a denser oceanic plate moves underneath (subducts) a less-dense continental plate, an oceanic trench is produced on the ocean side and a mountain range on the continental side. One example is the Cascade Mountain range, which extends north from California's Sierra Nevada Mountains and includes Mount Saint Helens.

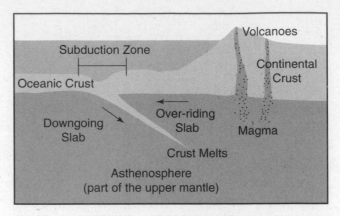

Figure 1.6 Subduction

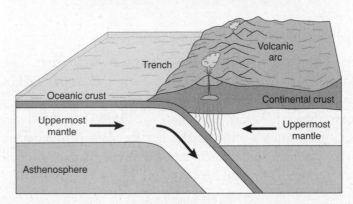

Figure 1.7 Oceanic-continental convergence

When *two oceanic plates converge*, they create an island arc—a curved chain of volcanic islands rising from the deep seafloor and near a continent. They are created by subduction processes and occur on the continent side of the subduction zone. Their curve is generally convex toward the open ocean. A deep undersea trench is located in front of such arcs where the descending plate dips downward. Examples include Japan and the Aleutian Islands in Alaska.

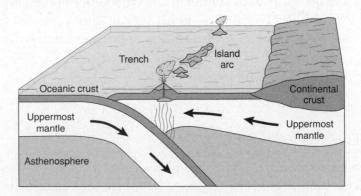

Figure 1.8 Oceanic-oceanic convergence

When *two continental plates collide,* mountain ranges are created as the colliding crust is compressed and pushed upward. An example is the northern margins of the Indian subcontinental plate being thrust under a portion of the Eurasian plate, lifting it and creating the Himalayas.

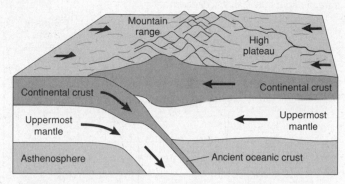

Figure 1.9 Continental-continental convergence

EARTHQUAKES

Earthquakes occur during abrupt movement on an existing fault, along tectonic plate boundary zones or along mid-oceanic ridges. A massive amount of stored energy, held in place by friction, is released in a very short period of time. The area where the energy is released is called the focus. From the focus, seismic waves travel outward in all directions. Directly above the focus, on Earth's surface, is the epicenter.

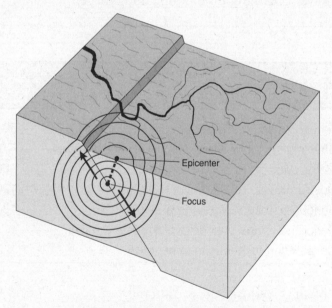

Figure 1.10 Relationship of an epicenter to a focus

The strength or magnitude of an earthquake is commonly measured by the logarithmic Richter scale and is recorded by a seismograph onto a seismogram.

However, the Richter magnitude scale is really comparing amplitudes of waves on a seismogram, not the strength (energy) of the quakes. Because of the logarithmic basis of the scale, each whole number increase in magnitude on the Richter scale represents a tenfold increase in measured amplitude. In terms of energy, each whole number increase corresponds to an increase of about 32 times the amount of energy released. It is the energy or strength of the earthquake that knocks down buildings and causes damage.

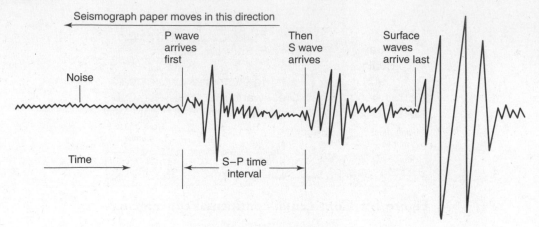

Seismograph paper moves in this direction

P wave arrives first

Then S wave arrives

Surface waves arrive last

Noise

Time

S–P time interval

Figure 1.11 Seismogram

There are two classes of seismic waves: body waves and surface waves. Body waves travel through the interior of Earth. There are two types of body waves: P waves and S waves. P waves travel through Earth and are caused by expansion and contraction of bedrock. S waves are produced when material moves either vertically or horizontally and travel only within the uppermost layers of Earth (along its surface). Surface waves produce rolling and/or swaying motion and are slower than P or S waves. Surface waves cause ground motion and damage.

The Richter Scale

Severity	Richter scale	Description	Occurrence (per year)
Minor	3.0–3.9	Rarely causes damage.	~ 50,000
Light	4.0–4.9	Shaking of indoor items. No significant damage.	~ 6,000
Moderate	5.0–5.9	Major damage to poorly constructed buildings.	~ 800
Strong	6.0–6.9	Destructive up to 100 miles (160 km).	~ 120
Major	7.0–7.9	Serious damage over large areas.	~ 18
Great	8.0–8.9	Serious damage over several thousand miles.	~ 1
Extreme	9.0+	Devastating for thousands of miles.	1 in 20 years

The severity of an earthquake depends upon:

1. The amount of potential energy that had been stored
2. The distance the rock mass moved when the energy was released
3. How far below the surface the movement occurred
4. The makeup of the rock material

Primary effects are due to the shaking and resulting damage to buildings and infrastructure and due to the loss of life or injury. Secondary effects include rock slides, flooding due to subsidence (sinking) of land, liquefaction of recent sediments, fires, and tsunamis. Damage due to earthquakes can be reduced through mapping of faults, preparing computer models and simulations, strengthening building codes, preparing emergency teams with adequate training, upgrading communication technology and availability, storing emergency supplies, and educating the public.

CASE STUDIES

- **Haiti, 2010:** The January 2010 earthquake in Haiti was a catastrophic, magnitude 7.0 quake that occurred at a depth of 8 miles (13 km) below the surface. About 3 million people were affected by the quake, with close to 200,000 people killed directly and more than 250,000 severely injured. The capital of Port-au-Prince and surrounding cities were entirely destroyed primarily due to inadequate building codes, making rescue attempts extremely difficult. The quake occurred in the vicinity of the northern boundary of the Caribbean tectonic plate, where it shifts eastward by about 0.79 inches (20 mm) per year in relation to the North American plate.

- **San Andreas Fault:** The San Andreas Fault, first discovered in 1895, is a continental transform fault that extends 800 miles (1,300 km) through California to Baja California in Mexico, forming the tectonic boundary between the Pacific plate and the North American plate. All land west of the fault on the Pacific plate is moving slowly to the northwest, while all land east of the fault is moving southwest. The rate of slippage averages about 1.5 inches (38 mm) per year. In 1906 a portion of the fault ruptured near San Francisco; the earthquake (estimated at 7.8) caused 3,000 deaths, many due to the resulting fires.

TSUNAMIS

Tsunamis are a series of waves created when a body of water is rapidly displaced usually by an earthquake. The effects of a tsunami can be devastating due to the immense volumes of water and energy involved. A tsunami can be generated when plate boundaries abruptly move and vertically displace the overlying water. Subduction-zone-related earthquakes generate the majority of all tsunamis. Tsunamis formed at divergent plate boundaries are rare since they do not generally disturb the vertical displacement of the water column. Tsunamis have a small wave height offshore, very long wavelength, and generally pass unnoticed at sea. Most tsunamis are generated in the Pacific and Indian Ocean basins. In 2004, the Indian Ocean 9.3 earthquake created tsunamis that killed ~300,000 people—one of the deadliest natural disasters in recorded history.

CASE STUDY

- **Tōhoku Earthquake and Tsunami:** In 2011, northern Japan was hit by a 9.0 magnitude earthquake that triggered a deadly 23-foot tsunami. The giant waves deluged cities and rural areas alike leaving a path of death and devastation in its wake. According to the official toll, the earthquake and resulting tsunami left 15,839 dead and 3,647 missing. In addition, cooling systems at a nearby nuclear power station failed shortly after the earthquake, causing a nuclear crisis. The initial nuclear reactor failure was followed by an explosion and eventual partial meltdowns and fires in other reactors, which resulted in the release of radioactivity directly into the atmosphere. More than 200,000 residents were evacuated from affected areas putting the disaster on par with the 1986 Chernobyl nuclear reactor explosion.

VOLCANOES

Active volcanoes produce magma (melted rock) at the surface. Other types of volcanoes are classified as intermittent, dormant, or extinct. The majority of volcanoes—95%—occur at subduction zones and mid-oceanic ridges. The remaining 5% occur at hot spots, areas where plumes of magma come close to the surface. Volcanoes may produce ejecta (lava rock and/or ash), molten lava, and/or toxic gases. The most common gases released by volcanoes are steam, carbon dioxide, sulfur dioxide, and hydrogen chloride. Volcanoes affect the climate by introducing large quantities of sulfur dioxide (SO_2) into the atmosphere that is later converted to sulfate ions (SO_4^{2-}) in the stratosphere. The sulfate particles reflect shorter wavelengths of solar radiation and serve as condensation nuclei for high clouds. In 1992, the year after the Mt. Pinatubo eruption, the effect of stratospheric sulfate particles decreased the average global temperature by as much as 1°F (0.5°C) by decreasing the amount of sunlight reaching Earth. The particles settle out of the atmosphere usually within two years and contribute to acid rain.

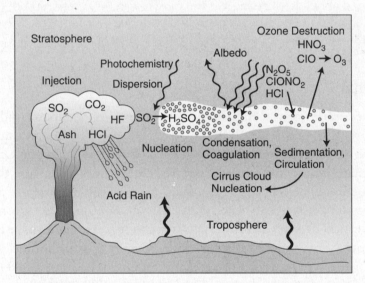

Figure 1.13 Atmospheric effects of volcanoes

Eruptions occur when pressure within a magma chamber forces molten magma up through a conduit (pipe) and out a vent at the top of the volcano. The type of eruption depends on the gases, the amount of silica in the magma (which determines viscosity), and how free the conduit is (whether the volcano flows or explodes). Correlation exists between seismic and volcanic activity. Benefits of volcanic eruptions include producing new landforms as seen in the Hawaiian Islands and increased soil nutrient levels produced from erosion of lava rock. Methods of dealing with threats from volcanoes include:

1. Modeling and data analysis for better volcanic activity prediction

2. Better evacuation plans

3. Study of precursors such as changes in the cone

4. Measuring changes in temperature and gas composition

5. Magnetic changes

6. Changes in seismic activity

Volcano Structure

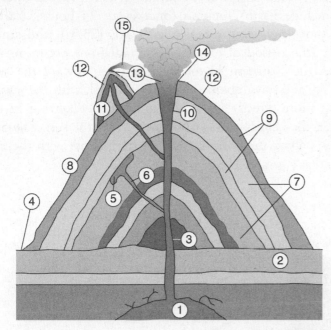

1. Magma chamber
2. Bedrock
3. Conduit (pipe)
4. Base
5. Sill
6. Branch pipe
7. Layers of ash emitted by the volcano
8. Flank
9. Layers of lava emitted by the volcano
10. Throat
11. Parasitic cone
12. Lava flow
13. Vent
14. Crater
15. Ash cloud

Figure 1.14 General volcano structure

CASE STUDIES

- **Mount Saint Helens:** Located in Washington State, Mount Saint Helens erupted in 1980. The earthquake removed trees, increased soil erosion, destroyed wildlife, and polluted the air with gases and ash. Other effects included mudflows, melting of glacial ice and snow, and clogged rivers that caused flooding. Fifty-seven people were killed.

- **Mount Pinatubo:** Mount Pinatubo is part of a chain of composite volcanoes on the west coast of the island of Luzon in the Philippines. In June 1991, Mount Pinatubo erupted for 9 hours, disgorged a cubic mile of volcanic debris, and vented 18 million metric tons of sulfur dioxide into the atmosphere which encircled Earth in three weeks after reaching the stratosphere. This was the largest sulfur dioxide cloud ever detected to date. The sulfate aerosols formed in the stratosphere increased reflection of solar radiation and within 3 years caused over a 2°F (1°C) overall cooling of Earth.

SEASONS, SOLAR INTENSITY, AND LATITUDE

Factors that affect the amount of solar energy at the surface of Earth (which directly affects plant productivity) include Earth's rotation (once every 24 hours), Earth's revolution around the sun (once per year), tilt of Earth's axis (23.5°), and atmospheric conditions. Summer (the period of greatest solar radiation) occurs in the Northern Hemisphere when the Northern Hemisphere is tilted toward the sun. The sun rises higher in the sky and stays above the horizon longer, with the rays of the sun striking the ground more directly (at less of an angle). Likewise, in the Northern Hemisphere winter, the hemisphere is tilted away from the sun. The Sun rises lower in the sky and stays above the horizon for a shorter period, with the rays of the sun striking the ground at a greater angle.

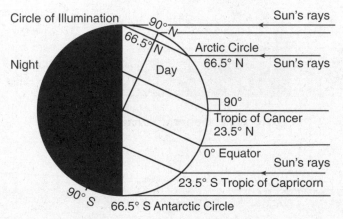

Figure 1.15 Summer solstice

Earth is closest to the sun during the Northern Hemisphere winter (December, January, and February) and farthest away during the Northern Hemisphere summer (June, July, and August). Seasons are NOT caused by Earth's distance from the sun, but rather by the angle of sunlight hitting the Earth.

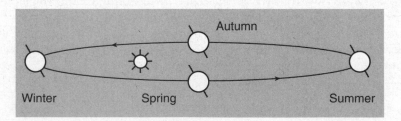

Figure 1.16 Seasons in the Northern Hemisphere

SOIL

Soils are a thin layer on top of most of Earth's land surface. This thin layer is a basic natural resource and deeply affects every other part of the ecosystem. For example, soils hold nutrients and water for plants and animals; water is filtered and cleansed as it flows through soils; and soils affect the chemistry of water and the amount of water that returns to the atmosphere to form rain. Soils are composed of three main ingredients: minerals of different sizes, organic materials from the remains of dead plants and animals, and open space that can be filled with water or air. A good soil for growing most plants should have about 45% minerals (with a mixture of sand, silt, and clay), 5% organic matter, 25% air, and 25% water.

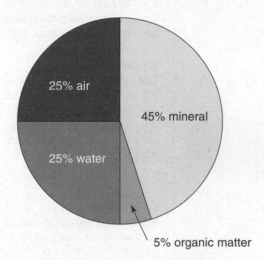

Figure 1.17 Soil components

Soils develop in response to several factors:

1. **Parent material.** This refers to the rock and minerals from which the soil derives. The nature of the parent rock, which can be either native to the area or transported to the area by wind, water, or glacier, has a direct effect on the ultimate soil profile.

2. **Climate.** This is measured by precipitation and temperature. It results in partial weathering of the parent material, which forms the substrate for soil.

3. **Living organisms.** These include the nitrogen-fixing bacteria *Rhizobium*, fungi, insects, worms, snails, etc., that help to decompose litter and recycle nutrients.

4. **Topography.** This refers to the physical characteristics of the location where the soil is formed. Topographic factors that affect a soil's profile include drainage, slope direction, elevation, and wind exposure.

With sufficient time, a mature soil profile reaches a state of equilibrium. Feedback mechanisms involving both abiotic and biotic factors work to preserve the mature soil profile. The relative abundance of sand, silt, and clay is called the soil texture.

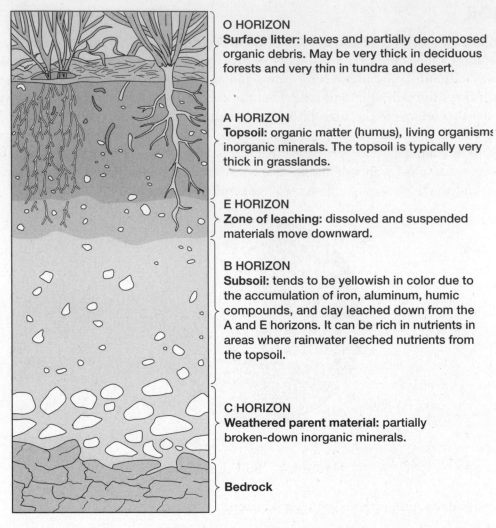

O HORIZON
Surface litter: leaves and partially decomposed organic debris. May be very thick in deciduous forests and very thin in tundra and desert.

A HORIZON
Topsoil: organic matter (humus), living organisms inorganic minerals. The topsoil is typically very thick in grasslands.

E HORIZON
Zone of leaching: dissolved and suspended materials move downward.

B HORIZON
Subsoil: tends to be yellowish in color due to the accumulation of iron, aluminum, humic compounds, and clay leached down from the A and E horizons. It can be rich in nutrients in areas where rainwater leeched nutrients from the topsoil.

C HORIZON
Weathered parent material: partially broken-down inorganic minerals.

Bedrock

Figure 1.18 Soil profile

Soil Components

Component	Description
Clay	Very fine particles. Compacts easily. Forms large, dense clumps when wet. Low permeability to water; therefore, upper layers become waterlogged.
Gravel	Coarse particles. Consists of rock fragments.
Loam	About equal mixtures of clay, sand, silt, and humus. Rich in nutrients. Holds water but does not become waterlogged.
Sand	Sedimentary material coarser than silt. Water flows through too quickly for most crops. Good for crops and plants requiring low amounts of water.
Silt	Sedimentary material consisting of very fine particles between the size of sand and clay. Easily transported by water.

Organic vs. Inorganic Fertilizers

Organic Fertilizer	Inorganic Fertilizer
Three common forms: animal manure, green manure, and compost.	Does not add humus to the soil, resulting in less ability to hold water and support living organisms (earthworms, beneficial bacteria, and fungi, etc.).
Improves soil texture, adds organic nitrogen, and stimulates beneficial bacteria and fungi.	Lowers oxygen content of the soil thereby keeping fertilizer from being taken up efficiently.
Improves water-holding capacity of soil.	Supplies only a limited number of nutrients (usually nitrogen and phosphorus).
Helps to prevent erosion.	Requires large amounts of energy to produce, transport, and apply.
	Releases nitrous oxide (N_2O)—a greenhouse gas.

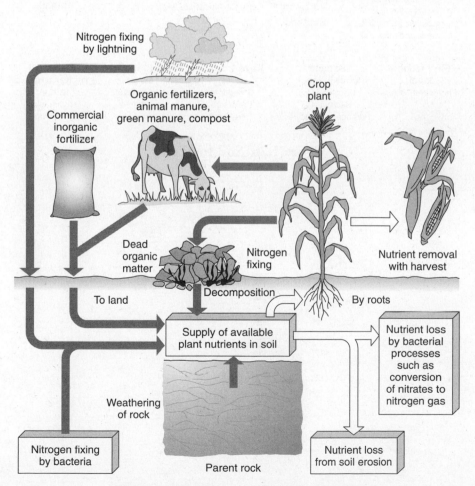

Figure 1.19 Plant nutrient pathways

SOIL FOOD WEB

The soil food web is the community of organisms living all or part of their lives in the soil. It describes a complex living system in the soil and how it interacts with the environments, plants, and animals.

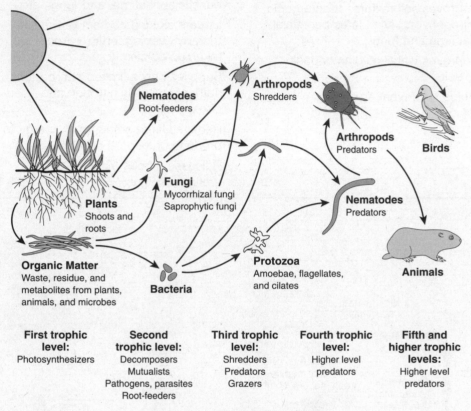

First trophic level:	Second trophic level:	Third trophic level:	Fourth trophic level:	Fifth and higher trophic levels:
Photosynthesizers	Decomposers Mutualists Pathogens, parasites Root-feeders	Shredders Predators Grazers	Higher level predators	Higher level predators

Figure 1.20 Soil food web

EROSION

Soil erosion is the movement of weathered rock or soil components from one place to another. It is caused by flowing water, wind, and human activity such as cultivating inappropriate land, burning of native vegetation, deforestation, and construction. Soil erosion destroys the soil profile, decreases the water-holding capacity of the soil, and increases soil compaction. Because water cannot percolate through the soil, it runs off the land, taking more soil with it (positive feedback loop). Because the soil cannot hold water, crops grown in areas of soil erosion frequently suffer from water shortage. In areas of low precipitation, erosion leads to significant droughts. Poor agricultural techniques that lead to soil erosion include monoculture, row cropping, overgrazing, improper plowing of the soil, and removing crop wastes instead of plowing the organic material back into the soil. There are three common types of soil erosion:

1. **Sheet erosion**—soil moves off as a horizontal layer

2. **Rill erosion**—fast-flowing water cuts small channels in the soil

3. **Gully erosion**—extreme case of rill erosion, where over time, channels increase in size and depth.

Soil erosion causes damage to agriculture, waterways (canals), and infrastructures (dams). It interferes with wetland ecosystems, reproductive cycles (as in salmon), oxygen capacity, and the pH of the water.

Soil Erosion

Desertification	Salinization	Waterlogging
Definition: Productive potential of arid or semiarid land falls by at least 10% due to human activity and/or climate change.	**Definition:** Water that is not absorbed into the soil and evaporates leaves behind dissolved salts in topsoil.	**Definition:** Saturation of soil with water resulting in a rise in the water table.
Symptoms: Loss of native vegetation; increased wind erosion; salinization; drop in water table; reduced surface water supply.	**Symptoms:** Stunted crop growth; lower yield; eventual destruction of plant life.	**Symptoms:** Saline water envelops deep roots killing plants; lowers productivity; eventual destruction of plant life.
Remediation: Reduce overgrazing; reduce deforestation; reduce destructive forms of planting, irrigation, and mining. Plant trees and grasses to hold soil.	**Remediation:** Take land out of production for a while; and/or install underground perforated drainage pipes; flush soil with freshwater into separate lined evaporation ponds; plant halophytes (salt-loving plants) such as barley, cotton, sugar beet and/or semi-dwarf wheat.	**Remediation:** Switch to less water-demanding plants in areas susceptible to waterlogging; utilize conservation-tillage farming; plant waterlogging-resistant trees with deep roots; take land out of production for a while; and/or install pumping stations with drainage pipes that lead to catchment-evaporation basins.

LANDSLIDES AND MUDSLIDES

Landslides occur when masses of rock, earth, or debris move down a slope. Mudslides, also known as debris flows or mudflows, are a common type of fast-moving landslide that tends to flow in channels. Landslides are caused by disturbances in the natural stability of a slope. They can happen after heavy rains, droughts, earthquakes, or volcanic eruptions. Mudslides develop when water rapidly collects in the ground and results in a surge of water-soaked rock, earth, and debris. Mudslides usually begin on steep slopes and can be triggered by natural disasters. Areas where wildfires or construction have destroyed vegetation on slopes are at high risk for landslides during and after heavy rains. Some areas are more likely to experience landslides or mudslides, including:

- Areas where wildfires or construction have destroyed vegetation
- Areas where landslides have occurred before
- Steep slopes and areas at the bottom of slopes or canyons
- Slopes that have been altered for construction of buildings and roads
- Channels along a stream or river
- Areas where surface runoff is directed

RELEVANT LAW

1935 Soil Erosion Act: As a result of the Dust Bowl, this act established the Soil Conservation Service. Mandates the protection of the nation's soil reserves. Deals with soil erosion problems, carries out soil surveys, and does research on soil salinity.

1977 Soil and Water Conservation Act: Provides for a continuing appraisal of U.S. soil, water, and related resources, including fish and wildlife habitats, and a soil and water conservation program to assist landowners and land users in furthering soil and water conservation.

THE ROCK CYCLE

There are three main categories of rocks: metamorphic, igneous, and sedimentary.

1. **Igneous**—formed by cooling and classified by their silica content. Intrusive igneous rocks solidify deep underground, cool slowly and have a large-grained texture (e.g., granite). Extrusive igneous rocks solidify on or near the surface, cool quickly, and have a fine-grained smooth texture (e.g., basalt). Igneous rocks are broken down by weathering and water transport. Most soils come from igneous rocks.

2. **Metamorphic**—formed by intense heat and pressure. Those with high quartz content form sandy soil (e.g., gneiss). Slate forms silty soil. Marble forms limestone clay. Common examples: diamond, marble, asbestos, slate, anthracite coal.

3. **Sedimentary**—formed by piling and cementing of various materials (diatoms, weathered chemical precipitates, fragments of older rocks) over time in low-lying areas. Fossils form only in sedimentary rock.

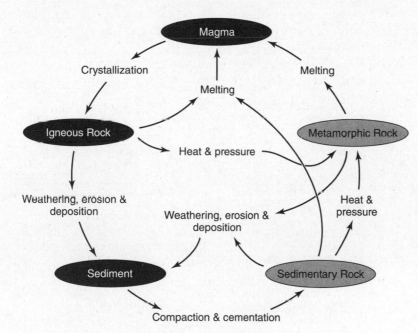

Figure 1.22 The rock cycle

QUICK REVIEW CHECKLIST

☐ **Geologic Time Scale**
    ☐ eon, era, periods, epochs, stages
    ☐ timeline of development

☐ **Earth Structure**
    ☐ layers of the Earth (crust-oceanic-continental, mantle, and core)
    ☐ Continental Drift Theory
    ☐ seafloor spreading
    ☐ plate tectonics
    ☐ boundaries (differences between transform, divergent, and convergent)

☐ **Earthquakes**
    ☐ epicenter, focus
    ☐ Richter scale vs. magnitude
    ☐ P wave vs. S wave
    ☐ primary vs. secondary effects
    ☐ mitigation
    ☐ tsunamis

☐ **Volcanoes**
    ☐ where they occur
    ☐ how they are formed
    ☐ atmospheric effects of volcanic eruptions
    ☐ basic structure
    ☐ mitigation

☐ **Seasons**
    ☐ angle of sun and how it affects season

☐ **Soil**
    ☐ components
    ☐ factors that affect soil development
    ☐ horizons (O, A, E, B, C)
    ☐ types (clay, gravel, loam, sand, silt)
    ☐ organic vs. inorganic fertilizers
    ☐ disadvantages of inorganic fertilizers
    ☐ soil cycle

☐ **Erosion**
    ☐ types (sheet, rill, and gully)

☐ **Rock Cycle**
    ☐ metamorphic—how it is formed and examples
    ☐ igneous—how it is formed and examples
    ☐ sedimentary—how it is formed and examples

MULTIPLE-CHOICE QUESTIONS

1. The majority of the rocks in the Earth's crust are:

   (A) igneous
   (B) metamorphic
   (C) sedimentary
   (D) basalt
   (E) volcanic

2. Which of the following is an example of an igneous rock?

   (A) Marble
   (B) Slate
   (C) Limestone
   (D) Granite
   (E) Sandstone

3. The smallest particle of soil is known as

   (A) clay
   (B) sand
   (C) silt
   (D) gravel
   (E) humus

4. Acid rain affects soil by

   (A) decreasing soil porosity
   (B) decreasing the pH
   (C) decreasing soil aeration
   (D) lowering nutrient capacity
   (E) All of the above

5. Which of the types of soil listed below contains the highest amount of nutrients?

   (A) clay
   (B) silt
   (C) sand
   (D) gravel
   (E) loam

6. An example of a volcano with broad, gentle slopes and built by the eruption of runny, fluid-type basalt lava would be *shield volcano* *P.255 ES book*

   (A) Mount Saint Helens
   (B) Krakatau
   (C) Kilauea
   (D) Vesuvius
   (E) Mount Rainier

7. Which of the following is at a convergent boundary where two continental plates are presently colliding?

    (A) The Appalachian Mountains
    (B) The Himalayas
    (C) The Andes Mountains
    (D) The Rocky Mountains
    (E) None of the above

8. The Dust Bowl of the 1930s resulted in the passage of what legislation?

    (A) Endangered American Wilderness Act
    (B) Soil and Water Conservation Act
    (C) Federal Land Management Act
    (D) Public Rangelands and Improvement Act
    (E) Soil Erosion and Conservation Act

9. Poor nutrient-holding capacity, good water infiltration capacity, and good aeration properties are examples of what type of particle found in soil?

    (A) Clay
    (B) Silt
    (C) Sand
    (D) Loam
    (E) Humus

10. An alkaline, dark soil, rich in humus, found in a semiarid climate would be most characteristic of   *p. 90 this book*

    (A) deserts
    (B) grasslands
    (C) tropical rain forests
    (D) deciduous forests
    (E) coniferous forests

11. Which period in geological time describes the following: "Development of flowering plants. Large diversity in dinosaurs but ending with their sudden extinction approximately 65 million years ago. Formation of the Andes Mountains. African and South American plates begin to separate. Climate is cooling. Shallow seas are prominent."?

    (A) Paleogene   *p. 78 this book*
    (B) Neogene   *p8 E5*
    (C) Jurassic
    (D) Cretaceous
    (E) Permian

12. The process of weathering produces what type of rock?

P.62 ESbook

    (A) Igneous
    (B) Metamorphic
    (C) Sedimentary
    (D) Volcanic
    (E) None of the above ·

13. A rock that would most likely contain a fossil would be

    (A) igneous
    (B) metamorphic
    (C) sedimentary
    (D) volcanic
    (E) All of the above

14. The most common element found in Earth's crust is

P. 26 ES

    (A) oxygen
    (B) hydrogen
    (C) iron
    (D) silicon
    (E) aluminum

15. The horizon of soil also known as the topsoil layer, that contains humus, minerals, and roots, and that is rich in living organisms is known as the

    (A) A layer
    (B) B layer
    (C) C layer
    (D) D layer
    (E) O layer

16. Earth is closest to the sun in the Northern Hemisphere during

    (A) winter
    (B) summer
    (C) spring
    (D) fall
    (E) all seasons

17. An earthquake of Richter magnitude 5 is how many times larger on the Richter scale than an earthquake of Richter magnitude 3?

    (A) One-fourth
    (B) One-half
    (C) Twice
    (D) Four times
    (E) One hundred times

Refer to the following table to answer Question 18.

	Most Stable	Least Stable
A	Bedrock	Sand
B	Unconsolidated sand	Bedrock
C	Clay	Bedrock
D	Sand and mud	Clay
E	Water-saturated sand	Sand

18. Which of the choices above represents the most stable and the least stable foundation material to build upon in areas that are frequented by earthquakes?

   (A) A
   (B) B
   (C) C
   (D) D
   (E) E

19. The San Andreas Fault in California occurs at

   (A) a convergent boundary
   (B) a divergent boundary
   (C) a transform boundary
   (D) a subduction zone
   (E) an oceanic ridge

20. Earth's surface is part of the

   (A) asthenosphere
   (B) lithosphere
   (C) benthosphere
   (D) troposphere
   (E) stratosphere

FREE-RESPONSE QUESTION

By: Dr. Ian Kelleher, Brooks School, North Andover, MA
B.S. University of Manchester, England;
Ph.D. Cambridge University

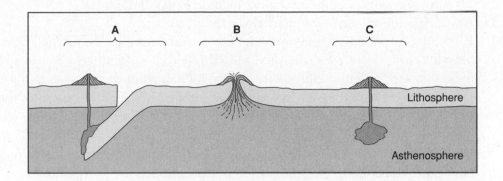

(a) Geological features A–C above are formed as a consequence of plate tectonics. For EACH feature above

(i) Identify AND describe the geological action that is occurring there
(ii) For each of the phenomena described in (i), give an actual example occurring on Earth

(b) Charles Darwin was the geologist, botanist, and zoologist on the research vessel *Beagle* when he made observations that lead to his book, *The Origin of Species*, in which the theory of evolution and natural selection was first introduced. A century later, scientists developed the theory of plate tectonics, describing how the solid Earth formed. Describe two ways in which evolution and/or speciation may occur as a consequence of plate tectonics.

(c) Mount Pinatubo in the Philippines erupted in 1991. Examine the temperature graph below and answer the following questions.

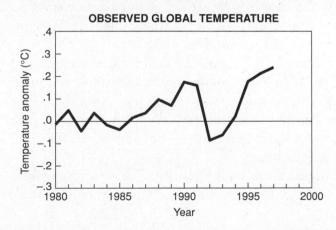

(i) Compare Earth's climate before and after the eruption of Mount Pinatubo.
(ii) Explain how the eruption of Mount Pinatubo might affect short-term and long-term climate change.

MULTIPLE-CHOICE ANSWERS AND EXPLANATIONS

1. **(A)** Igneous rocks are solidified from magma. If the magma cools slowly, the rocks are coarser in nature. Had the question been worded "the majority of the rocks on the surface of Earth," the answer would have been sedimentary.

2. **(D)** Other examples of igneous rocks include basalt and quartz.

3. **(A)** Clay is the smallest-sized particle found in soil.

4. **(E)** Acid rain causes calcium and magnesium compounds to be leached from the soil, which decreases the natural buffering effect and reduces the soil pH (makes it more acidic). Acids release toxic materials from compounds and are absorbed by vegetation (mercury, lead, cadmium, etc.). Acid rain promotes the growth of mosses that tend to retain water in the soil, thereby decreasing soil aeration (waterlogged). Mosses decrease the abundance of mycorrhizal fungi that help plants absorb nutrients. Acid rain decreases plants' resistance, making them more susceptible to disease, insects, drought, etc.

5. **(E)** Loam is soil composed of sand, silt, and clay in relatively even concentrations. Loam soils generally contain more nutrients and humus than sandy soils, have better infiltration and drainage than silty soils, and are easier to till than clay soils.

6. **(C)** Kilauea is a shield volcano that is characterized by basalt lava building enormous, low-angle, gently sloping cones. The fluid nature of the lava prevents it from piling in steep mounds. Shield volcanoes occur along the mid-oceanic ridge, where seafloor spreading is in progress and along subduction zones related to volcanic arcs. The largest volcanoes on Earth are shield volcanoes.

7. **(B)** Notice the word "presently." The Appalachian and Rocky mountains were formed at *ancient* convergent plate boundaries. The Andes lie at a convergent boundary where oceanic lithosphere is being subducted under the South American continent.

8. **(B)** Refer to the case study presented in this chapter.

9. **(C)** Water flows through sandy soils too fast for many crops and requires frequent irrigation.

10. **(B)** Soils found in grasslands are rich in organic nutrients.

11. **(D)** Refer to the section "Geologic Time Scale" in this chapter.

12. **(E)** Weathering does not produce rock.

13. **(C)** Sedimentary rock is formed by the piling of material over time. If conditions are right, organisms that die may be covered by this material and become fossilized. Fossils are impressions made up of minerals.

14. **(A)** Eight elements make up 99% of Earth's crust. In order of decreasing abundance, they are oxygen, silicon, aluminum, iron, calcium, sodium, potassium, and magnesium.

15. **(A)** If the topsoil is brown or black, it is rich in nitrogen and is good for crops. If the topsoil is gray, yellow, or red, it is low in organic matter and poor for crops.

16. **(A)** The angle of sunlight determines the season, not how close Earth is to the sun.

17. **(E)** The Richter scale is a $\log_{10}$ scale and measures the magnitude of an earthquake. $5 - 3 = \underline{2}$ and $10^2 = 100$. However, the Richter scale does NOT measure the energy of an earthquake. The energy of a Richter magnitude 5 has 32 times more energy than a Richter magnitude of 4.

18. **(A)** Soil type can substantially increase earthquake risk. The worst soils to build upon include deep, loose sand; silty clays; sand and gravel; and soft, saturated granular soils. Earthquake forces are amplified on water-saturated soils, changing the soil from a solid to a liquid, a process known as liquefaction. Liquefaction makes the ground incapable of supporting a foundation. During liquefaction, the ground can crack or heave, causing uneven settling or building collapse. The best soils to build upon to reduce damage from earthquakes are bedrock (deep and unbroken rock formations) and stiff soils. These soil types are best, since much less vibration is transferred through the foundation to the structure above.

19. **(C)** Places where plates slide past each other are called transform boundaries. The most famous transform boundary in the world is the San Andreas fault.

20. **(B)** The asthenosphere is the region below the lithosphere, estimated as being from 50 to several hundred miles (85 to several hundred km) thick.

IMPORTANT NOTE TO STUDENT

Your answers to the Free Response Questions (FRQs) will most likely be <u>much</u> shorter than what you see in this book and that is perfectly fine and expected. The answers presented in this book are very comprehensive and try to cover a wide range of possible student responses. When comparing your essays with the essays presented in this book, try to see if there are any major topics that you may have left out or if you have made any factual errors in what you did write about. Remember, you only have on average 23 minutes to write each essay.

FREE-RESPONSE ANSWER

(a) Feature *A* is a subduction zone. One lithospheric plate is subducting (sinking) below another, largely due to differences in density (the denser plate sinks). This is an example of a convergent plate boundary. As the subducted plate sinks to greater depths, the temperature increases to the point where it begins to melt. This molten magma is less dense than the solid rock around it, so it rises up and forms a chain of volcanic mountains parallel to the plate boundary. The Cascade Mountains in Washington State are examples of a volcanic arc. When two oceanic plates converge, they create an island arc—a curved chain of volcanic islands rising from the deep seafloor and near a continent. They are created by subduction processes and occur on the continent side of the subduction zone. Japan is an example of an island arc.

Feature *B* is a divergent plate boundary. Lithospheric plates are moving apart. The space created between them is filled by hot molten magma coming up from the asthenosphere that then cools and adds to the crust. In oceanic crust, they are known as a mid-oceanic ridge. The Mid-Atlantic Ridge is an example. When they form on continental crust, they are known as rift valleys, such as the African Rift Valley.

Feature *C* is a hot spot. This is a place in the asthenosphere where the temperature is higher than average such that localized melting occurs. This molten rock, being less dense due to its temperature and state of matter, rises up. It forms a volcano on Earth's surface. Over geologic time, the location of the hot spot remains constant, whereas the lithospheric plate moves over it. This causes a chain of volcanoes to form over time from a single hot spot. The Hawaiian Islands are an example of the consequences of hot spots.

(b) Evolution may occur as a consequence of geographic separation of one population of a species into two or more populations. Plate tectonics may cause this separation by either of two methods.

First, a divergent (constructive) plate boundary could cause one landmass to be divided into two or more distinct parts, perhaps even separated by an ocean. For example, identical fossils can be found on the east cost of South America and the west coast of Africa, indicating that these were once the same connected landmass. After these two continents diverged, different species would evolve from this common ancestor as a reaction to the different environments and consequent environmental pressures on the different landmasses.

Second, faulting occurring as a consequence of plate tectonics may cause a river to be diverted. The new path of the river could divide a population into two and serve as a geographic barrier preventing gene flow. Different conditions in geographically separated regions would eventually lead to the evolution of different species as each population adapted to its environment in different ways.

Plate tectonics may result in climatic change in one of the following ways: First, Earth's atmosphere has changed considerably throughout geological history, largely as a consequence of gases emitted through volcanic activity caused by plate tectonics. These changes in the atmosphere have caused climate change. For example, there is evidence that the Earth was much warmer hundreds of millions of years ago.

Second, lithospheric plates move over the surface of Earth at speeds of a few centimeters a year. Individual plates have moved thousands of miles over geologic history. As the latitude of the plate changes, so will its climate. For example, some rocks in Alaska indicate that they were originally deposited at a time when the plate had a tropical climate and so must have been closer to the equator.

Species evolve in reaction to these climate changes. For example, animals will adapt to shifting food sources as different plants grow in different climates. A species of animal may develop fur over time (through natural selection) as temperatures decrease over time.

As lithospheric plates move to latitudes further from the equator, climates will have greater seasonal variations. This could lead to evolutionary adaptations such as plants shedding their leaves to conserve resources or animals hibernating during winter months to conserve energy.

Evidence suggests that some mass extinctions in Earth's history may have been caused by large-scale volcanic activity, such as flood basalts, which occur as a consequence of plate tectonics. These mass extinctions tend to be followed by periods with high rates of evolution (punctuated) and increase in species diversity (adaptive radiation) as ecological niches are filled.

(c)

(i)   After the eruption of Mount Pinatubo, Earth's temperature was approximately 0.3°C lower for the next two years. Earth's temperature rose by approximately 0.1°C during the third year. By the fourth year after the eruption, Earth's temperature had returned to the pre-eruption level.

(ii)   In the short term, dust and other particulates released into the atmosphere from the eruption would block the Sun's rays and tend to decrease temperatures. For example, sulfur dioxide gas that is injected into the stratosphere from volcanic activity chemically reacts with oxygen to form sulfate aerosols (solid particles), which tend to reflect solar radiation back into space. The aerosol remains in suspension long after solid ash particles have fallen to Earth. Without replenishment, the sulfuric acid aerosol layer is gradually depleted. This decrease in energy reaching the Earth would result in lower global temperatures. Fine ash particles from an eruption column fall out too quickly to significantly cool the atmosphere over an extended period of time.

In the long term, gases such as carbon dioxide released during the eruption would accumulate in the stratosphere. There they would contribute to the greenhouse effect. They would absorb energy radiated back from the Earth, leading to an increase in global temperature. The degree of temperature change from the Mount Pinatubo eruption via this mechanism would be much less than that by dust blocking the sun, but the effect would last much longer.

The Atmosphere

CHAPTER 2

COMPOSITION

Earth's atmosphere is composed of seven primary compounds:

Component

NITROGEN (N_2)   78%

Fundamental nutrient for living organisms. Deposits on Earth through nitrogen fixation and reactions involving lightning and subsequent precipitation. Returns to the atmosphere through combustion of biomass and denitrification.

OXYGEN (O_2)   21%

Oxygen molecules are produced through photosynthesis and are utilized in cellular respiration.

WATER VAPOR (H_2O)   0–4%

Largest amounts occur near equator, over oceans, and in tropical regions. Areas where atmospheric water vapor can be low are polar areas and deserts.

CARBON DIOXIDE (CO_2)   <<1%

Volume of CO_2 has increased about 25% in the last 300 years due to the burning of fossil fuels and deforestation. CO_2 is produced during cellular respiration and the decay of organic matter. It is a reactant in photosynthesis. CO_2 is also a major greenhouse gas. Humans are responsible for about 5,500 million tons of CO_2 per year. The average time of a CO_2 molecule in the atmosphere is approximately 100 years.

METHANE (CH_4)   <<<1%

Methane contributes to the greenhouse effect. Since 1750, methane has increased about 150% due to the use of fossil fuels, coal mining, landfills, grazers, and flooding of rice fields. Human activity is responsible for about 400 million tons per year as compared with approximately 200 million tons per year produced naturally. Average cycle of a methane molecule in the atmosphere is approximately 10 years.

NITROUS OXIDE (N_2O)   <<<1%

Concentration increasing about 0.3% per year. Sources include burning of fossil fuels, use of fertilizers, burning biomass, deforestation, and conversion to agricultural land. Humans are responsible for about 6 million tons per year. N_2O is a contributor to the greenhouse effect. Average time of a N_2O molecule in the atmosphere is approximately 170 years.

OZONE (O_3)   <<<1%

97% of ozone is found in the stratosphere (ozone layer) 9–35 miles (15–55 km) above Earth's surface. Ozone absorbs UV radiation. Ozone is produced in the production of photochemical smog. A "hole" in the ozone layer occurs over Antarctica. Chlorofluorocarbons (CFCs) are the primary cause of the breakdown of ozone.

STRUCTURE

The atmosphere consists of several different layers:

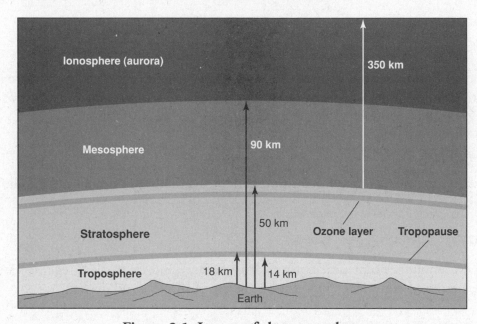

Figure 2.1 Layers of the atmosphere

Layers

TROPOSPHERE

0–7 miles (0–11 km) above surface. 75% of atmosphere's mass is in the troposphere. Temperature decreases with altitude, reaching –76°F (–60°C) near the top. Weather occurs in this zone.

STRATOSPHERE

Temperature increases with altitude due to absorption of heat by ozone. Ozone is produced by UV radiation and lightning. Contains the ozone layer.

MESOSPHERE

Temperature decreases with altitude. Coldest layer. Ice clouds occur here. Meteors (shooting stars) burn up in this layer.

TIP

Don't confuse weather and climate. Weather describes whatever is happening outdoors in a given place at a given time. Weather is what happens from minute to minute. Climate describes the total of all weather occurring over a period of years in a given place. This includes average weather conditions, regular weather sequences (like winter, spring, summer, and fall), and special weather events (like tornadoes and floods.)

THERMOSPHERE (IONOSPHERE)

Temperature increases with height due to gamma rays, X-rays, and UV radiation. Molecules are converted into ions which results in the aurora borealis (northern lights) in the northern hemisphere and the aurora australis (southern lights) in the southern hemisphere. The aurora borealis most often occurs from September to October and from March to April.

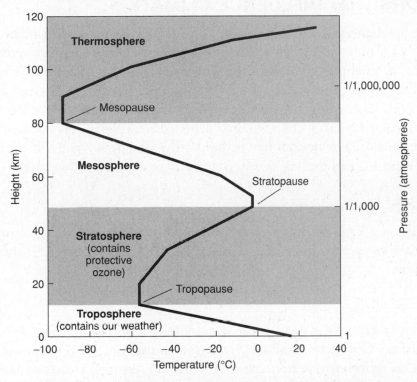

Figure 2.2 Changes in temperature in the atmosphere

WEATHER AND CLIMATE

Weather is caused by the movement or transfer of heat energy, which influences the following physical properties: temperature, air pressure, humidity, precipitation, available sunshine determined by cloud cover, wind speed, and wind direction. Climate describes the total of all weather occurring over a period of years in a given place. Energy can be transferred wherever there is a temperature difference between two objects. Energy can be transferred through radiation, conduction, and convection.

Radiation is the flow of electromagnetic radiation. It is the method by which Earth receives solar energy.

Conduction involves the transfer of heat through a substance that results from a difference in temperature between different parts of the substance.

Convection is the primary way energy is transferred from hotter to colder regions in Earth's atmosphere and is the primary determinant of weather patterns. Convection involves the movement of the warmer and therefore more energetic molecules in air. Convection takes place both vertically and horizontally. When air near the ground becomes warmer and therefore less dense than the air above it, the air rises. Pressure differences that develop because of temperature differences result in wind or horizontal convection.

TIP

Several different factors influence climate:

- Air mass
- Air pressure
- Albedo
- Altitude
- Angle of sunlight
- Clouds
- Distance to oceans
- Fronts
- Heat
- Land changes
- Latitude
- Location
- Humidity or moisture content of air
- Mountain ranges
- Pollution
- Rotation
- Wind patterns
- Human activity

Regions nearer the equator receive much more solar energy than regions nearer the poles and are consequently much warmer. These latitudinal differences in surface temperature create global-scale flows of energy within the atmosphere, giving rise to the major weather patterns of the world. Without convection and the transfer of energy, the equator would be about 27°F (15°C) warmer and the Arctic would be about 45°F (25°C) colder than they actually are.

FACTORS THAT INFLUENCE CLIMATE

Evidence for changes in the climate come from data used to measure climate (which is available for only the last few hundred years), written accounts (subjective), and data from material present at the time. These materials consists of tree rings, fossilized plants, insect and pollen samples, gas bubbles trapped in glaciers, deep ice core samples, lake sediments, stalactites and stalagmites, marine fossils including coral analysis, sediments including rafted debris, dust analysis, and isotope ratios in fossilized remains. The bottom line is that Earth's climate has gone through many cycles of warming and cooling trends. Many different factors influence the climate.

AIR MASS

An air mass is a large body of air that has similar temperatures and moisture content. Air masses can be categorized as equatorial, tropical, polar, arctic, continental, or maritime.

AIR PRESSURE

Most of the total mass of the atmosphere—99%—is within 20 miles (32 km) of Earth's surface. Gravity on an air mass results in air pressure and is measured in millibars, inches of mercury, or hectopascals (hPa). Air pressure decreases with altitude. Low pressure usually produces cloudy and stormy weather. High pressure masses contain cool, dense air that descend toward Earth's surface and becomes warmer. High pressure is usually associated with fair weather.

ALBEDO

Albedo is reflectivity. Materials like ocean water have low albedo, whereas land masses have moderate albedo. The highest albedo is snow and ice. Hence, periods when polar ice is highly extended will promote further cooling. This is a positive-feedback mechanism. Dust in the atmosphere has the same effect. It forms a high albedo veil around Earth so that a significant amount of solar radiation is reflected before it reaches the surface. The dust may come from dry climate periods, volcanic eruptions, meteor impacts, and so on.

ALTITUDE

For every 1,000 feet (300 m) rise in elevation, there is a 3°F (1.5°C) drop in temperature. Every 300 feet (90 m) rise in elevation is equivalent to a shift of 62 miles (100 km) north in latitude and biome similarity.

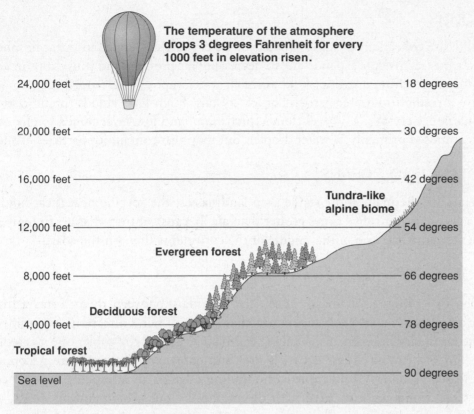

The temperature of the atmosphere drops 3 degrees Fahrenheit for every 1000 feet in elevation risen.

24,000 feet	18 degrees
20,000 feet	30 degrees
16,000 feet	42 degrees
	Tundra-like alpine biome
12,000 feet	54 degrees
Evergreen forest	
8,000 feet	66 degrees
Deciduous forest	
4,000 feet	78 degrees
Tropical forest	
Sea level	90 degrees

Figure 2.3 Change in temperature in response to change in altitude

ANGLE OF SUNLIGHT

In the Northern Hemisphere winter, Earth is closest to the sun. The angle of sunlight reaching Earth affects the climate. Areas closest to the equator receive the most sunlight and therefore higher temperatures.

CARBON CYCLE

The consumption of carbon in the form of carbon dioxide (CO_2) results in cooling. Two different processes consume carbon dioxide: carbonate rock weathering and silicate rock weathering.

Carbonate rock weathering: $\mathbf{CO_2} + H_2O + CaCO_3 \rightarrow Ca^{2+} + 2HCO_3^-$

Silicate rock weathering: $2\mathbf{CO_2} + H_2O + CaSiO_3 \rightarrow Ca^{2+} + 2HCO_3^- + SiO_2$

The production of carbon in the form of carbon dioxide results in warming. Both carbonate formation in the oceans and metamorphic breakdown of carbonate yield carbon dioxide.

Carbonate formation in the oceans: $Ca^{2+} + 2HCO_3^- \rightarrow \mathbf{CO_2} + H_2O + CaCO_3$

Metamorphic breakdown of carbonate: $SiO_2 + CaCO_3 \rightarrow \mathbf{CO_2} + CaSiO_3$

CLOUDS

Clouds are collections of water droplets or ice crystals suspended in the atmosphere. As warmer air rises, it expands due to decreasing air pressure and thus drops in temperature; therefore, it cannot hold as much water vapor. The vapor begins to condense forming tiny water particles or ice crystals. High-level clouds (prefix *cirr*) are primarily ice crystals. Midlevel clouds (prefix *alto*) and low-level clouds (prefix *strat*) are composed primarily of water droplets but may also contain ice particles or snow.

DISTANCE TO OCEANS

Oceans are thermally more stable than landmasses; the specific heat (heat-holding capacity) of water is five times greater than air. Because of this, changes in temperature are more extreme in the middle of the continents than on the coasts.

FRONTS

When two different air masses meet, the boundary between them forms a front. The air masses can vary in temperature, dew point, or wind direction. A warm front is the boundary between an advancing warm air mass and the cooler one it is replacing. Since warm air is less dense, it rises and cools, and the moisture it contains is released as rain. A cold front is the leading edge of an advancing mass of cold air. Cold fronts are associated with thunderhead clouds, high surface winds, and thunderstorms. After a cold front passes, the weather is usually cool with clear skies.

GREENHOUSE EFFECT

The most important greenhouse gases are water (H_2O), carbon dioxide (CO_2), and methane (CH_4). Without this effect, Earth would be cold and inhospitable. If taken too far, Earth could evolve into a hothouse.

HEAT (CONVECTION)

Climate is influenced by how heat energy is exchanged between air over the oceans and the air over land.

LAND CHANGES

Climate is influenced by urbanization and deforestation.

LANDMASS DISTRIBUTION

Materials absorb and reflect solar radiation to different extents. Ocean water is much more absorbent than landmasses so that continents reflect a lot more solar energy back into space than the oceans. Earth receives more solar radiation at low latitudes (near the equator) than near the poles. An Earth with landmasses clustered at low latitudes would reflect more solar energy into space, resulting in a cooler planet than one with more equatorial ocean area. Approximately 600–800 million years ago, there were significant glacial deposits in North America, Australia, and Africa. At this time, paleomagnetism of rocks suggests that these continents were near the south pole and that the equatorial Earth was largely ocean.

LATITUDE

The higher the latitudes, the less solar radiation. This affects the climate.

LOCATION

Climate is influenced by the location of high and low air pressure zones and where landmasses are distributed.

MOISTURE CONTENT OF AIR (HUMIDITY)

The moisture content of air is a primary determinant of plant growth and distribution and is a major determinant of biome type (desert vs. tropical forest). Atmospheric water vapor supplies moisture for clouds and rainfall, and it plays a role in energy exchanges within the atmosphere. Water vapor is also a greenhouse gas as it traps heat energy leaving Earth's surface. The dew point is the temperature at which condensation takes place.

MOUNTAIN RANGES

The presence or absence of mountain ranges affects the climate. Mountains influence whether one side of the mountain will receive rain or not (rain shadow effect). The side facing the ocean is the windward side and receives the most rain; the side of the mountain opposite the ocean is the leeward side and receives little rain. Temperatures decrease as the altitude increases. Orographic lifting occurs when an air mass is forced from a low elevation to a higher elevation as it moves over rising terrain. As the air mass gains altitude, it expands and cools, which can raise the relative humidity and create clouds and, under the right conditions, precipitation.

PLATE TECTONICS AND VOLCANOES

Plate tectonics affect atmospheric carbon dioxide, which factors into climate changes through the greenhouse effect. Volcanoes produce carbon dioxide. If global volcanism slows, as would be the case when supercontinents stabilize, less atmospheric carbon dioxide released into the atmosphere would trigger global cooling. Increased volcanism puts more carbon dioxide in the atmosphere and results in more greenhouse warming.

POLLUTION

Greenhouse gases are emitted from both natural sources (e.g., volcanoes) and anthropogenic (human) sources (e.g., industry, transportation, etc.).

PRECESSION

The wobble of Earth on its axis changes the amount of energy received by the sun. Changes in the orientation of Earth in space (tilt and obliquity) also have an effect on climate.

ROTATION

Daily temperature cycles are primarily influenced by Earth's rotation on its axis (once every 24 hours). At night, heat escapes from the surface. Daily minimum temperatures occur just before sunrise.

SOLAR OUTPUT

Changes in solar output of only 1% per 100 years would change Earth's temperature by up to 1°F (0.5°C). Times of sunspot activity (every 11, 90, and 180 years) correspond to decreases in solar radiation reaching Earth. The sun's magnetic field reverses every 22 years.

VOLCANOES

Sulfur-rich volcanic eruptions can eject material into the stratosphere, potentially causing tropospheric cooling and stratospheric warming. Volcanic aerosols exist in the atmosphere for an average of one to three years, and may result in tropospheric cooling. Volcanic aerosols injected into the stratosphere can also provide surfaces for ozone-destroying reactions. Over the course of millions of years, large volumes of volcanic ash deposited in the oceans can increase the iron content in seawater. This additional iron can promote biotic activity, which can lower the CO_2 concentration of seawater, and hence atmospheric CO_2 levels, resulting in global cooling. Over the course of weeks to years, ongoing production of ash from volcanoes may locally change climate by modifying the local atmosphere. Recent research also suggests that large eruptions may trigger El Niño climatic events.

WIND PATTERNS

Wind patterns are influenced by temperature, pressure differences (gradients), and the Coriolis effect.

- The Sun heats the atmosphere unevenly.
- The air closest to the surface is warmer and rises.
- Air at high elevations is cooler and sinks.
- This rising and falling sets up convection processes and is the primary cause of winds.
- Global air circulation is caused and affected by:
  - uneven heating of the Earth's surface.
  - seasons.
  - the Coriolis effect.
  - the amount of solar radiation reaching the Earth's surface over a period of time.
  - convection cells created by areas of warm ocean water which in turn are caused by differences in water density, winds, and the Earth's rotation.

During relatively calm, sunny days, the land warms up faster than the sea. This causes the air above it to become less dense than the air over the sea, which results in a sea breeze. A land breeze occurs during relatively calm, clear nights when the land cools down faster than the sea. This results in the air above the land becoming denser than the air over the sea. As a result, air moves from the land towards the coast.

Once air has been set in motion by the pressure gradient force, it undergoes an apparent deflection from its path. This apparent deflection is called the "Coriolis force" and is a result of the Earth's rotation. As air moves from high to low pressure in the Northern Hemisphere, it is deflected to the right by the Coriolis force. In the Southern Hemisphere, air moving from high to low pressure is deflected to the left by the Coriolis force. The amount of deflection the air makes is directly related to both the speed at which the air is moving and its latitude. Therefore, slowly blowing winds will be deflected only a small amount, while stronger winds will be deflected more. Likewise, winds blowing closer to the poles will be deflected more than winds at the same speed closer to the equator. The Coriolis force is zero at the equator.

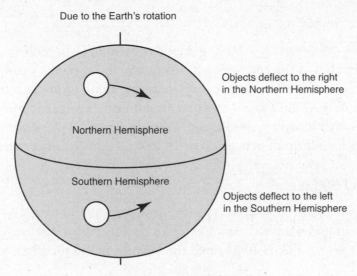

Figure 2.4 Coriolis effect

HUMAN ACTIVITY

Climate can also be influenced by human activity. Deforestation, urbanization, heat island effects, release of pollutants including greenhouse gases and the burning of fossil fuels, and the production of acid rain are examples of how humans have altered climatic patterns. Increased pollution alone, combined with an increase in convectional uplift in urban areas, tends to increase the amount of rainfall in urban areas as much as 10% when compared with undeveloped areas.

Major Climatic Periods

Several major climatic periods have occurred. They are described below.

2,000,000 B.C.E. TO 12,000 B.C.E. (PLEISTOCENE ICE AGE)

Large glacial ice sheets covered much of North America, Europe, and Asia. The Pleistocene had periods when the glaciers retreated (interglacial) because of warmer temperatures and advanced because of colder temperatures (glacial). During the coldest periods of the Pleistocene Ice Age, average global temperatures were probably 7°F–9°F (4–5°C) colder than they are today.

12,000 B.C.E. TO 3000 B.C.E.

This warming of Earth and subsequent glacial retreat began about 14,000 years ago. The warming was shortly interrupted by a sudden cooling period between 10,000–8500 B.C.E. Scientists speculate that this cooling may have been caused by the release of fresh water trapped behind ice on North America draining into the North Atlantic Ocean. The release altered vertical currents in the ocean, which exchange heat energy with the atmosphere. The warming resumed by 8500 B.C.E. By 5000 to 3000 B.C.E., average global temperatures reached their maximum level and were 2°F–4°F (1–2°C) warmer than they are today, a period known as the Climatic Optimum. During the Climatic Optimum, many of Earth's great ancient civilizations began and flourished. In Africa, the Nile River had three times its present volume, indicating a much larger tropical region.

3000 B.C.E. TO 750 B.C.E.

From 3000 to 2000 B.C.E., a cooling trend occurred. This cooling caused large drops in sea levels and the emergence of many islands (Bahamas) and coastal areas that are still above sea level today. A short warming trend took place from 2000 to 1500 B.C.E., followed once again by colder conditions. Colder temperatures from 1500 to 750 B.C.E. caused renewed ice growth in continental glaciers and alpine glaciers and a sea level drop of between 6 to 10 feet (2–3 m) below present-day levels.

750 B.C.E. TO 900 C.E.

The period from 750 B.C.E. to 900 C.E. saw warming up to 150 B.C.E. During the time of the Roman Empire (150 B.C.E. to 300 C.E.) a cooling began that lasted until about 900 C.E. At its height, the cooling caused the Nile River and the Black Sea to freeze.

900 C.E. TO 1200 C.E. (LITTLE CLIMATIC OPTIMUM)

During this warm period, the Vikings established settlements on Greenland and Iceland. The snow line in the Rocky Mountains was about 1,200 feet (370 m) above current levels. A period of cool and more extreme weather followed the Little Climatic Optimum. There are records of floods, great droughts, and extreme seasonal climate fluctuations up to the 1400s.

1550 C.E. TO 1850 C.E. (LITTLE ICE AGE)

From 1550 to 1850 C.E., global temperatures were at their coldest since the beginning of the Holocene. During the Little Ice Age, the average annual temperature of the Northern Hemisphere was about 2°F (1.0°C) lower than today.

1850 C.E. TO PRESENT

The period from 1850 to the present is one of general warming.

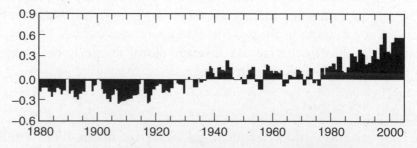

Figure 2.5 Average changes in ocean and land temperatures (°C) from 1880 to 2000

ATMOSPHERIC CIRCULATION AND THE CORIOLIS EFFECT

Due to the rotation of Earth on its axis, rotation around the sun, and the tilt of Earth's axis, the sun heats the atmosphere unevenly. Air closer to Earth's surface is the warmest and rises. Air at higher elevations is cooler and, as such, more dense and sinks. This sets up convection processes and is the primary cause for winds. Global air circulation is also affected by uneven heating of Earth's surface, seasons, the Coriolis effect, the amount of solar radiation reaching the Earth over long periods of time, convection cells created by warm ocean waters that commonly leads to hurricanes, and ocean currents, which are caused by differences in water density, winds, and Earth's rotation.

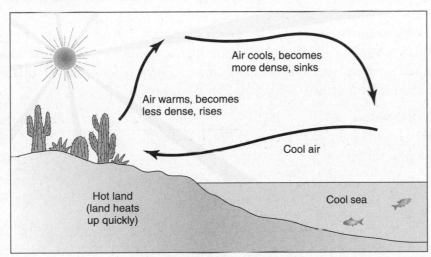

Figure 2.6 Convection cell

The trade winds are the prevailing pattern of easterly surface winds found in the tropics, within the lower portion of the Earth's atmosphere, in the lower section of the troposphere near the Earth's equator. The trade winds blow predominantly from the northeast in the Northern Hemisphere (northeast trade winds) and from the southeast in the Southern Hemisphere (southeast trade winds), strengthening during the winter. Historically, the trade winds have been used by captains of sailing ships to cross the world's oceans for centuries; they also enabled European empire expansion into the Americas and helped trade routes to become established across the Atlantic and Pacific oceans. The trade winds act as the steering flow for tropical storms that form over the Atlantic, Pacific, and south Indian oceans and make landfall in North America, Southeast Asia, and India, respectively. Trade winds also steer African dust westward across the Atlantic Ocean into the Caribbean Sea, as well as portions of southeast North America.

Horizontal winds move from areas of high pressures to areas of low pressures. Wind speed is determined by pressure differences between air masses. The greater the pressure difference is, the greater the wind speed. Wind direction is based upon from where the wind is coming. A wind coming from the east is called an easterly. Wind speed is measured with an anemometer, and wind direction is measured with a wind vane.

Earth's rotation on its axis causes winds not to travel straight. This is called the Coriolis effect. It causes prevailing winds in the Northern Hemisphere to spiral clockwise out from high-pressure areas and spiral counterclockwise in toward low-pressure areas.

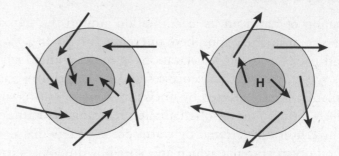

Figure 2.7 Circulation wind patterns of high- and low-pressure systems in the Northern Hemisphere. The pattern reverses in the Southern Hemisphere

The worldwide system of winds, which transports warm air from the equator where solar heating is greatest toward the higher latitudes where solar heating is diminished, gives rise to Earth's climatic zones. Three types of air circulation cells associated with latitude exist—Hadley, Ferrel, and polar.

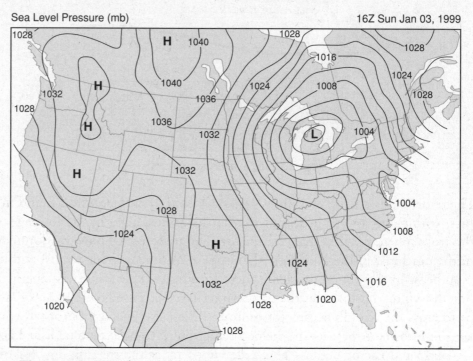

Figure 2.8 Wind speed is directly related to the distance between the isobars. The closer together they are, the stronger the wind

Hadley Air Circulation Cells

Air heated near the equator rises and spreads out north and south. After cooling in the upper atmosphere, the air sinks back to Earth's surface within the subtropical climate zone (between 25° and 49° north and south latitudes). Surface air from subtropical regions returns toward the equator to replace the rising air. The equatorial regions of the Hadley cells are characterized by high humidity, high clouds, and heavy rains. The monthly average temperatures are around 90°F (32°C) at sea level, and there is no winter. The vegetation is tropical rain forest. Temperature variation from day to night (diurnal) is greater than from season to season.

Subtropical regions of the Hadley cell are characterized by low relative humidity, little cloud formation, high ocean evaporation due to low humidity, and many of the world's deserts. The climate is characterized by warm to hot summers and mild winters.

The tropical wet and dry (or savanna) climate has a dry season more than two months long. Annual losses of water through evaporation in this region exceed annual water gains from precipitation.

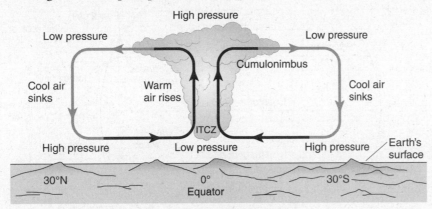

Figure 2.9 Hadley Cell

Ferrel Air Circulation Cells

Ferrel cells develop between 30° and 60° north and south latitudes. The descending winds of the Hadley cells diverge as moist tropical air moves toward the poles in winds known as the westerlies. Midlatitude climates can have severe winters and cool summers due to midlatitude cyclone patterns. The western United States is drier in summer than the eastern United States due to oceanic high pressures that brings cool, dry air down from the north. The climate of this area is governed by both tropical and polar air masses. Defined seasons are the rule, with strong annual cycles of temperature and precipitation. The seasonal fluctuation of temperature is greater than the change in temperature occurring in a 24-hour cycle. Climates of the middle latitudes have a distinct winter season. The area of Earth controlled by Ferrel cells contains broadleaf deciduous and coniferous evergreen forests.

Polar Air Circulation Cells

The polar cells originate as icy-cold, dry, dense air that descends from the troposphere to the ground. This air meets with the warm tropical air from the midlatitudes. The air then returns to the poles, cooling and then sinking. Sinking air suppresses precipitation; thus, the polar regions are deserts (deserts are defined by moisture, not temperature). Very little water exists in this area because it is tied up in the frozen state as ice. Furthermore, the amount of snowfall per year is relatively small.

In general, climates in the polar domain are characterized by low temperatures, severe winters, and small amounts of precipitation, most of which falls in summer. The annual fluctuation of temperature is greater than the change in temperature occurring in a 24-hour cycle. In this area where summers are short and temperatures are generally low throughout the year, temperature rather than precipitation is the critical factor in plant distribution and soil development. Two major biomes exist—the tundra and the taiga.

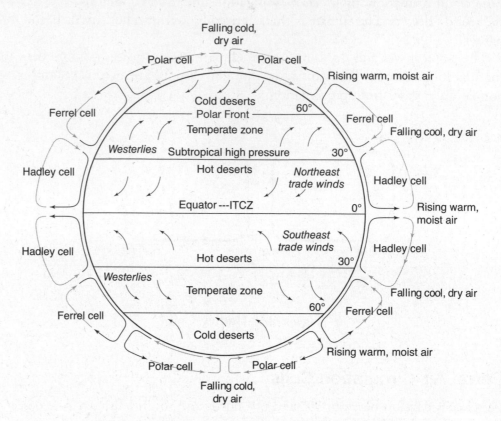

Figure 2.10 The Hadley, Ferrel, and polar cells

Hurricanes, Cyclones, and Tornadoes

Hurricanes are the most severe weather phenomenon on the planet. Hurricane Katrina that hit New Orleans, Louisiana, in 2005 was responsible for about $81 billion in damage and approximately 1,830 deaths. Hurricanes begin over warm oceans in areas where the trade winds converge. A subtropical high-pressure zone creates hot daytime temperatures with low humidity that allows for large amounts of evaporation. The Coriolis effect initiates the cyclonic flow.

The stages of hurricane development include the presence of separate thunderstorms that have developed over tropical oceans, and cyclonic circulation that begins to cause these thunderstorms to move in a circular motion. This cyclonic circulation allows them to pick up moisture and latent heat energy from the ocean. In the center of the hurricane is the eye, an area of descending air and low pressure. The energy of a hurricane dissipates as it travels over land or moves over cooler bodies of water. Rainfall can be as much as 24 inches (0.6 m) in 24 hours. A storm surge, which results from the increase in the height of the ocean near the eye of a hurricane, can cause extensive flooding.

Tornadoes are swirling masses of air with wind speeds close to 300 miles per hour (485 kph). Like hurricanes, the center of the tornado is an area of low pressure. In the United States, tornadoes are frequent from April through July and occur in the center of the United States in an area known as "Tornado Alley." Due to advances in weather forecasting, modeling, and warning systems, the death rate due to tornadoes has decreased significantly.

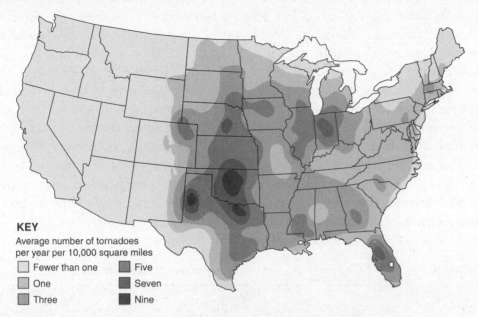

KEY

Average number of tornadoes
per year per 10,000 square miles

- ☐ Fewer than one
- ☐ One
- ☐ Three
- ☐ Five
- ☐ Seven
- ☐ Nine

Figure 2.11 Average number of tornadoes per year

While both tornadoes and tropical cyclones are spinning, turbulent vortices of wind, they have little in common. Tornadoes have diameters on the scale of hundreds of meters and are produced from a single convective storm, such as a thunderstorm. Tropical cyclones, on the other hand, have diameters of hundreds of kilometers and are comprised of many convective storms. Tornadoes occur primarily over land, as solar heating of the land surface usually contributes the development of the thunderstorm that spawns the vortex. In contrast, tropical cyclones are an oceanic phenomenon and die out over land due to the loss of a moisture source. Additionally, while tornadoes require substantial vertical shear of the horizontal winds (i.e., change of wind speed and/or direction with height) to form, tropical cyclones require very low values of vertical shear in order to form and grow. Finally, tropical cyclones have lifetimes that are measured in days, while tornadoes typically last for less than an hour.

CASE STUDY

Hurricane Katrina in 2005 was one of the deadliest and most destructive hurricanes in the history of the United States. Nearly 2,000 people died in the actual hurricane and in the subsequent floods. Total property damage was estimated at $81 billion. Katrina caused severe destruction along the Gulf coast from central Florida to Texas, much of it due to the storm surge. The largest number of deaths occurred in New Orleans, Louisiana, which flooded as the levee system failed and 80% of the city and the surrounding areas became flooded. The worst property damage occurred in coastal areas, such as Mississippi beachfront towns with storm surge waters reaching 6–12 miles (10–19 km) inland.

Monsoons

Monsoons are strong, often violent winds that change direction with the season. Monsoon winds blow from cold to warm regions because cold air takes up more space than warm air. Monsoons blow from the land toward the sea in winter and from the sea toward land in the summer. India's climate is dominated by monsoons. During the Indian winter, which is hot and dry, the monsoon winds blow from the northeast and carry little moisture. The temperature is high because the Himalayas form a barrier that prevents cold air from passing onto the subcontinent. Furthermore, most of India lies between the Tropic of Cancer and the equator, so the sun's rays shine directly on the land. During the summer the monsoons move onto the subcontinent from the southwest. The winds carry moisture from the Indian Ocean and bring heavy rains from June to September. Farmers in India rely on these torrential summer rainstorms to irrigate their land. Additionally, a large amount of India's electricity is generated by water power provided by the monsoon rains.

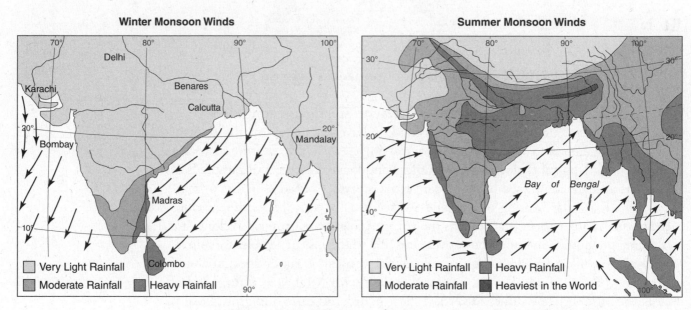

Figure 2.12 **Winter and summer monsoon wind patterns**

Rain Shadow Effect

A rain shadow is a dry area on the mountainside facing away from the direction of the wind. The mountains block the passage of rain-producing weather systems, casting a "shadow" of dryness behind them. Warm, moist air rises through orographic lifting to the top of a mountain range or large mountain, where, due to decreasing atmospheric pressure with increasing altitude, it expands and cools to reach its dew point. At the dew point, moisture condenses onto the mountain, and it precipitates on the top and windward sides of the mountain. The air descends on the leeward side, but due to the process of precipitation, it has lost much of its initial moisture. Typically, descending air also gets warmer down the leeward side of the mountain, creating an arid region.

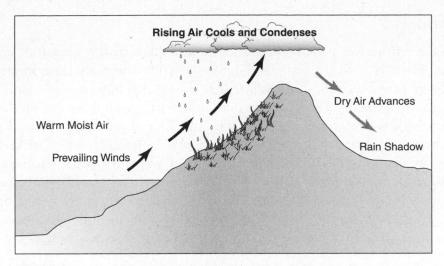

Figure 2.13 Rain shadow effect

EL NIÑO AND LA NIÑA
Normal State

During normal (non-El Niño or "La Nada") conditions, easterly trade winds move water and air warmed by the sun toward the west (Walker circulation). The ocean is generally around 24 inches (60 cm) higher in the western Pacific and the water about 14°F warmer. The trade winds, in piling up water in the western Pacific, make a deep—450 feet (150 m)—warm layer in the west that pushes the thermocline down while it rises in the east. The shallow—90 feet (30 m)—eastern thermocline allows the winds to pull up nutrient-rich water from below, which increases fishing stocks. The western side of the equatorial Pacific is characterized by warm, wet low-pressure weather, as the collected moisture is released in the form of typhoons and thunderstorms.

TIP

Questions about El Niño and La Niña are very common on the APES exam. Be sure you know these two processes!

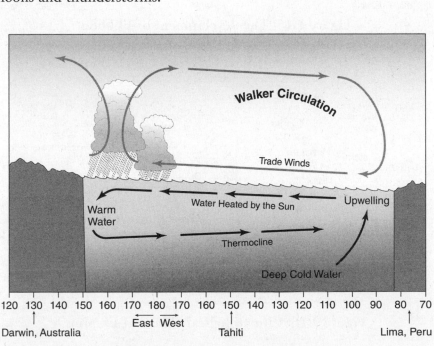

Figure 2.14 Normal conditions

El Niño

When the air pressure patterns in the South Pacific reverse direction (the air pressure at Australia is higher than at Tahiti), the trade winds decrease in strength and can reverse direction. The result is that the normal flow of water away from South America decreases and ocean water piles up off South America. This pushes the thermocline deeper and decreases the upwelling of nutrient-rich deep water, which results in extensive fish kills off the South American coast. With a deeper thermocline and decreased westward transport of water, the sea surface temperature increases in the eastern Pacific. This is the warm phase of El Niño-Southern Oscillation (ENSO) called El Niño. The net result is a shift of the prevailing rain pattern from the normal western Pacific to the central Pacific; rainfall is more common in the central Pacific while the western Pacific becomes relatively dry.

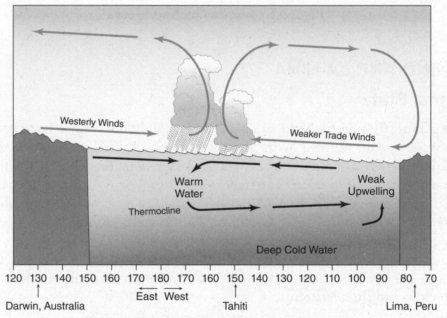

Figure 2.15 The development of El Niño

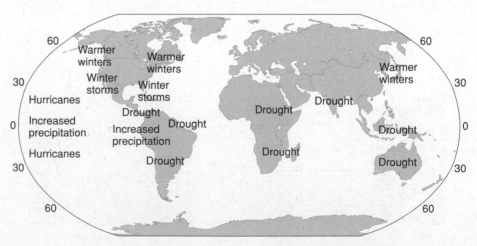

Figure 2.16 Climatological effects of El Niño

La Niña

There are occasions when the trade winds that blow west across the tropical Pacific are stronger than normal, leading to increased upwelling off South America and hence *cooler-than-normal* sea surface temperatures. The prevailing rain pattern also shifts farther west than normal. These winds pile up warm surface water in the western Pacific. This is the cool phase of ENSO called La Niña. La Niña is characterized by unusually cold ocean temperatures in the eastern equatorial Pacific. La Niña tends to bring nearly the opposite effects of El Niño to the United States, with wetter-than-normal conditions across the Pacific Northwest and both dryer and warmer-than-normal conditions in the southern states. Winter temperatures are warmer than normal in the southeastern United States and cooler than normal in the northwest. The increased temperatures in the southeast during La Niña years correlate with the substantial increase in hurricanes that occurs during the same time period. La Niña is also responsible for heavier-than-normal monsoons in India and Southeast Asia.

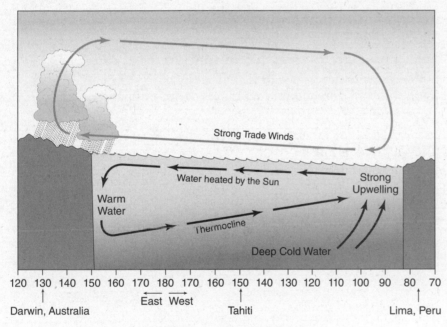

Figure 2.17 La Niña

QUICK REVIEW CHECKLIST

☐ **Atmosphere Composition**
  ☐ Nitrogen, oxygen, water vapor, carbon dioxide, methane, nitrous oxide, ozone

☐ **Structure of the Atmosphere**
  ☐ Ionosphere, mesosphere, stratosphere, troposphere

☐ **Weather and Climate**
  ☐ Differences between weather and climate
  ☐ Factors that affect climate: air mass, air pressure, albedo, altitude, angle of sunlight, clouds, distance to oceans, fronts, heat, land changes, latitude, location, humidity, mountain ranges, pollution, rotation, wind patterns
  ☐ Human activity and influence on climate

☐ **Atmospheric Circulation**
  ☐ Coriolis effect
  ☐ Types of winds (breezes): sea breeze, land breeze
  ☐ Hadley, Ferrel, and polar cells
  ☐ Hurricanes and Tornadoes: what causes them, mitigation, case studies (Katrina)

☐ **El Niño and La Niña**
  ☐ How they develop
  ☐ Environmental effects

MULTIPLE-CHOICE QUESTIONS

1. The zone of the atmosphere in which weather occurs is known as the

   (A) ionosphere
   (B) mesosphere
   (C) troposphere
   (D) thermosphere
   (E) stratosphere

2. 99% of the volume of gases in the lower atmosphere, listed in descending order of volume, are

   (A) O_2, N_2, CO_2, H_2O
   (B) H_2O, N_2, O_2, CO_2
   (C) O_2, CO_2, N_2, H_2O
   (D) CO_2, H_2O, O_2, N_2
   (E) N_2, O_2, H_2O, CO_2

3. Regional climates are most influenced by

   (A) latitude and altitude
   (B) prevailing winds and latitude
   (C) altitude and longitude
   (D) latitude and longitude
   (E) Coriolis effect and trade winds

4. A low-pressure air mass is generally associated with

   (A) hot, humid weather
   (B) fair weather
   (C) tornadoes
   (D) cloudy or stormy weather
   (E) hurricanes

5. La Niña would produce all the following effects EXCEPT

   (A) more rain in southeast Asia
   (B) wetter winters in the Pacific Northwest region of the United States
   (C) warmer winters in Canada and northeast United States
   (D) warmer and drier winters in the southwest and southeast United States
   (E) more Atlantic hurricanes

6. On the leeward side of a mountain range, one would expect

   (A) more clouds and rain than on the windward side
   (B) more clouds but less rain than on the windward side
   (C) colder temperatures
   (D) less clouds and less rain than on the windward side
   (E) no significant difference in climate compared with the windward side

7. The ozone layer exists primarily in what section of the atmosphere?

(A) Troposphere
(B) Stratosphere
(C) Mesosphere
(D) Thermosphere
(E) Ionosphere

8. Along the equator,

(A) warm, moist air rises
(B) warm, moist air descends
(C) warm, dry air descends
(D) cool, dry air descends
(E) cool, moist air descends

9. The gas that is responsible for trapping most of the heat in the lower atmosphere is

(A) water vapor
(B) ozone
(C) carbon dioxide
(D) oxygen
(E) nitrogen

10. Characteristics or requirements of a monsoon include all of the following EXCEPT

(A) a seasonal reversal of wind patterns
(B) large land areas cut off from continental air masses by mountain ranges and surrounded by large bodies of water
(C) different heating and cooling rates between the ocean and the continent
(D) extremely heavy rainfall
(E) heating and cooling rates between the oceans and the continents that are equal

11. An atmospheric condition in which the air temperature rises with increasing altitude, holding surface air down and preventing dispersion of pollutants, is known as (a)

(A) temperature inversion
(B) cold front
(C) warm front
(D) global warming
(E) upwelling

12. The surface with the lowest albedo is

    (A) snow
    (B) ocean water
    (C) forest
    (D) desert
    (E) black topsoil

13. Jet streams over the U.S. travel primarily

    (A) north to south
    (B) south to north
    (C) east to west
    (D) west to east
    (E) in many directions

14. The correct arrangement of atmospheric layers, arranged in order from the most distant from Earth's surface to the one closest to Earth's surface, is

    (A) troposphere, stratosphere, mesosphere, thermosphere
    (B) thermosphere, mesosphere, stratosphere, troposphere
    (C) stratosphere, troposphere, mesosphere, thermosphere
    (D) thermosphere, mesosphere, troposphere, stratosphere
    (E) None of the above

TIP

Be careful when answering questions that involve placing items in order that you have the order going in the proper direction. A common trick is to have the correct choices in the opposite direction.

15. Areas of low pressure are typically characterized by _____ air and move toward regions where the pressure is _____ with time.

    (A) sinking, falling
    (B) rising, falling
    (C) sinking, rising
    (D) rising, rising
    (E) None of the above

16. The global circulation pattern that dominates the tropics is called the

    (A) Ferrel cell
    (B) Polar cell
    (C) Bradley cell
    (D) Hadley cell
    (E) Tropical cell

17. The date with the shortest amount of daylight in the Southern Hemisphere is

    (A) June 22
    (B) January 1
    (C) December 21
    (D) September 1
    (E) March 18

18. Jet streams follow the sun in that as the sun's elevation _____ (increases, decreases) each day in the spring, the jet stream shifts by moving _____ (north, south) during the Northern Hemisphere spring.

   (A) increases, north
   (B) increases, south
   (C) decreases, north
   (D) decreases, south
   (E) None of the above

19. Usually, fair and dry/hot weather is associated with high pressure around _____ latitude with rainy and stormy weather associated with low pressure around _____ latitude.

   (A) 0 degrees N/S, 90 degrees E/W
   (B) 90 degrees E/W, 90 degrees N/S
   (C) 30 degrees N/S, 50–60 degrees N/S
   (D) 50–60 degrees N/S, 30 degrees N/S
   (E) 45 degrees N/S, 45 degrees E/W

20. The three necessary ingredients for thunderstorm formation are

   (A) moisture, lifting mechanism, instability
   (B) lifting mechanism, mountains, oceans
   (C) stability, moisture, heat
   (D) lifting mechanism, fronts, moisture
   (E) deserts, mountains, clouds

TIP

Before writing your essays, be sure to map out or brainstorm what you are going to write about. A few minutes planning and organizing your essays will get you a much higher score.

FREE-RESPONSE QUESTION

(a) Choose ONE of the following: Hadley cell, Ferrel cell, or polar cell. In your description, include the following:

   (i)   A description of the type of cell
   (ii) An explanation of how that cell develops
   (iii) The cell's location with respect to the equator

(b) Describe the characteristics of that cell in terms of climatic conditions. In your description, include temperature, relative humidity, and prevailing winds.

   (i) Identify and describe ONE biome that would exist at sea level within the specific latitudes of that cell. In your description, give examples of both plants and animals that would be found within that biome.

MULTIPLE-CHOICE ANSWERS AND EXPLANATIONS

1. **(C)** The troposphere is the atmospheric layer closest to Earth and extends for about 11 miles (18 km) above Earth at the equator and about 5 miles (8 km) above Earth at the poles. Temperature declines as altitude increases.

2. **(E)** Nitrogen (78%), oxygen (21%), water vapor (about 0–4%), and the rest below 1%.

3. **(A)** Latitude expresses how far north or south of the equator a location is. The equator is 0° latitude, and the poles are at 90°. For every 1,000 feet (300 m) in altitude, there is a 3°F (1.5°C) drop in temperature.

4. **(D)** A low-pressure air mass (low) occurs when warm air, which is less dense than cooler air, spirals inward toward the center of a low-pressure area. Since the center of the low-pressure area is of even less density and pressure, the air in this section rises and the warm air cools as it expands. The temperature begins to fall and may go below the dew point—the point at which air condenses into water droplets. These water droplets make up clouds. If the droplets begin to coalesce on condensation nuclei, rain follows.

5. **(C)** During La Niña, large portions of central North America experience increased storminess, increased precipitation, and an increased frequency of significant cold-air outbreaks, while the southern states experience less storminess and precipitation. Also, there tends to be considerable month-to-month variations in temperature, rainfall, and storminess across central North America during the winter and spring seasons.

6. **(D)** The rain shadow effect occurs on the leeward side of a mountain, the side away from the ocean. Moist air from the ocean rises when it hits mountains, cools, and loses its moisture on the windward side. On the leeward side, air is much drier. For example, the western side of the Sierra Nevada Mountain Range in California is much wetter than the eastern side.

7. **(B)** 97% of ozone (O_3) is found in the lower stratosphere, which is 9 to 35 miles (15–55 km) above Earth's surface. Temperature increases with altitude in the stratosphere due to absorption of heat energy by ozone molecules.

8. **(A)** Hadley cells occur between 0° and 25° north and south latitudes (equatorial region). In this area, there is upward air motion, cooling of the air due to uplift, high humidity, high clouds, and heavy rains.

9. **(A)** Water vapor is present in such abundance throughout the atmosphere that it acts like a blanket of insulation, trapping heat and forcing surface temperatures higher than they would be otherwise. Water vapor is roughly eight times more effective than carbon dioxide as a greenhouse gas.

10. **(E)** During monsoon season, winds blow from cooler ocean areas (higher pressure) to warmer landmasses (lower pressure). As the air rises over the land masses, it cools and is unable to retain water, producing great amounts of precipitation. In winter, the ocean is now warmer and the cycle reverses. Drier air travels from the land out to the ocean. Monsoons exist in Australia, Africa, and North and South America.

11. **(A)** Temperature inversions are atmospheric conditions in which the air temperature rises with increasing altitude, holding surface air down and preventing dispersion of pollutants.

12. **(E)** Albedo is a measure of reflection of sunlight from a surface. Of the choices, dark topsoil absorbs the most energy and therefore reflects the least amount of energy, resulting in the lowest albedo.

13. **(D)** Jet streams are large-scale upper air flows that travel from west to east and are produced by differences in temperature. They can travel as fast as 250 miles per hour (400 kph) and travel between 3 and 8 miles (5–13 km) above Earth's surface.

14. **(B)** Remember, the question required you to place the layers in order from the most distant to the closest.

15. **(B)** In a high-pressure system, air pressure is greater than the surrounding areas. This difference in air pressure results in wind. In a high-pressure area, air is denser than in areas of lower pressure. The result is that air will move from the high-pressure area to an area of lower pressure. Clear skies and fair weather usually occur in these regions. On the other hand, winds tend to blow into a low-pressure system because air moves from areas of higher pressure into areas of lower pressure. As winds blow into a low-pressure system, the air moves up. This upward flow of air can cause clouds, strong winds, and precipitation to form.

16. **(D)** Hadley cells dominate the tropics.

17. **(A)** The seasons in the Southern Hemisphere are opposite that of the Northern Hemisphere. For example, spring in the Northern Hemisphere is fall in the Southern Hemisphere.

18. **(A)** The position of the jet stream also determines where the storm track is. As the jet stream moves north during spring, the storm track moves north, leaving the southern plains of Texas and Oklahoma and moving into the northern plains near the Dakotas.

19. **(C)** Except for a few locations, most of the world's deserts are located along 30 degrees N/S latitude with lush forests from abundant rains located around 50–60 degrees N/S latitude.

20. **(A)** Moisture, a lifting mechanism, and instability are all needed for thunderstorms to form. The moisture is needed for rain. The lifting mechanism is needed to get the air moving initially in an upward direction, and the unstable atmosphere insures the upward-moving air continues to do so.

FREE-RESPONSE ANSWER

Let's do this essay together, using it as a teaching tool rather than just providing an answer and rubric. Let's choose the Hadley cell for this example.

The first step is to brainstorm. Write a list of key words that would apply to the question. Remember that the order of the keywords is not important, we will put them into the correct order later. Here is a sample of key words: Hadley, Ferrel, polar, temperature, solar insolation, humidity, biomes, plants, and animals.

Now that we have around 10 key words, let's expand the list by adding details—items that we will discuss in our essay. We can also begin to map out the order in which we will discuss the items.

- Hadley: 0° to 30°, deserts, equatorial regions, tropical rain forests, subtropical areas, savannas
- Location
- Temperature (heat moves from equator to colder areas)
- Relative humidity

- Prevailing winds
- Solar insolation
- Biomes
- Animals
- Plants

Remember, you have 23 minutes to spend on this essay, so do not spend more than 5 minutes in organization. Now look over the list and add anything that you have missed. Now it is time to begin writing.

The first step is to write a thesis statement: **"The world's biomes are primarily determined by climatic conditions. Deserts are characterized as areas of low precipitation, while tropical rain forests are characterized by areas of high precipitation."**

The next step is to begin describing what determines climatic conditions that, in turn, affect the type of life present within that zone. **"Solar insolation, that is the amount of sunlight received on Earth, is greatest at the equator and diminishes toward the poles. Since heat flows from warmer regions to cooler regions, the warmer air produced at the equator moves through major worldwide wind patterns that distribute this energy worldwide. As one moves from the equator to the poles, there are three major air circulation cells—Hadley, Ferrel, and polar."** At this point, a labeled sketch would be helpful.

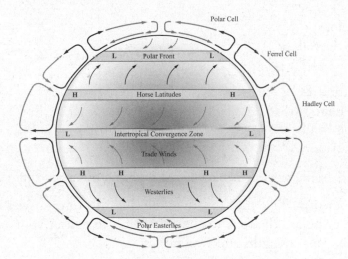

TIP

When writing your FRQ essays, a picture is worth a thousand words! If you can sketch out a labeled diagram of what you are describing it will go a long way in improving your score.

Now, for the remaining time, we can refer to the detailed essay outline and complete the essay, adding details and examples where necessary.

From the equator to 30° north and south latitudes, Hadley cells exist. Since this area of Earth receives the greatest solar radiation due to Earth's axis tilt, this area of Earth is the warmest. Near the equator, this warm, moist air rises. As the warm air rises, it begins to cool and become denser. Since cooler air cannot hold as much water vapor as warmer air, the humidity of the air increases to the point where clouds are produced. This, in turn, causes great amounts of rainfall. Monthly average temperatures are quite high at sea level, and there is no winter. Vegetation near sea level is tropical rain forest. In these tropical systems, temperature variations from day to night are greater than

from season to season. Tropical rain forests, which extend about 1,500 miles (2,400 km) north and south of the equator, are found in South and Central America, Africa, and southeast Asia. Climatic conditions can include rain throughout the year, monsoons—a short, dry season followed by a heavy, rainy period, and tropical savanna with characteristic wet and dry seasons.

Tropical rain forests have characteristically high-species plant and animal diversity. Vegetation is dense. Bromeliads, orchids, ferns, and palms are present. Leaves are large in an effort to absorb sunlight, and there is little need to conserve water lost through transpiration. Soils are characteristically low in nutrients with the nutrients being stored in vegetation. Soil is characteristically acidic. Decomposition of organic material is high due to temperature and moisture. Leaching of soil nutrients is high; therefore, soil quality is very low. Abundant insects and animal biodiversity are characteristic. Examples of animals that one might find in a tropical rain forest biome might include numerous species of butterflies, ants, mosquitoes, millipedes, bats, monkeys, sloths, tarsiers, hippopotamuses, macaws, toucans, parrots, anacondas, alligators, and numerous species of frogs.

At this point, we think we are done. Our last job is to be sure that we have answered all questions. Let's put a check beside each topic that we answered and that we were required to address:

- Describe the cell ✔
- How it develops ✔
- Location of the cell ✔
- Characteristics of the cell

  1. Temperature ✔
  2. Relative humidity ✔
  3. Prevailing winds ✔
  4. Solar insolation ✔

- List examples of plants and animals living within a biome in that cell ✔

Now it is your turn. Take the same question, but this time answer it in terms of either a Ferrel or polar cell. Your teacher may wish to collect your essay and give you pointers on your strengths and weaknesses.

Global Water Resources and Use

CHAPTER 3

FRESHWATER AND SALTWATER

Over 70% of Earth's surface is covered by water. Oceans hold about 97% of all water on Earth, while freshwater constitutes about 3%. Of the freshwater that is available, most of it is trapped in glaciers and ice caps. The rest is found (in descending order) in groundwater, lakes, soil moisture, atmospheric moisture, rivers, and streams.

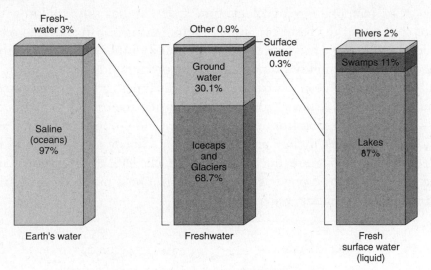

Figure 3.1 Distribution of Earth's water

Water has many unique properties:

1. Strong hydrogen bonds hold water molecules to each other.

2. The temperature of water changes slowly due to its high specific heat capacity.

3. Water has a high boiling point.

4. A lot of energy is needed to evaporate water.

5. Water dissolves many compounds.

6. Water filters out harmful UV radiation in aquatic ecosystems.

7. Water adheres to many solid surfaces.

8. Water expands when it freezes.

Most human settlements are determined by the availability of freshwater. The highest per capita supplies of freshwater are in countries with high precipitation and small populations (Iceland, Norway, and so on). Lowest per capita freshwater supplies are in areas with low rainfall and high populations (Egypt, Israel, and so on).

The use of freshwater, a limited resource, is growing at twice the rate of population growth. In the United States, the average amount of freshwater allocated per person for all purposes is approximately 500,000 gallons (1,900,000 l) per year.

LAKES

Most lakes on Earth are located in the Northern Hemisphere at higher latitudes and are generally found in mountainous areas, rift zones, areas with ongoing or recent glaciations, or along the courses of mature rivers. Processes that form lakes include: (1) tectonic uplift of a mountain range that creates a depression that accumulates water; (2) advance and retreat of glaciers that scrape depressions in the Earth's surface where water accumulates (e.g., the Great Lakes of North America); (3) salt or saline lakes that form where there is no natural outlet or where the water evaporates rapidly and the drainage surface of the water table has a higher-than-normal salt content (e.g., the Great Salt Lake, the Aral Sea, and the Dead Sea); (4) oxbow lakes formed by erosion in river valleys; and (5) crater lakes formed in volcanic craters and calderas that fill up with water more rapidly than they empty (e.g., Crater Lake in Oregon).

All lakes are temporary over geologic time scales, as they slowly fill in with sediments or spill out of the basin containing them. Changes in the level of a lake are controlled by the difference between the input and output compared to the total volume of the lake. Significant input sources are precipitation onto the lake, runoff carried by streams and channels from the lake's catchment area, groundwater channels and aquifers, and artificial sources from outside the catchment area. Output sources include evaporation from the lake, surface and groundwater flows, and any extraction of lake water by humans. Variations in climate conditions and human water requirements will create fluctuations in the lake level. Artificial lakes are constructed for hydroelectric power generation, recreational purposes, industrial use, agricultural use, or domestic water supply.

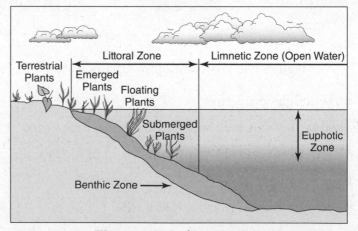

Figure 3.2 Lake zonation

Lakes have three zones: the littoral zone, which is a sloped area close to land; the photic or open-water zone, where sunlight is abundant; and the deep-water benthic zone. The depth to which light can reach in lakes depends on turbidity, or the amount and type of suspended particles in the water. These particles can be either sedimentary (i.e., silt) or biological (e.g., algae or detritus) in origin.

The material at the bottom of a lake can be composed of a wide variety of inorganic material, such as silt or sand, and organic material, such as decaying plant or animal matter. The composition of the lake bed has a significant impact on the flora and fauna found near the lake, as it contributes to the amount and the types of nutrients available. A lake may be in-filled with deposited sediment and gradually become wetland.

Oligotrophic lakes are generally clear due to low nutrient levels and have little plant life. Mesotrophic lakes have good clarity and an average level of nutrients. Eutrophic lakes are enriched with nutrients, resulting in large amounts of plant growth with possible algal blooms. Hypertrophic lakes have been excessively enriched with nutrients, have poor water clarity, and are subject to devastating algal blooms. These lakes usually result from human activities, such as heavy use of fertilizers or sewage outlets in the lake catchment area. Such lakes are of little use to humans and have a poor ecosystem due to decreased amounts of dissolved oxygen. Oligotrophic lakes are lakes with low primary productivity, the result of low nutrient content. These lakes have low algal production, and consequently, often have very clear waters, with high drinking water quality.

Because of the high specific heat capacity of water, lakes moderate the surrounding region's temperature and climate. In the daytime, a lake can cool the land beside it with local winds, resulting in a sea breeze; at night, it can warm it with a land breeze.

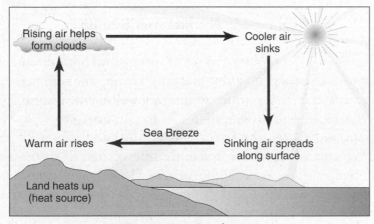

Figure 3.3 Sea breeze

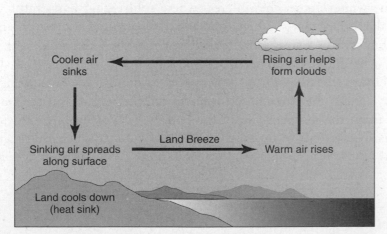

Figure 3.4 Land breeze

The stratification or layering of water in lakes is due to density changes caused by changes in temperature. The density of water increases as temperature decreases until it reaches its maximum density at about 39°F (4°C), causing thermal stratification—the tendency of deep lakes to form distinct layers in the summer months. Deep water is insulated from the sun and stays cool and denser, forming a lower layer called the hypolimnion. The surface and water near the shore are warmed by the sun, making them less dense, so that they form a surface layer called the epilimnion.

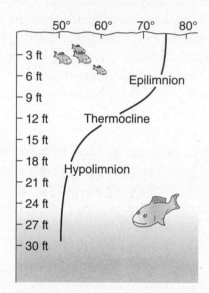

Figure 3.5 Thermal stratification

Seasonal turnover refers to the exchange of surface and bottom water in a lake or pond that happens twice a year (spring and fall). During the summer, the sun heats water near the surface of lakes, which results in a well-defined warm layer of water occurring over a cooler one (stratification). As the summer progresses, temperature differences increase between the layers, and a thin middle layer, or thermocline, develops, where a rapid transition in temperature occurs. With the arrival of fall and cooler air temperatures, water at the surface of lakes begins to cool and becomes heavier. During this time, strong fall winds move the surface water around, which promotes mixing with deeper water—a condition known as *fall turnover*. As the mixing continues, lake water becomes more uniform in temperature and oxygen level. As the winter approaches in areas where subfreezing temperatures are common, the lake surface temperatures approach the freezing mark (water is most dense at 4°C). Thus, as lake waters move toward freezing, the water sinks to the lake bottom when it reaches 4°C. Colder water remains above, perhaps eventually becoming capped by an ice layer, which further prevents the winds from stirring the water mass. With spring, the surface ice begins to melt, and cold surface waters warm until they reach the temperatures of the bottom waters, again producing a fairly uniform temperature distribution throughout the lake. When this occurs, winds blowing over the lake again set up a full circulation system known as *spring turnover*.

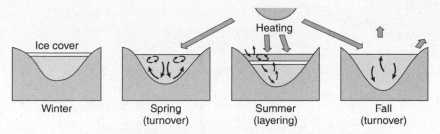

Figure 3.6 Seasonal lake turnover

WETLANDS

Wetlands include swamps, estuaries, marshes, and bogs. Wetlands are characterized by a water table that stands at or near the land surface for a long enough season each year to support aquatic plants. They are considered the most biologically diverse of all ecosystems and occur where the soil is either permanently or seasonally saturated with moisture, often covered partially or completely by shallow pools of water. The water found in wetlands can be saltwater, freshwater, or brackish (water that has more salinity than freshwater, but not as much as seawater). Plant life found in wetlands includes mangrove, water lilies, cattails, sedges, tamarack, black spruce, and cypress. Animal life includes many different amphibians, reptiles, birds, and mammals.

Wetlands have historically been drained for real estate development or flooded for use as recreational lakes. By 1993, half of the world's wetlands had been drained. Wetlands provide a valuable flood control function and are very effective at filtering and cleaning water.

AQUIFERS

An aquifer is a geologic formation that contains water in quantities sufficient to support a well or spring. Unconfined aquifers have as their upper boundary the water table. Typically (but not always), the shallowest aquifer at a given location is unconfined, meaning it does not have a confining layer between it and the surface. Unconfined aquifers usually receive recharge water directly from the surface, from precipitation or from a body of surface water (e.g., a river, stream, or lake). Confined aquifers have the water table above their upper boundary and are typically found below unconfined aquifers. The term "perched" refers to groundwater accumulating above an area of low permeability such as clay. The unsaturated zone is directly below the surface and contains some water. In the unsaturated zone, water and air fill the voids between soil or rock particles. Deeper in the ground is the zone of saturation. In the zone of saturation, the subsurface is completely saturated with water. The point where the zone of aeration meets the zone of saturation is known as the water table. Water table levels fluctuate naturally throughout the year based on seasonal variations. In addition, the depth of the water table varies.

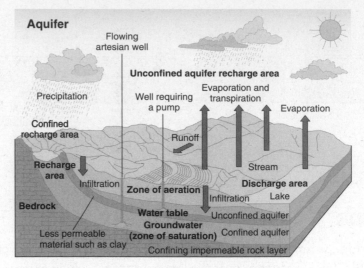

Figure 3.7 A typical aquifer

OCEANS

Approximately 71% of the Earth's surface is covered by the oceans. More than half of this area is below 10,000 feet (3,000 m) deep, with the average salt content of seawater being around 3.5%. The oceanic crust is composed of a dense, thin layer of solidified volcanic basalt as compared to the thicker but less dense continental crust, which is composed primarily of granite. The ocean floor spreads from mid-ocean ridges, where two tectonic plates adjoin. Where two plates move toward each other, one plate subducts or moves under another plate (oceanic or continental), leading to an oceanic trench.

Oceans have a significant effect on the biosphere, as oceanic evaporation is the primary source for precipitation and ocean temperatures affect climate and wind patterns. Approximately 250,000 marine life-forms are currently known, with many times that number yet to be discovered. Oceans are divided into specific zones:

Oceanic Zones	
Aphotic	The depths beyond which less than 1% of sunlight penetrates.
Benthic	The ecological region at the lowest level of a body of water.
Disphotic	The zone that is dimly lit and does not have enough light penetrating from the surface to carry out photosynthesis.
Neritic	Extends from the low tide mark to the edge of the continental shelf, with a relatively shallow depth extending to about 650 feet (200 m). Generally well-oxygenated water, low water pressure, available light for photosynthesis, and relatively stable temperature and salinity levels. High biodiversity. Also known as sublittoral or photic zone.
Oceanic	The region of open sea beyond the edge of the continental shelf; includes 65% of the ocean's open water.
Pelagic	Includes all open ocean regions.
Photic (Euphotic)	The depth of the water that is exposed to sufficient sunlight for photosynthesis to occur. Biologically diverse.

Ocean Zonation (not to scale)

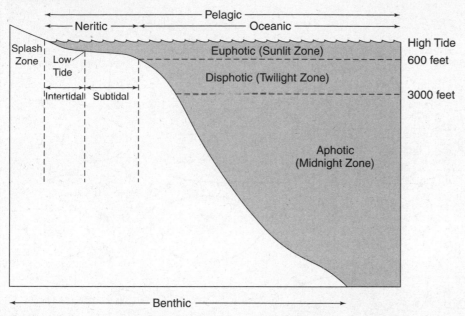

Figure 3.8 Ocean zonation

OCEAN CIRCULATION

The Northern Hemisphere is dominated by land and the Southern Hemisphere by oceans. Temperature differences between summer and winter are more extreme in the Northern Hemisphere because the land warms and cools more quickly than water. Heat is transported from the equator to the poles mostly by atmospheric air currents but also by oceanic water currents. The warm waters near the surface and colder waters at deeper levels move by convection. Changes in ocean temperatures have a direct bearing on ocean currents. During summers, a thermocline develops in ocean waters between the warm surface water and the cooler bottom water.

Surface ocean currents are driven by wind patterns that result from the flow of high thermal energy sources generated at the tropics (higher pressure) to low-energy sources in polar areas (lower pressure). They serve to distribute the heat generated near the tropics. Deep-water, density-driven currents are controlled primarily by differences in temperature and salt content. Denser, saltier water sinks, and less-dense water rises. About 90% of the ocean volume circulates due to density differences in temperature and salinities, while the remaining 10% is involved in wind-driven surface currents. In the Northern Hemisphere, north-flowing currents are warm (originating near the equator), and south-flowing currents are colder (originating from the Arctic area).

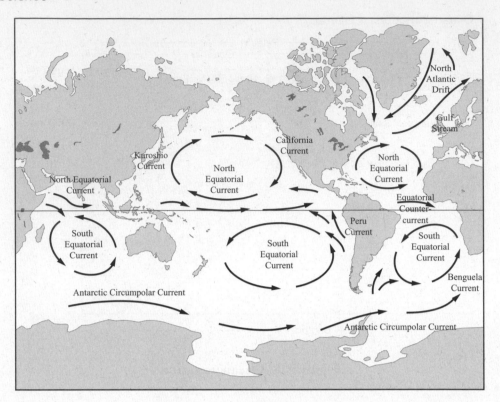

Figure 3.9 Ocean surface currents

Ocean water has warmed significantly during the past 50 years. The greatest amount of warming has occurred in the top surface layers of the ocean. The temperature of the Antarctic Southern Ocean rose by 0.31°F (0.17°C) between the 1950s and the 1980s—twice the rate for the world's oceans as a whole. Since the 1950s, the California Current that runs southward along the west coast of the United States has risen about 2.7°F (1.5°C) and has resulted in a significant decrease in plankton with resulting rippling effects within the food web. Possible reasons for dramatic increases in ocean temperatures include:

1. Significant slowing of the ocean circulation that transports warm water to the North Atlantic

2. Large reductions in the Greenland and West Antarctic ice sheets

3. Accelerated global warming due to carbon cycle feedbacks in the terrestrial biosphere

4. Decreases in upwelling

5. Releases of terrestrial carbon from permafrost regions and methane from hydrates in coastal sediments.

The Gulf Stream transports warm water from the Caribbean northward. A branch of the Gulf Stream known as the North Atlantic Drift is responsible for bringing warmer temperatures to Europe. Evaporation of ocean water in the North Atlantic results in a cooling effect and a higher salt concentration, both of which increase the density of the water. As the denser water sinks, it creates a southern circulation pattern. As glaciers in Greenland melt due to the effects of global warming, the density of this ocean water decreases due to more freshwater. This, in effect, could stall the North Atlantic Drift and bring colder temperatures and flooding to Europe.

The Great Ocean Conveyor Belt

There is constant motion in the ocean in the form of a global ocean "conveyor belt" driven by thermohaline currents. These currents are density driven and are affected by both temperature and salinity. Cold, salty water is dense and sinks to the bottom of the ocean, while warm water is less dense and rises to the surface. Warm water from the Gulf Stream enters the Norwegian Sea and provides heat to the atmosphere in the northern latitudes. The loss of heat by the water in this area makes the water cooler and denser, causing it to sink. As more warm water is transported north, the cooler water sinks and moves south, making room for the incoming warm water. This cold bottom water flows south to Antarctica. Eventually, the cold bottom waters warm and rise to the surface in the Pacific and Indian oceans. It takes water about 1,600 years to move through the entire conveyor belt. The ocean conveyor belt plays an important role in supplying heat to the polar regions, and thus in regulating the amount of sea ice in these regions. Insofar as thermohaline circulation governs the rate at which deep waters are exposed to the surface, it may also play an important role in determining the concentration of carbon dioxide in the atmosphere. For more information on global warming and its effect on thermohaline circulation, refer to Chapter 11.

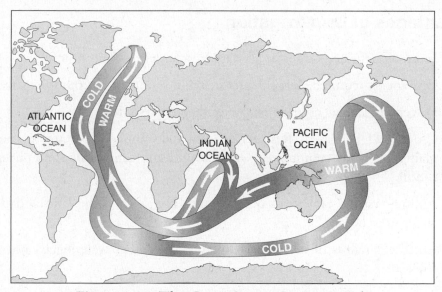

Figure 3.10 The Great Ocean Conveyor Belt

Upwellings

Upwellings occur when prevailing winds produced through the Coriolis effect, and moving clockwise in the Northern Hemisphere, push warmer, nutrient-poor surface waters away from the coastline. This surface water is then replaced by cooler, nutrient-rich deeper waters. The deeper waters contain high levels of nitrates and phosphates, which result from the decomposition and sinking of surface water plankton. When these nutrients are brought to the surface through upwelling, they supply necessary nutrients for phytoplankton, which form the base of the oceanic food chain.

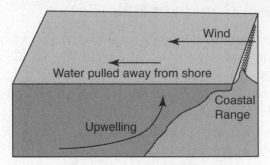

Figure 3.11 Upwelling

AGRICULTURAL, INDUSTRIAL, AND DOMESTIC USE

About 70% of freshwater is used for agriculture. Use of water for agriculture depends upon national wealth, climate, and degree of industrialization. Canada uses about 10% of its freshwater resources for agriculture, whereas India uses about 90%. Up to 70% of water intended for agriculture in developing countries may not reach crops due to seepage, evaporation, and leakage. Drip irrigation, the most efficient type of irrigation, is used on less than 1% of crops worldwide.

Advantages of Drip Irrigation

1. Increased efficiency. Almost all water reaches crops—no runoff.

2. Less energy required. Lower water demand results in less pumping costs.

3. Lower demand on aquifers or depleted water resources.

4. Crop yield increases—fertilizer is accurately applied directly to roots of plants and can be monitored. Reduces salinization and nitrate and phosphate runoff.

5. Tubing systems can be adapted to meet contours of the land and can be changed as needed.

6. Correct amounts of water means plants are neither waterlogged nor water stressed.

Industry uses about 25% of all freshwater, ranging from a high of about 75% in Europe to less than 5% in developing countries. Water used for cooling power plants is the largest sector. Water returns 60 times its economic value when used for industrial purposes rather than for agriculture.

Domestic uses of freshwater include water being used for flushing toilets, bathing, drinking, and so on. People living in developed countries use about 10 times more water for personal use than people living in less-developed countries.

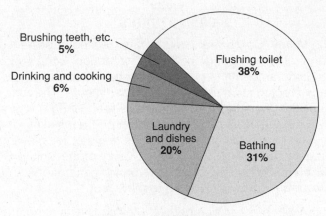

Figure 3.12 Water use

SURFACE AND GROUNDWATER ISSUES

Surface water infiltrates and percolates through the soil into aquifers—layers of porous rock, sand, and gravel where water is trapped above a nonporous layer or bedrock. The surface area in which water infiltrates into the aquifer is called the recharge zone. If pollution enters an aquifer, the aquifer is no longer a source of safe drinking water. Movement of water through aquifers is very slow. Artesian wells occur where the water breaks through to the surface. Aquifers in the United States hold 30 times more water than all U.S. lakes and rivers combined, with groundwater supplying almost 40% of all freshwater in the United States. When removal of water exceeds the recharge rate, the land sinks (subsidence). Depletion of water in aquifers also leads to sinkholes and saltwater intrusion—a condition in which seawater replaces the freshwater in the aquifer, making it unusable for human use. The region where water-saturated soil meets water-unsaturated soil is called the water table. The water table is unique to a region and can rise and fall with rainfall variations, depletion rates, etc.

Groundwater is considered to be one of the last "free resources," as anyone who can afford to drill can draw up water. Thus, the extraction of groundwater becomes a "Tragedy of the Commons," with high economic externalities.

Water-Renewal Rates

Source of Water	Average Renewal Rate
Groundwater (deep)	~ 10,000 years
Groundwater (near surface)	~ 200 years
Lakes	~ 100 years
Glaciers	~ 40 years
Water in the soil	~ 70 days
Rivers	16 days
Atmosphere	8 days

CASE STUDIES

San Joaquin Valley, California: Groundwater-related subsidence is the sinking of land resulting from groundwater extraction. Land subsidence occurs when large amounts of ground water have been withdrawn from certain types of rocks, such as fine-grained sediments. The rock compacts because the water is partly responsible for holding the ground up. When the water is withdrawn, the rocks fall in on themselves. The desert areas of the world are requiring more and more water for growing populations and agriculture. In the San Joaquin Valley of the United States, groundwater pumping for crops has gone on for generations and has resulted in the entire valley sinking up to thirty feet.

Mexico City: A city of 22 million people, Mexico City is almost entirely dependent on exploiting groundwater for its needs. The water table in Mexico City is dropping almost six feet (2 meters) per year. Such a dramatic change in land elevation causes massive impacts on buildings and infrastructure, such as cracking and tilting.

GLOBAL PROBLEMS

Both water shortages and rising sea levels are global problems.

Water Shortages

The rate of water consumption is growing twice as fast as the population growth rate. Freshwater shortages that result from this demand can be due to natural weather patterns that reduce rainfall, rivers changing course, flooding that contaminates existing supplies, competition for available water, overgrazing and the resulting erosion, pollution of existing supplies, and competing interests that reduce water conservation programs. Water is a limiting factor as it limits the amount of food that can be produced in a region. If food cannot be grown locally due to water shortages, then food must be imported at additional costs.

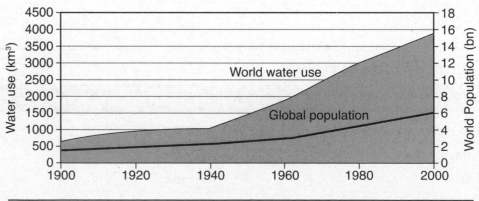

Source: FAO Aquastat, UN

Figure 3.13 Global Water Use vs. Population Growth

Rising Sea Levels

Rising sea level is primarily due to two factors: thermal expansion of water and the melting of ice caps and glaciers. Thermal expansion of seawater involves increasing the distance between neighboring water molecules, and this distance increases with increasing temperature. Translated over the mean depth of the ocean 2.4 miles (3.8 km), a 1-degree increase in temperature will cause a sea level rise of about 28 inches (70 cm).

During the end of the last ice age about 18,000 years ago, when global temperatures were about 10°F (5°C) warmer than they are today, sea level was about 430 feet (130 m) higher than it is today. Much of the rise was due to ice that was on land melting and filling the oceans. When ice that is floating on the water melts, it does not contribute to a rise in sea level. However, it does affect climate by changing the albedo. With higher sea levels and more water covering Earth's surface, more heat energy is absorbed by the water and less is reflected back into space, resulting in higher temperatures.

During the 20th century, sea levels rose 6–8 inches (15–25 cm). Approximately 1–2 inches (2–5 cm) of the rise resulted from the melting of mountain glaciers. Another 1–2 inches (2–5 cm) resulted from the expansion of ocean water that resulted from warmer ocean temperatures. Best scientific estimates indicate that sea levels will rise 7 inches (18 cm) by the year 2030 and 23 inches (58 cm) by the year 2090. Climate models have suggested that temperatures in polar regions will increase more and at a faster pace than in other areas of the world. Since 1995, more than 5,400 square miles, an area equal to Connecticut and Rhode Island combined, have broken off the Antarctic ice shelves and melted.

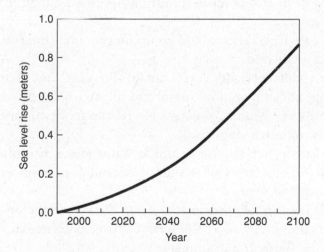

Figure 3.14 Projected rise in sea level

Wetlands are the most-impacted areas affected by rising sea levels. A 1-foot (30 cm) increase in sea level would result in up to 40% of the U.S. wetlands being destroyed. Other impacts would include: erosion of beaches and bluffs, salt intrusion into aquifers and surface waters, inundation of seawater into low-lying areas, and increased flooding and storm damage.

Much of the world's population—20%—lives in coastal regions, and half of the world's population lives within 120 miles (200 km) of the coast. Asia, Latin America, and the Caribbean have the highest percentages of people living near the coast.

FRESHWATER CONSERVATION

Several conservation methods can be used to increase the quantity of available freshwater:

Methods to Increase Supplies of Freshwater: Description and Drawbacks

- **Changes in personal habits:** Turn off shower while washing. Wash only full loads of clothes and dishes; turn off water when brushing teeth. Check for and repair leaks around the home. Use a broom when cleaning driveways and patios. Adjust sprinklers to prevent runoff. Water at night or early morning.
- **Construct dams and reservoirs:** Interferes with fish migration and destroys natural rivers. Leakages, earthquakes, evaporation, sediment buildup, and displacement of people are also consequences. Up to 60 million people worldwide (many being indigenous minorities) have been relocated due to dam and reservoir construction.
- **Desalinate water:** Rate of production is low and is expensive (three to four times more expensive than any other process). Issues of brine disposal.
- **Drip irrigation:** Drip irrigation conserves water by reducing evaporation, but it is expensive. Most large agricultural corporations can afford it, but small, independent farmers cannot. Not suitable for annual crops.
- **Education:** Informing and educating the public on water conservation.
- **Encourage the use of recycled products that require less water to produce:** Costs of collecting products may be outweighed by savings.
- **Engineer systems to collect more runoff:** Water may be high in pollutants and expensive to reprocess.
- **Levy taxes or user fees:** Prices would go up on products. International competitiveness would be affected.
- **Line irrigation channels and cover canals:** Prevents loss of irrigation water through seepage and evaporation. Initial and maintenance costs are expensive.
- **Meter all water used:** Municipalities would recoup money on meter installation with paying less to water suppliers.
- **Plant crops that do not require as much water and xeriscaping:** Xeriscaping reduces urban runoff. Issues of market economies, crop prices and demand, weather patterns, and so on.
- **Rebates or legislation of low-flush toilets, shower restrictors, etc.:** Rebates would be offered by water companies, which, over time, would recoup costs by having to buy less water from suppliers.
- **Reduce government subsidies:** Increased water costs would be passed on to consumers and result in greater personal conservation.
- **Reprocess (recycle) water:** The public is not supportive of using reprocessed toilet water. Reprocessed water could be used for irrigation but would require separate pipeline systems. Reuse of gray water is becoming popular in new developments.
- **Seed clouds:** Water availability to other areas would be affected.
- **Tiered price scale:** Would reduce effective family income for larger families. Could be remedied through exemptions or allocated share of water per family member.

- **Use of icebergs:** Expensive and most of it would be lost before it reached final destination.
- **Use more groundwater:** If rate of use exceeds rate of recharge, then subsidence, sinkholes, and saltwater intrusion could occur.

CASE STUDIES

- **Aswan High Dam, Egypt:** Completed in the 1970s, the Aswan High Dam in Egypt was built to supply irrigation water. The water that is available is only half of what was expected due to evaporation and seepage losses in unlined canals. Several other problems were encountered. First, the elimination of nutrients onto farmlands now requires the use of expensive fertilizers. Second, the depletion of nutrients into the Mediterranean caused a decline in certain fish catches. Third, large amounts of standing water caused the proliferation of snails and ultimately resulted in a debilitating disease known as schistosomiasis, with some areas having infection rates of 80%.

- **Bangladesh:** In the 1960s, thousands of wells were dug in Bangladesh by foreign governments and humanitarian organizations in an effort to supply freshwater to the population. Shortly thereafter, arsenic compounds from the soil began to leach into the groundwater. Arsenic poisoning began to appear among the population, with millions of people showing symptoms.

- **Colorado River Basin:** Diversion of water from the Colorado River has led to water right disputes between California, Arizona, and Mexico. Dams on the Colorado River trap large quantities of silt (over 10 million metric tons per year) and reduce nutrient levels in farmlands below the dam. As a result, more fertilizer is required. Farm irrigation has resulted in high levels of salts in the alkaline soils to become incorporated in agricultural runoff. Millions of acres of once-valuable farmland are now useless due to the salt buildup in soil, a process known as salinization.

- **James Bay, Canada:** Diversion of rivers into Hudson Bay to generate electrical power has resulted in massive flooding. During one flood, up to 10,000 caribou drowned. In addition, mercury has leached out of rocks and into water, with nearby residents showing signs of mercury poisoning. The project also created expensive legal battles and created many issues with indigenous people whose land was flooded.

- **Ogallala Aquifer:** The Ogallala Aquifer underlies eight states from Texas to North Dakota. The Ogallala Aquifer used to hold more freshwater than all freshwater lakes, streams, and rivers on Earth. Due to pumping of this groundwater for agricultural, domestic, and industrial uses, many locations are experiencing water shortages.

- **Three Gorges Dam, China:** In 1949, China had no large reservoirs and only 40 small hydroelectric stations. By 1985, there were 80,000 reservoirs and 70,000 hydroelectric stations. The Three Gorges Dam required relocation of 1.2 million people.

TIP

Case studies can be brought into your essay answers to bring a historical connection to the question. Try to bring at least one case study into your essays—your score will go higher. You will also find several multiple-choice questions on the test that focus on case studies. Only the most relevant case studies are included in this book.

QUICK REVIEW CHECKLIST

- ☐ **Properties of Water**
  - ☐ name five physical properties of water
  - ☐ various forms and amounts of freshwater available (lakes, glaciers, ice caps, etc.)

- ☐ **Ocean Circulation**
  - ☐ causes of currents (winds, densities, temperature, thermoclines, etc.)
  - ☐ differences between Northern Hemisphere and Southern Hemisphere currents
  - ☐ causes and consequences of increases in ocean temperatures

- ☐ **Water Usage**
  - ☐ agricultural usage (various forms, advantages and disadvantages of irrigation)
  - ☐ industrial usage (relative amount compared to other uses, recycling methods)
  - ☐ domestic usage (relative amount compared to other uses)

- ☐ **Surface Water vs. Groundwater**
  - ☐ relative amount of freshwater available
  - ☐ relative renewal rates
  - ☐ pollution (current status, mitigation)

- ☐ **Rising Sea Level**
  - ☐ historical perspective
  - ☐ contributing factors
  - ☐ impact (financial and cultural)

- ☐ **Conservation**
  - ☐ name five conservation methods for industrial, agricultural, and domestic water usage

- ☐ **Case Studies**
  - ☐ Aswan High Dam
  - ☐ Bangladesh
  - ☐ Colorado River Basin
  - ☐ James Bay
  - ☐ Ogallala Aquifer
  - ☐ Three Gorges

MULTIPLE-CHOICE QUESTIONS

1. Water vapor returning to the liquid state is called

   (A) evaporation
   (B) transpiration
   (C) boiling
   (D) condensation
   (E) vaporization

2. The temperature at which air becomes saturated and produces liquid is called

   (A) the saturation point
   (B) the dew point
   (C) the condensation point
   (D) relative humidity
   (E) absolute humidity

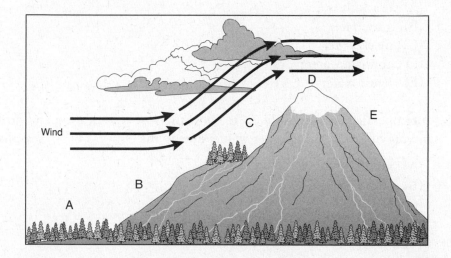

3. The area that would receive the most precipitation would be

   (A) *A*
   (B) *B*
   (C) *C*
   (D) *D*
   (E) *E*

4. The rain shadow effect would be located at point

   (A) *A*
   (B) *B*
   (C) *C*
   (D) *D*
   (E) *E*

5. Of the freshwater on Earth that is not trapped in snow packs or glaciers, most of it (95%) is trapped in

    (A) lakes
    (B) rivers
    (C) aquifers
    (D) dams
    (E) estuaries, marshes, and bogs

6. The primary use of freshwater is for

    (A) industry
    (B) domestic use
    (C) fishing
    (D) agriculture
    (E) landscaping

7. A mixture of freshwater and saltwater is known as
    (A) brackish water
    (B) gray water
    (C) black water
    (D) connate water
    (E) lentic water

8. A temperate lake is most likely to show thermal stratification and limited mixing of surface and deeper waters during the _____ season.

    (A) winter
    (B) spring
    (C) summer
    (D) fall
    (E) None of the above

9. Of the following methods of irrigation, the one that currently conserves the most water is

    (A) flooding fields
    (B) irrigation channels
    (C) sprinklers
    (D) drip irrigation
    (E) misters

10. The largest use for industrial water is for

    (A) cooling electrical power plants
    (B) automobile manufacturing
    (C) mining
    (D) the food and beverage industry
    (E) aquaculture

11. A country that would represent large per capita water use would be

    (A) China
    (B) India
    (C) Israel
    (D) United States
    (E) Iceland

12. When compared with the rate of population growth, the worldwide demand rate for water is

    (A) about half
    (B) about the same
    (C) about two times
    (D) about three times
    (E) about ten times

13. The U.S. per capita use of water on a daily basis is closest to
    (A) 50 gallons
    (B) 100 gallons
    (C) 1,500 gallons
    (D) 5,000 gallons
    (E) 10,000 gallons

14. Countries that are more likely to suffer from water stress would be located in

    (A) North America
    (B) South America
    (C) western Europe
    (D) the Middle East
    (E) Asia

15. What fraction of the world's population does not have access to adequate amounts of safe drinking water?

    (A) 1/2
    (B) 1/3
    (C) 1/4
    (D) 1/6
    (E) 1/10

16. The oceanic zone that extends from the low tide mark to the edge of the continental shelf is known as the

    (A) aphotic zone
    (B) benthic zone
    (C) neritic zone
    (D) pelagic zone
    (E) photic zone

17. Rising sea levels due to global warming would be responsible for all of the following EXCEPT

    (A) destruction of coastal wetlands
    (B) beach erosion
    (C) increased damage due to storms and floods
    (D) increased salinity of estuaries and aquifers
    (E) all would be the result of rising sea levels

18. What does La Niña bring to the southeastern United States?

    (A) Warm winters
    (B) Extremely cold winters
    (C) Hot summers
    (D) Cooler than normal summers
    (E) None of the above

19. Which of the following ocean currents flows without obstruction or barriers around Earth?
    (A) Gulf Stream
    (B) California Current
    (C) Antarctic Circumpolar Current
    (D) Aghulas Current
    (E) They all flow unimpeded

20. Saltwater intrusion into groundwater occurs most often when

    (A) the water table near the coast drops
    (B) surface salts from irrigation water seep into the ground
    (C) storms at sea create unusually low tides
    (D) less surface water reaches the water table
    (E) salt water is pumped into aquifers to replenish water tables

FREE-RESPONSE QUESTION

An APES class visited a local freshwater stream near their high school and performed several tests to determine the water quality of the stream.

(a) Choose ONE test of water quality. For that test, provide

   (i)   A description of exactly what that water quality test measures
   (ii) A brief explanation of how that test is performed
   (iii) How the results of that test are interpreted and they relate to overall water quality

(b) Assume that the results of the water quality test that you chose in (a) are such that they are negatively impacting stream water quality. Describe ONE remediation technique that could be used to reduce the environmental impact.

(c) Sketch a food web of a freshwater or riparian ecosystem. In your diagram

   (i)   Include at least one producer, one primary consumer, one secondary consumer, and one tertiary consumer and label one of each type as to their role in the pyramid
   (ii) Use arrows to show the relationships that exist in your diagram between the producers and the consumers

MULTIPLE-CHOICE ANSWERS AND EXPLANATIONS

1. **(D)** Evaporation is water changing from a liquid state to a gaseous state below the boiling point. Transpiration is water moving through a plant. Vaporization is moving from a liquid state to a gaseous state.

2. **(B)** The saturation point is the maximum amount of water vapor that a particular volume of air at a given temperature can hold. The condensation point is the temperature and pressure at which water vapor turns into liquid water. Absolute humidity is the mass of water vapor in a given volume of air. Relative humidity is the ratio of the actual amount of water vapor held in the atmosphere compared with the maximum amount that the air could hold and is influenced by temperature and atmospheric pressure.

3. **(C)** As the air lifts (orographic lifting), it becomes cooler. Cooler air holds less water vapor. At location *C*, the air is holding the maximum amount of water vapor. Given the fact that the temperature has decreased, it would receive the maximum amount of rain. At the top of the mountain at location *D*, much of the water vapor has been depleted from the air.

4. **(E)** Point *E* is on the leeward side of the mountain. This side receives little precipitation because most of the rain has been deposited on the windward side. The leeward side is experiencing the rain shadow effect.

5. **(C)** The oceans hold 97% of all water on Earth. Freshwater only makes up 3%. Of that 3%, 90% of it is trapped in ice and snow, which is rapidly melting due to global warming. Of the freshwater left, the majority is found in groundwater, with the remaining 3% of freshwater found in lakes, rivers, and

streams. Of the total amount of water on Earth, only 0.01% is located in lakes, rivers, and streams.

6. **(D)** Agriculture uses about 70% of all freshwater. Use for agriculture depends upon national wealth, climate, and degree of industrialization. Industry uses about 25% of all freshwater, with Europe using the most and developing countries using the least. Water used for cooling of power plants is the largest sector.

7. **(A)** Gray water is sewage water that does not contain toilet wastes. Black water is sewage that does contain toilet wastes. Connate water is also known as fossil water and is water that has been trapped within sediment or rock structures at the time the rock was formed. Lentic water is the standing water of lakes, marshes, ponds, and swamps.

8. **(C)** During the summer, the surface water warms up much faster than the deep water. The warmer surface water is less dense than the cooler, deep water, so it stays on the surface. The wind mixes the surface water but only near the surface. The lake tends to become stratified, with a warmer upper layer or epilimnion and a cooler lower layer or hypolimnion. The boundary between these layers is called the thermocline.

9. **(D)** Drip irrigation can increase yields and decrease water requirements and labor. It provides the plant with continuous, near-optimal soil moisture by conducting water directly to the plant. It saves water because only the plant's root zone receives moisture.

10. **(A)** Industry used about 25% of all freshwater, ranging from 75% in Europe to less than 5% in developing countries.

11. **(D)** Highest per capita supplies of freshwater are in countries with high rainfall and low populations (Iceland, Norway, and so on). These water-rich countries have low water withdrawals. Remember, *per capita* means "per person."

12. **(C)** Since populations are increasing and increased populations result in higher levels of pollution and since freshwater sources are finite, the amount of freshwater per person is decreasing each year.

13. **(C)** In the United States, renewable or replacement water averages 2.4 million gallons (9 million L) per person per year. The average amount withdrawn from water supplies in the United States is about 500,000 gallons (1.9 million L) per person per year (1,500 gallons [5,700 L] per day).

14. **(D)** Areas that do not receive as much precipitation include polar regions (cold air cannot hold as much water as warmer air), midcontinental areas (they are too far from oceans and the clouds have released much of their moisture before they reach inland), subtropical deserts (air masses are subsiding), and the leeward sides of mountains near coastal regions (rain shadow effect).

15. **(D)** It is estimated that over 1 billion people lack access to safe drinking water. A child dies every 8 seconds worldwide from contaminated water sources (over 5 million children each year).

16. **(C)** The neritic zone, also called coastal waters or the sublittoral zone, is the part of the ocean extending from the low tide mark to the edge of the continental shelf, with a relatively shallow depth extending to about 650 feet (200 m). The neritic zone has generally well-oxygenated water, low water pressure, and relatively stable temperature and salinity levels. These, combined with presence of light and the resulting photosynthetic life, such as phytoplankton and floating seaweed, make the neritic zone the location of the majority of sea life.

17. **(E)** Sea levels are rising and are caused by both natural and human factors. A 0.5-inch (1 cm) rise in sea level erodes beaches about 3 feet (1 m) horizontally. It is predicted that within 100 years, there will be a net loss of up to 43% in coastal wetlands due to rising sea levels.

18. **(A)** La Niña can bring warm winters to the southeast and cooler-than-normal winter temperatures to the northwest United States. It is the cold counterpart of El Niño. La Niña's strong easterly winds bring cold ocean water to the surface in the eastern Pacific and causes increased rainfall in the western Pacific. The jet stream rather than coming through the Pacific Northwest is diverted over Alaska and into the Great Lakes region.

19. **(C)** The Antarctic Circumpolar Current is the most powerful ocean current system on Earth and exerts a strong influence on climate. It circles Earth in the southern hemisphere and connects the three great ocean basins—Atlantic, Indian, and Pacific. Unlike in the Northern Hemisphere, there are no landmasses to break up this large, continuous stretch of water.

20. **(A)** Normally, the groundwater underlying coastal regions has an upper layer of freshwater with salt water beneath it. The layering occurs because rain, falling as freshwater, is less dense than salt water. When freshwater is withdrawn at a faster rate than it can be replenished, a drawdown of the water table occurs, with a resulting decrease in the overall hydrostatic pressure. When this happens near an ocean coastal area, saltwater from the ocean intrudes into the freshwater aquifer.

FREE-RESPONSE ANSWER

(a) Choose ONE test of water quality. For that test, provide

    (i)   A description of exactly what that water quality test measures

        Biological oxygen demand (BOD) is a measure of the oxygen used by microorganisms to decompose organic wastes. If there is a large quantity of organic waste in the water supply, bacterial counts will be high. Therefore, the demand for oxygen will be high, resulting in a high BOD level. As the waste is consumed or dispersed through the water, BOD levels will begin to decline. Nitrates and phosphates in a body of water can also contribute to high BOD levels. Nitrates and phosphates are plant nutrients and can cause plant life and algae to grow quickly. When plants grow quickly, they also die quickly. This contributes to the organic waste in the water, which is then decomposed by bacteria—resulting in a high BOD level. When BOD levels are high, dissolved oxygen (DO) levels decrease because the oxygen that is available in the water is being consumed by the bacteria. Because less dissolved oxygen is available in the water, fish and other aquatic organisms may not survive.

    (ii)   A brief explanation of how that test is performed

        The BOD test takes five days to complete and is performed using a dissolved oxygen test kit. The BOD level is determined by comparing the DO level of a water sample taken immediately with the DO level of a water sample that has been incubated in the dark for five days. The difference between the two DO levels represents the amount of oxygen required for the decomposition of any organic material in the sample and is a good approximation of the BOD level.

(iii) How the results of that test are interpreted and how they relate to overall water quality

BOD levels of 1–2 ppm are indicative of good water quality without much organic waste present in the water supply. A water supply with a BOD level of 3–5 ppm is considered moderately clean. In water with a BOD level of 6–9 ppm, the water is considered somewhat polluted because there is usually organic matter present, and bacteria are decomposing this waste. At BOD levels of 10 ppm or greater, the water supply is considered very polluted with organic wastes. At these BOD levels, organisms that are more tolerant of lower dissolved oxygen may appear and become numerous (such as leeches and sludge worms). Organisms that need higher oxygen levels (like caddis fly larvae and mayfly nymphs) will not survive.

(b) Assume that the results of the water quality test that you chose in (a) are such that they are negatively impacting stream water quality. Describe ONE remediation technique that could be used to reduce the environmental impact.

High BOD usually indicates the presence of organic waste(s) in the water. The first step in reducing the environmental impact of low dissolved oxygen content in the water is to identify the source of the waste. By carefully testing various sites along the stream, it may be possible to identify exactly the source of the organic pollution; i.e., leaking sewer line, leaking septic tank, discharge from a factory, runoff from a cattle feedlot, etc. Once the source has been identified, several options are available: (1) contact the polluter and let them know your results and/or (2) contact local, state, or national authorities. If the source of pollution is a point-source such as a leak in a sewer line, this can generally be easily corrected. However, non-point pollution sources such as agricultural runoff may be more difficult to locate and identify. However, remediation techniques could involve changes in: the type of fertilizer being used and its application; erosion and sediment control techniques; changes in animal feeding operations; changes in cattle grazing management; and/or changes in irrigation water management.

A sample food web is included here ONLY for reference. The student does NOT need to include any of the organisms listed here to receive full credit. However, the food web that is drawn must be correct.

(c) Sketch a food web of a freshwater or riparian ecosystem. In your diagram

(i) include at least one producer, one primary consumer, one secondary consumer, and one tertiary consumer and label each as to its role in the pyramid

(ii) use arrows to show the relationships that exist in your diagram between the producers and the consumers

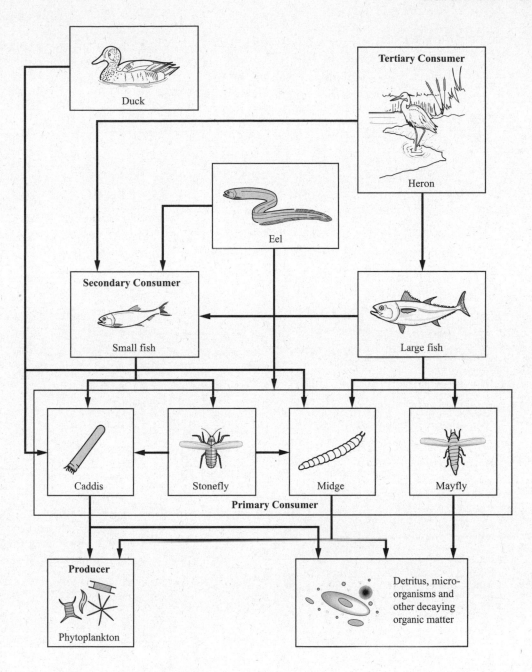

UNIT II: THE LIVING WORLD (10–15%)

Areas on Which You Will Be Tested

A. **Ecosystem Structure**—biological populations and communities, ecological niches, interactions among species, keystone species, species diversity and edge effects, major terrestrial and aquatic biomes.

B. **Energy Flow**—photosynthesis and cellular respiration, food webs and trophic levels, and ecological pyramids.

C. **Ecosystem Diversity**—biodiversity, natural selection, evolution, and ecosystem services.

D. **Natural Ecosystem Change**—climate shifts, species movement, and ecological succession.

E. **Natural Biogeochemical Cycles**—carbon, nitrogen, phosphorus, sulfur, water, and conservation of matter.

Ecosystems

CHAPTER 4

ECOLOGY

Ecology is the branch of biology that deals with the relations of organisms to one another and to their physical surroundings. Understanding ecosystems requires having an understanding of ecology.

Biological Populations and Communities

Organisms that resemble each other, that are similar in genetic makeup, chemistry, and behavior, and that are able to interbreed and produce fertile offspring belong to the same species. Organisms of the same species (intraspecific) that interact with each other and occupy a specific area form a population. Populations of different species (interspecific) living and interacting within an area create communities. A community is made up of all the populations of different species that live together within a particular area. An ecosystem is a system formed by the interaction of a community of organisms with their physical environment. Organisms make up populations that make up communities that make up ecosystems that make up the biosphere: biosphere → ecosystems → communities → populations → species → organisms.

Members of a population can be dispersed in an area in three ways:

> **Remember**
>
> "Intra" means within. "Inter" means between.

1. **Clumped.** Some areas within the habitat are dense with organisms, while other areas contain few members. This type of distribution is found in environments that are characterized by patchy resources. Clumped distribution is the most common type of dispersion found in nature because animals need certain resources to survive, and when these resources become scarce during certain parts of the year, animals tend to "clump" together around these crucial resources; e.g., elephants in Africa clumped together near water holes during the dry season. Individuals might also be clustered together in an area due to social factors such as family groups; e.g., wildebeests. Organisms that usually serve as prey also form clumped distributions in areas where they can hide and detect predators easily; e.g., ducks. Other causes of clumped distributions are the inability of offspring to independently move from their habitat. This is seen in juvenile animals that are immobile and strongly dependent upon parental care; e.g., chimpanzees. Clumped distribution in species also acts as a mechanism against predation as well as an efficient mechanism to trap or corner prey; e.g., lions. Threatened or endangered species are also more likely to be clumped in their distribution.

2. **Random.** Little interaction between members of the population leading to random spacing patterns. Random distribution usually occurs in habitats where environmental conditions and resources are consistent. This pattern of dispersion is characterized by the lack of any strong social interactions between species; e.g., dandelion seeds being dispersed by wind.

3. **Uniform.** Fairly uniform spacing between individuals. Uniform distributions are found in populations in which the distance between neighboring individuals is maximized. The need to maximize the space between individuals generally arises from competition for a resource such as moisture or nutrients, or as a result of direct social interactions between individuals within the population, such as territoriality; e.g., penguins often exhibit uniform spacing by aggressively defending their territory among their neighbors or creosote bushes releasing chemicals that inhibit the growth of other plants around them (allelopathy).

Ecosystem (Community) Characteristics

(1) Physical appearance	Relative size; stratification; distribution of populations and species
(2) Species diversity	Number of different species
(3) Species abundance	Number of individuals of each species
(4) Niche structure	Number of ecological niches; how they resemble or differ from each other; species interactions

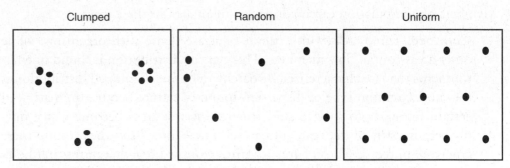

Figure 4.1 Clumped, random, uniform distribution

Ecological Niches

An ecological niche is the particular area within a habitat occupied by an organism and includes the function of that organism within an ecological community. The physical environment influences how organisms affect and are affected by resources and competitors. The niche reflects the specific adaptations that a species has acquired through evolution. To describe an organism's niche involves a description of the organism's adaptive traits, habitat, and place in the food web. A niche also takes into account the types and amounts of resources the species uses and its interactions with both living (biotic) and nonliving (abiotic) factors in its habitat.

Generalist species (*K*) live in broad niches. They are able to withstand a wide range of environmental conditions. Examples of generalist species include cockroaches, mice, and humans. Specialist species (*r*) live in narrow niches and are sensitive to environmental changes. Since they cannot tolerate change, they are more prone to extinction. An example of a specialist species is the giant panda, which only eats a certain type of bamboo. When environmental conditions are stable, specialist species have an advantage since there are few competitors as each species occupies its own unique niche (competitive exclusion principle). However, when habitats are subjected to rapid changes, the generalist species usually fare better since they are more adaptable.

Law of Tolerance and Limiting Factors

Earth's ecosystems are affected by both biotic (living) and abiotic (nonliving) factors, and are regulated by the Law of Tolerance, which states that the existence, abundance, and distribution of species depends on the tolerance level of each species to both physical and chemical factors. The abundance or distribution of an organism can be controlled by certain factors (e.g., the climatic, topographic, and biological requirements of plants and animals) where levels of these exceed the maximum or minimum limits of tolerance of that organism. A limiting factor is any abiotic factor that limits or prevents the growth of a population. Limiting factors in terrestrial ecosystems may include the level of soil nutrients, amount of water, light, and temperature. In aquatic ecosystems, major limiting factors include pH, the amount of dissolved oxygen in the water, and salinity.

Resource Partitioning

Resources in an environment are limited. Some species have evolved to "share" a specific resource. This sharing may take the form of two species eliminating direct competition by utilizing the same resource at different times (temporal partitioning)—e.g., one species of spiny mouse feeds on insects during the day while a second species of spiny mouse feeds on the same insects at night or when competing species use the same resource by occupying different areas or habitats within the range of occurrence of the resource (spatial partitioning)—e.g., different species of fish feeding at different depths in a lake or different species of monkey feeding at different heights in a tree. A third form of resource partitioning occurs when two species share the same resource but have evolved slightly different structures to utilize the same resource (morphological partitioning)—e.g., two different species of bees have evolved different proboscis lengths to utilize various sized flowers of the same species.

Interactions Among Species

Various types of interactions occur among species. They can benefit one or both species, harm one or both species, or not affect one of the species involved.

Interactions Among Species

Interaction	Description
Amensalism 	The interaction between two species whereby one species suffers and the other species is not affected. Usually this occurs when one organism releases a chemical compound that is detrimental to another organism. Examples: The bread mold *Penicillium* secretes penicillin, which is a chemical that kills bacteria. The black walnut tree releases a chemical that kills neighboring plants. Amensalism is common in chaparral and desert communities as it stabilizes the community by reducing competition for scarce nutrients in the water. This chemical interaction is known as allelopathy.
Commensalism 	The interaction between two species whereby one organism benefits and the other species is not affected. Forms of commensalism include: (1) using another organism for transportation such as the remora on a shark or mites on dung beetles; (2) using another organism for housing such as epiphytic plants like orchids growing on trees or birds living in the holes of trees; and (3) using something that another organism created such as hermit crabs using the shells of marine snails for protection.
Competition 	Competition can be either intraspecific (competition between members of the same species) or interspecific (competition between members of different species). Competition is the driving force of evolution whether it is for food, mating partners, or territory. Intraspecific competition results in organisms best suited for surviving in a changing environment. Competition is prominent in predator-prey relationships with the predator (seeking food) and the prey (seeking survival). Examples of types of competition include: (1) interference—occurs directly between individuals by interfering with foraging, survival, or reproduction or by preventing a species to establish itself within a habitat; (2) exploitation—occurs indirectly through a common limiting resource that acts as an intermediate; by using the resource it depletes the amount available to others; and (3) apparent—occurs indirectly between two species, which are both sought after by the same predator.
Mutualism 	The interaction between two species whereby both species benefit. Symbiosis is a lifelong positive interaction that involves close physical and/or biochemical contact such as the relationship between trees and mycorrhizal fungi, or the relationships can be shorter as in the case of bees pollinating flowers. Mutualism can be obligatory such as mycorrhizal fungi being totally dependent on their plant hosts, or it can be nonobligatory as seen in *Rhizobia* bacteria reproducing either in the soil or in a symbiotic relationship with legumes.
Parasitism 	The interaction between two species whereby one species is benefited at the expense of the other. If the parasite lives on the host, it is known as an ectoparasite (mosquito) and often has elaborate mechanisms and strategies for acquiring a host (leeches that locate their host by sensing movement and confirming the identity through skin temperature and chemical cues before attaching). If the parasite lives within the host, it is known as an endoparasite (tapeworm) and acquires its hosts by passive mechanisms such as the ingestion of egg cells. Epiparasites feed on other parasites. Biotrophic parasites must keep their hosts alive and represent a successful mode of life; many viruses are examples of biotrophic parasites because they use the host's genetic and cellular processes to multiply. Necrotrophs are parasites that eventually kill their hosts. Social parasites involve behaviors that benefit the parasite and harm the host (e.g., cuckoo birds that use other birds to raise their young or the nectar-robbing behavior of insects and birds). Hosts have evolved defense mechanisms (immune systems, plant toxins) to diminish parasitism.

Interactions Among Species (continued)

Interaction	Description
Predation	Predators hunt and kill prey through the act of predation. Whereas predators kill their prey, parasites have evolved mechanisms to keep their host alive, since their survival depends on a viable host. Predators can be opportunistic and kill and eat almost anything, or they may be specialists and only prey upon certain organisms. Predators that eat only meat are called carnivores; those that eat both meat and vegetation are known as omnivores.
Saprotrophism	Saprotrophs obtain their nutrients from dead or decaying plants or animals through absorption of soluble organic compounds. Saprotrophs include many fungi, bacteria, and protozoa. Vultures and dung beetles are also saprotrophs.

Keystone Species

A keystone species is a species whose very presence contributes to a diversity of life and whose extinction would lead to the extinction of other forms of life. Through various interactions, a small number of individuals from a keystone species have a very large and disproportionate impact on how ecosystems function. An ecosystem may experience a dramatic shift if a keystone species is removed even though that keystone species was a small part of the ecosystem as measured by biomass or productivity. A classic keystone species is a small predator that prevents a particular herbivorous species from decimating a dominant plant species. Since the prey numbers are low, the keystone predator numbers could be even lower and still be effective. Yet without the predators, the herbivorous prey would explode in numbers, wipe out the dominant plants, and dramatically alter the character of the ecosystem. Two examples of keystone species are listed below:

STARFISH (SEA STARS)

Starfish prey on sea urchins, mussels, and other shellfish that have no other natural predators. If starfish are removed from the ecosystem, mussel populations explode and drive out other species, and sea urchin populations rise to the point where they begin to decimate coral reefs

SEA OTTERS IN KELP FORESTS

Sea otters prey on sea urchins. Kelp roots serve as anchors and not as nutrient-absorbing systems as found in land plants. If left unchecked, the sea urchins destroy the kelp forests by foraging on kelp roots.

Species Diversity

Organisms that live in different environments are specifically adapted to their particular biome.

AQUATIC ORGANISMS

Water provides buoyancy and reduces the need for support structures such as legs and trunks. Water has high thermal capacity, so most aquatic organisms do not spend energy on temperature regulation. Many organisms obtain nutrients directly from the water, thereby reducing energy spent on searching for food. These include filter feeders such as barnacles, clams, and oysters. Water allows dispersal of gametes and larvae to new areas. Water screens out UV radiation. Intertidal organisms have evolved methods of not being swept away by waves and prevent water loss during low tide through shells and exoskeletons, salt-removing mechanisms, and lower metabolic rates due to cooler ambient temperatures.

DESERT ORGANISMS

Desert plants are spaced apart due to limiting factors and consist primarily of succulents (cactus) and short-lived annuals (wildflowers). Succulents store water, have small surface areas exposed to sunlight, have vertical orientation to minimize exposure to the sun, open their stomata at night, have waxy leaves to minimize transpiration, have deep roots to tap groundwater (mesquite and creosote), and have shallow roots to collect water after short rainfalls (prickly pear and saguaro cactus). Sharp spines on cacti reflect sunlight, create shade, and discourage herbivores. Cacti secrete toxins into the soil to prevent interspecific competition (allelopathy). Desert plants store biomass in seeds. Wildflowers have short life spans and are dependent on water for germination.

Desert animals are small and have small surface areas. They spend time in underground burrows where it is cooler, and they are often nocturnal. Aestivation is common. Some animals are able to metabolize dry seeds. Kangaroo rats produce their own water and secrete concentrated urine. Insects and reptiles have thick outer coverings to minimize water loss.

GRASSLAND ORGANISMS

Grasses grow out from the bottom (basal meristem) so they can grow again after being nibbled by grazing animals. Grass species are drought resistant. Deciduous trees and shrubs shed leaves during the dry season to conserve water. Grazers and browsing animals eat vegetation at different heights so as to not compete. Giraffes eat near the top of trees, elephants eat lower down, zebras eat taller grasses, and gazelles eat shorter grasses. Some animals migrate to find water. Others become dormant. Still others survive on seeds during the dry season. Some animals live in burrows to hide and escape predators. In addition, their fur color sometimes matches the color of the surroundings to act as camouflage.

FOREST ORGANISMS

In the tropics, some animals live in tree canopies where shelter and available food supplies (leaves, flowers, and fruits) are abundant and where they can escape from predators that live closer to the ground. Epiphytes (orchids and bromeliads) live on trunks and branches of trees, and they catch organic matter falling from the canopy. Epiphytes do not root in the soil. Instead, they obtain their moisture from the air or from dampness on the surface of their host. Some plants have very large leaves to capture scarce light. Roots of trees are shallow and spread out to capture nutrients in poor soil. To compensate for little support by shallow roots, trees may have buttresses. Flowers have elaborate devices to attract pollinators since wind is minimal in dense growth.

In temperate deciduous forests, broadleaf deciduous trees lose their leaves in winter and become dormant to conserve water and energy. Deciduous trees shift their metabolism from a photosynthesis-based system when light and temperature are favorable to one utilizing glucose and amino acids during the winter. This also helps keep the tree from freezing and acts as a kind of antifreeze.

In evergreen coniferous forests, small, waxy-coated needles are able to withstand the cold and drought of winter and have low surface area to reduce transpiration. Furthermore, conifers are always replacing their needles, unlike deciduous trees that replace their leaves only once a year. Decomposed needles make soil acidic, preventing many competing species from surviving in the soil environment. Some animals hibernate to conserve energy during winter when resources are scarce.

TEMPERATE SCRUB FOREST LAND ORGANISMS

Chaparral plants have small, waxy-coated leaves to reduce transpiration. Many plants produce toxins that leach into the soil to prevent competition (allelopathy). Vegetation becomes dormant during the dry season. Leaves do not fall during the dry season due to the stress of replacing leaves without an adequate water supply. Plant thorns are common for protection. Vegetation is adapted to fires and is common due to the high oil content in the brush. Fires reduce competition and allow seeds to germinate. Rodents are common and store seeds in underground burrows.

TUNDRA ORGANISMS

Polar grassland plants are adapted to low sunlight, low amounts of free water, high winds, and low temperatures. Tundra plants primarily grow during summer months. Leaves on plants have waxy outer coatings. Many plants survive winter as roots, stems, bulbs or tubers. Lichens dehydrate during winter to avoid frost damage.

Animals have adapted in several ways. They have extra layers of fat, chemicals in the blood to keep it from freezing, and compact bodies to conserve heat. They also have thick skin, thick fur, and waterproof feathers above downy insulating feathers. Additionally, animals either migrate during the coldest months or live underground (lemmings).

Edge Effects

An edge effect refers to how the local environment changes along some type of boundary or edge. Forest edges are created when trees are harvested, particularly when they are clear-cut. Tree canopies provide the ground below with shade and maintain a cooler, moister environment below. In contrast, a clear-cut allows sunlight to reach the ground, making the ground warmer and drier—environments not suitable for many forest plants. As time passes and a stand of young trees emerges on a clear-cut, the environment in the young stand changes and the edge begins to fade. As the mature forest develops, the edge fades away.

The edge effect is the result of two different conditions influencing the plants and animals that live on the edge. Some animal species (deer, elk, white-tailed deer, and pheasants) survive well in a forest edge since they are able to find food in the clearing, are able to benefit due to various habitats near one another, are able to hide in the nearby trees, or are well-adapted to human disturbances. These animals are called "edge species." Other animals, such as the spotted owl, do not do well in edges. Negative edge effects become most extreme when forest lands share the edge with agriculture or suburbs. Species composition and diversity can change dramatically if organisms adapted to one set of conditions cannot adapt to the change. If the edge effect is gradual or has indistinct boundaries and over which many species cross, it is called an open community. A community that is sharply divided from its neighbors is called a closed community. Preservation of large blocks of habitat and linking smaller habitat blocks together with migration corridors are techniques to protect rare and endangered species.

Major Terrestrial and Aquatic Biomes

Biomes are a major regional or global biotic community characterized by the dominant forms of plant life and the prevailing climate. Temperature and precipitation are the most important determinants of biomes. Biomes are classified by the type of dominant plant and animal life. Species diversity within a biome is directly related to net productivity, availability of moisture, and temperature.

Biomes

ANTARCTIC

Area surrounding south pole. Rainfall <2 inches (5 cm) per year.

BENTHOS (HADAL)

Bottom of oceans. No sunlight, therefore no plant life. Primary input of energy comes from dead organic matter settling and chemosynthesis.

COASTAL ZONES

Includes estuaries, wetlands, and coral reefs. High diversity and counts of animal and plant species due to runoff from land.

CORAL REEFS

Warm, clear, shallow ocean habitats near land and generally in the tropics. There are three types of coral reefs. Fringing reefs grow on continental shelves near the coastline. Barrier reefs are parallel to shoreline but farther from shore. Coral atolls are rings of coral that grow on top of sunken oceanic volcanoes. Coral reefs are disappearing at an alarming rate due to increase in sea temperature, pollution, dredging, and sedimentation.

DESERTS

Occur between 15° and 25° north and south latitude and generally occur in the interior of continents. Occupy about 20% of all landmass. Rainfall less than 20 inches (50 cm) per year. Air currents are descending, which generally diminishes the formation of rain (rain requires ascending air—orographic lifting). Soils often have abundant nutrients but lack organic matter and consist of sand and closely packed boulders. Little humus in the soil profile.

FRESHWATER WETLANDS

Freshwater wetlands include freshwater swamps, marshes, bogs, prairie potholes (important stopping grounds for migrating birds), ponds, peat bogs, and riparian areas (areas near rivers and streams). Ground is saturated with freestanding water, although standing water can be seasonal. Soil is low in oxygen. Important breeding areas rich in insects, amphibians, and reptiles. Prime areas for human development and recreation resulting in large amounts of habitat destruction. Critical for freshwater supplies. Easily polluted. Dominant plants are floating plants (phytoplankton). Animal life is abundant. Estuaries are rich in nutrients and are breeding grounds for fish. Water input includes runoff, groundwater flow, and streams. A lake is a body of water of considerable size surrounded by land. The vast majority of lakes on Earth are freshwater and most, like in the Northern Hemisphere, are at higher altitudes. Lakes can be classified on the basis of their richness of nutrients, which affects plant growth and animal diversity. Oligotrophic lakes are clear and low in nutrients, which results in a small amount of plant life and other forms of biomass. Eutrophic lakes are rich in nitrogen and phosphorus with large and diverse populations of phyto- and zooplankton, which supports a large diversity of fish. During warm weather, the oxygen content of the water may decrease resulting in die-offs. Hypertrophic lakes are excessively enriched with nutrients and subject to algal blooms. These lakes result from human activity such as the heavy use of fertilizers in the lake catchment area.

GRASSLANDS

Found in areas too dry for forests and too wet for deserts. Rainfall is seasonal. Temperatures are moderate. Grasslands occupy approximately 25% of all land area on Earth. Few trees and shrubs due to frequent seasonal fires and water availability. Dominant plants are grasses and perennials with extensively developed roots. Soils are rich in organic matter. Upper soil horizons are alkaline, dark, and rich in humus. Extensively used by humans for agriculture.

HYDROTHERMAL VENTS

Occur in the deep ocean where hot-water vents rich in sulfur compounds are found, which provide energy for chemosynthetic bacteria.

INTERTIDAL

The area of the shoreline exposed to water during high tide and air during low tide. Water movement brings in nutrients and removes waste products. Extremely rich in biodiversity. Susceptible and sensitive to pollution from land runoff and ocean pollution.

OCEAN

Oceans occupy 75% of Earth's surface. Areas of low diversity and low productivity except near the shoreline. Low in nitrogen and phosphorus, which limits plant growth and the smaller organisms that feed on them. Large animals occur but in low density.

SAVANNAS

Warm year-round. Scattered trees. Environment is intermediate between grassland and forest. Extended dry season followed by a rainy season. Consists of grasslands with stands of deciduous shrubs and trees. Trees and shrubs generally shed leaves during the dry season, which reduces the need for water. Food is limited during the dry season, requiring many animals to migrate. Soils are rich in nutrients. Contain large herds of grazing animals and browsing animals that provide resources for predators.

TAIGA (CONIFEROUS OR BOREAL FORESTS)

Generally found between 45° and 60° north latitude. Occupies approximately 17% of land surface. Forests of cold climates of high latitudes and altitudes. Two types of taiga: open woodland and dense forest. More precipitation than the tundra. Generally precipitation occurs during the summer months due to midlatitude cyclones. Soils are generally poor in nutrients because of large amounts of leaching caused by rainfall. The soils are acidic due to the decomposition of needles (an adaptation to conserve water). Deep layer of litter on the surface, and decomposition is slow because of low temperatures. Dense stands of small trees cause understory to be low in light. Low biodiversity due to harshness of environment. Disturbances include fires, storms, and insect infestations.

TEMPERATE DECIDUOUS FORESTS

Forests found in milder temperatures than boreal forests of the taiga. Rapid decomposition due to mild temperatures and precipitation, which results in a small amount of litter on surface. Because of the mild climate, this biome has been greatly exploited by humans for agriculture, lumber, and urban development. Soil is generally poor in nutrients. Tall deciduous trees. Rich and diverse understory. Low density of large mammals due to shade that prevents much ground vegetation.

TEMPERATE RAIN FORESTS

Moderate temperatures and rainfall exceeding 100 inches (250 cm) per year. Low biodiversity because of limited light, which limits food for herbivores. Major resource for timber.

TEMPERATE SHRUBLAND (CHAPARRAL)

Hot, dry summers with mild, cool, and rainy winters. During summer and spring, a subtropical high-pressure zone exists over the area. Rain falls during the winter due to midlatitude cyclones. Average rainfall is between 15–40 inches (30–75 cm) per year. Characterized by dense shrub growth. Decomposition is slow during dry months. Few large mammals. Erosion is common after fires. Because of climate, this area is being utilized for urbanization. Found in select coastal regions.

TEMPERATE WOODLANDS

Drier climate than deciduous forests. Dominated by small trees. Stands of trees are open, allowing light to reach ground. Fires are common.

TROPICAL RAIN FORESTS

High and constant temperature (average approximately 80°F [27°C]). Rainfall ranges between 75–100 inches (200–250 cm) per year. Found within Hadley cells. High species diversity for both plants and animals (up to 100 different tree species per square kilometer as opposed to three to five in temperate zones). Vegetation is dense. Soils are low in nutrients, and most nutrients are stored in vegetation. Soil is acidic. Decomposition of organic material is very fast due to temperature and moisture. Leaching is high. Abundant insect and animal biodiversity. Humans are clearing tropical rain forests for agriculture and cattle raising through a technique called slash and burn. Due to poor soil nutrient levels, agricultural activity lasts for only a limited time and can be sustained only by applying expensive fertilizer, which causes farmers to destroy even more forest.

TROPICAL SEASONAL FOREST

Occurs in areas of seasonal rainfall (monsoon) that is followed by long, dry season. Warm temperatures year-round. Contains mixture of deciduous and drought-tolerant evergreen trees.

TUNDRA

60° north latitude and above. Influenced by polar cells. Alpine tundra is located in mountainous areas, above the tree line with well-drained soil. Dominant animals include small rodents and insects. Arctic tundra is frozen, treeless plains, low rainfall, and low average temperatures with poor drainage due to the frozen ground. Growing season lasts about 2 months. Soil has few nutrients because of low vegetation and little decomposition. Permafrost is permanently frozen ground and a barrier for roots.

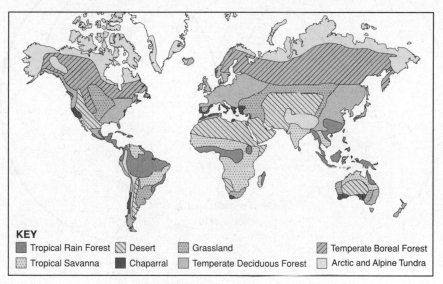

Figure 4.2 Major biomes of the world

ENERGY FLOW

The ultimate source of energy is the sun. Plants are able to use this light energy to create food. The energy in food molecules flows to animals through food webs.

Photosynthesis and Cellular Respiration

$$6CO_2 + 6H_2O + \text{sunlight} \rightarrow C_6H_{12}O_6 + 6O_2$$

Plants remove carbon dioxide from the atmosphere by a chemical process called photosynthesis, which uses light energy to produce carbohydrates and other organic compounds. Plants capture light primarily through the green pigment chlorophyll.

Chlorophyll is contained in organelles called chloroplasts. The glucose or the energy derived from its oxidation during cellular respiration is then used to form other organic compounds such as cellulose (for support), lipids (waxes and oils), and amino acids and then proteins. Oxygen gas is released into the atmosphere during photosynthesis. Plants also emit carbon dioxide during respiration. They produce less carbon dioxide than they absorb and therefore become net sinks of carbon.

Organisms that undergo photosynthesis are called photoautotrophs. Factors that affect the rate of photosynthesis are the amount of light and its wavelength, carbon dioxide concentration, the availability of water, and temperature.

Organisms dependent on photosynthetic organisms (autotrophs) are called heterotrophs. In general, cellular respiration is the opposite of photosynthesis. In respiration, glucose is oxidized by the cells to produce carbon dioxide, water and chemical energy. This energy is then stored in the molecule adenosine triphosphate (ATP).

$$C_6H_{12}O_6 + 6O_2 \rightarrow 6CO_2 + 6H_2O + energy$$

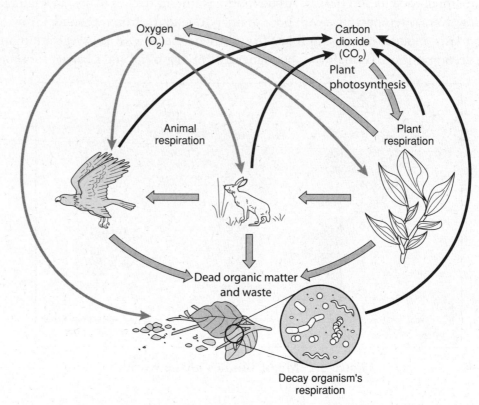

Figure 4.3 The carbon cycle viewed as an ecosystem cycle

Food Webs and Trophic Levels

Primary producers (autotrophs) are plants. They convert solar energy into chemical energy through photosynthesis. Primary consumers are heterotrophs (herbivores—plant eaters) and get their energy by consuming primary producers. Primary consumers have developed defense mechanisms against predation, some of which include speed, flight, quills, tough hides, camouflage, and horns and antlers.

Secondary (and higher) consumers are also heterotrophs and may be either strictly carnivores (meat eaters) or omnivores (eat both plants and animals).

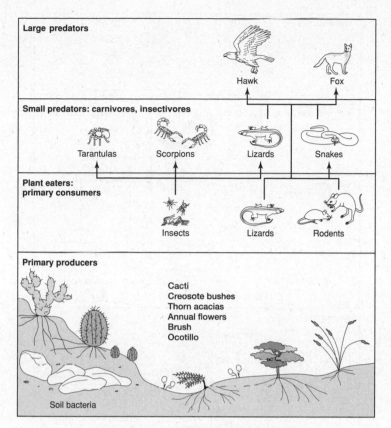

Figure 4.4 Desert food web

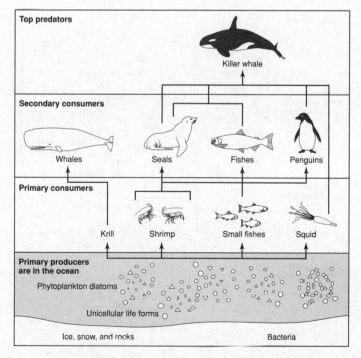

Figure 4.5 Oceanic food web

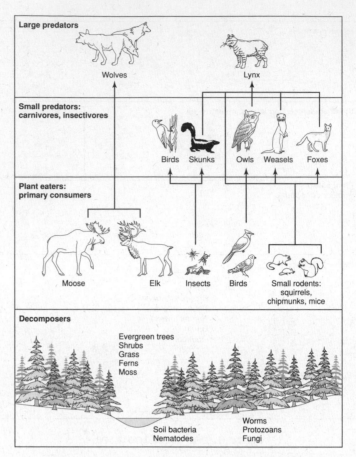

Figure 4.6 Coniferous forest food web

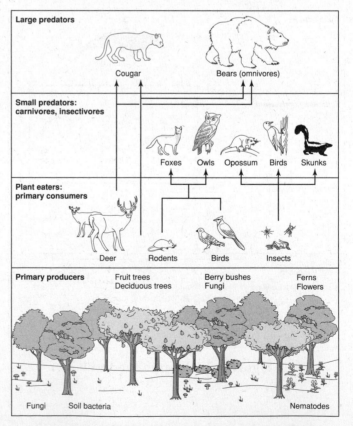

Figure 4.7 Deciduous forest food web

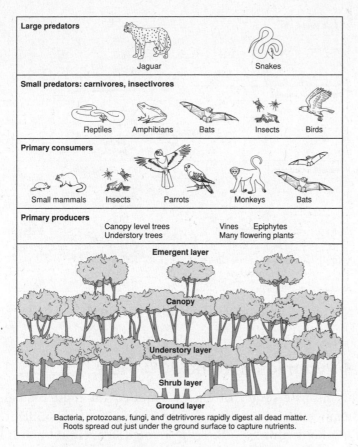

Figure 4.8 Tropical rain forest food web

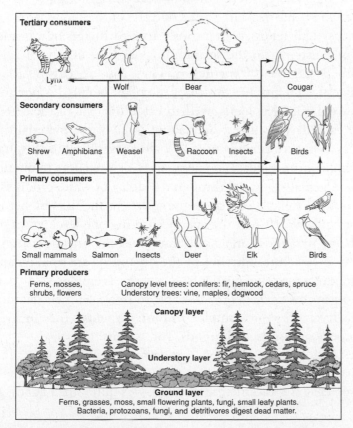

Figure 4.9 Temperate rain forest food web

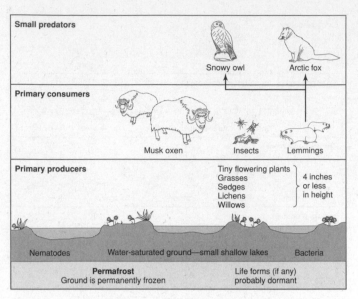

Small predators	Snowy owl	Arctic fox
Primary consumers	Musk oxen	Insects	Lemmings
Primary producers		Tiny flowering plants / Grasses / Sedges / Lichens / Willows	4 inches or less in height

Nematodes   Water-saturated ground—small shallow lakes   Bacteria

Permafrost
Ground is permanently frozen

Life forms (if any)
probably dormant

Figure 4.10 Tundra food web

Ecological Pyramids and the 10% Rule

Sunlight is the ultimate source of energy required for most biological processes (except for chemosynthetic organisms). Less than 3% of all sunlight that reaches Earth is used in photosynthesis. The energy stored in chemical bonds is released to animals and plants through cellular respiration.

Losses of potential energy (in the form of heat energy) occur as one moves up an energy pyramid and conforms with the second law of thermodynamics, which states that any closed system tends spontaneously toward increasing disorder (entropy); some energy is transferred to the surroundings as heat in any energy conversion; and no real process can be 100% efficient. These losses occur in various ways. For example, in digestive inefficiency, much of the plant material is not able to be broken down. For example, elephants need to eat about 5% of their body weight in plant material each day but digest only about 40% of the material that they consume. Other losses occur in energy used by predators for cellular respiration, energy required for temperature regulation, energy used by predators to obtain food or for reproduction, and energy released through the decay of waste products. There is an average 90% loss in available energy as one moves to the next-higher trophic level. Likewise, approximately 10% of the energy entering one level passes to the next.

Detritus energy pyramids (organisms that consume organic wastes) are significantly different in structure. The size of the organisms are smaller, and organisms exist in environments rich in nutrients so energy is not needed to obtain food. Additionally, organisms are generally not able to move on their own to any degree. Finally, trophic levels are more complex and interrelated as they include algae, bacteria, fungi, protozoa, insects, arthropods, and worms.

A notable exception in the scheme of a pyramid of biomass occurs in aquatic ecosystems. In this ecosystem, the producers are mainly microscopic algae. Although the total number of algae is great, their total biomass is quite small at any given time.

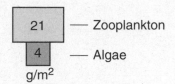

Figure 4.11 Inverted pyramid of biomass—aquatic ecosystem

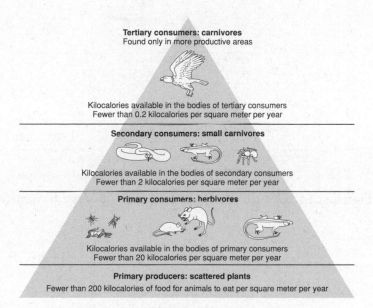

Figure 4.12 Desert biomass—energy pyramid

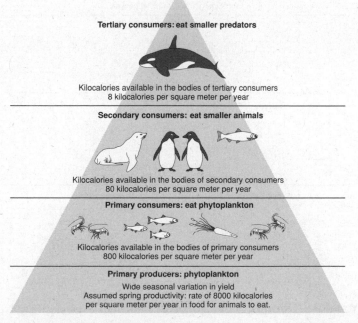

Figure 4.13 Marine biomass—energy pyramid

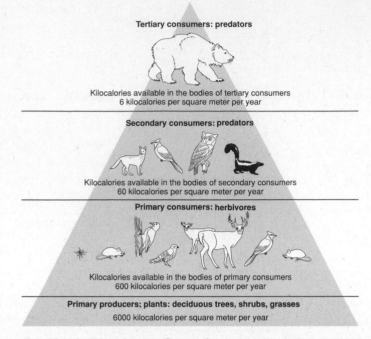

Tertiary consumers: predators

Kilocalories available in the bodies of tertiary consumers
6 kilocalories per square meter per year

Secondary consumers: predators

Kilocalories available in the bodies of secondary consumers
60 kilocalories per square meter per year

Primary consumers: herbivores

Kilocalories available in the bodies of primary consumers
600 kilocalories per square meter per year

Primary producers: plants: deciduous trees, shrubs, grasses
6000 kilocalories per square meter per year

Figure 4.14 Deciduous forest biomass—energy pyramid

TIP

As energy flows
through systems,
more of it becomes
unusable at each
step.

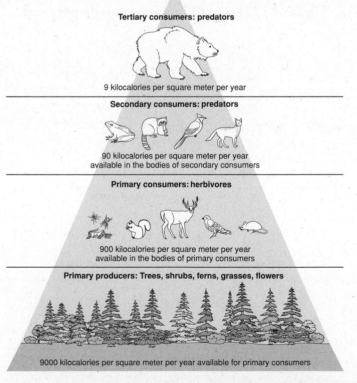

Tertiary consumers: predators

9 kilocalories per square meter per year

Secondary consumers: predators

90 kilocalories per square meter per year
available in the bodies of secondary consumers

Primary consumers: herbivores

900 kilocalories per square meter per year
available in the bodies of primary consumers

Primary producers: Trees, shrubs, ferns, grasses, flowers

9000 kilocalories per square meter per year available for primary consumers

Figure 4.15 Temperate rain forest biomass—energy pyramid

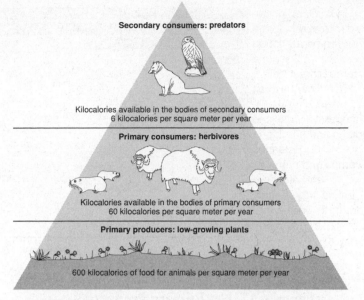

Secondary consumers: predators

Kilocalories available in the bodies of secondary consumers
6 kilocalories per square meter per year

Primary consumers: herbivores

Kilocalories available in the bodies of primary consumers
60 kilocalories per square meter per year

Primary producers: low-growing plants

600 kilocalories of food for animals per square meter per year

Figure 4.16 Tundra biomass—energy pyramid

ECOSYSTEM PRODUCTIVITY: GPP AND NPP

Of all available sunlight that reaches the Earth, less than 3% for land plants and less than 1% for aquatic plants is used for photosynthesis. This relatively low efficiency of the conversion of solar energy into energy stored in carbon compounds sets the overall amount of energy available to heterotrophs at all other trophic levels. Gross primary production (GPP) is the rate at which plants capture and fix (store) a given amount of chemical energy as biomass in a given length of time. Some fraction of this fixed energy is used by primary producers for cellular respiration and the maintenance of existing tissues. The remaining fixed energy is referred to as net primary production (NPP) and is the rate at which all the plants in an ecosystem produce net useful chemical energy and is equal to the difference between the rate at which the plants produce useful chemical energy (GPP) and the rate at which they use some of that energy during respiration.

$$NPP = GPP - \text{plant respiration}$$

Some net primary production goes toward growth and reproduction of primary producers, while some is consumed by herbivores. Open oceans, due to their large proportion of the Earth's surface, collectively have the highest new productivity. However, when compared on a one-to-one per square meter basis, estuaries have the highest net primary productivity.

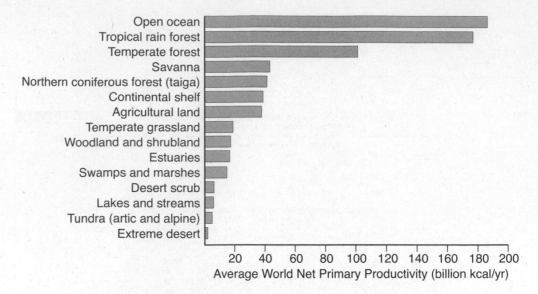

Figure 4.17 Net Primary Productivity—compared by total surface area on Earth

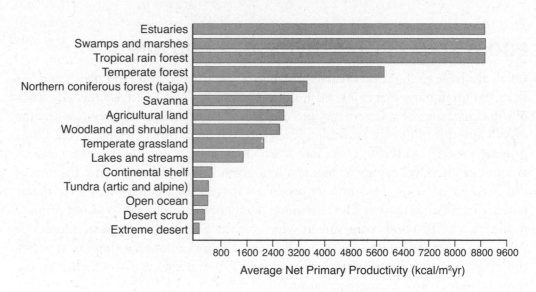

Figure 4.18 Net Primary Productivity—compared per square meter

ECOSYSTEM DIVERSITY

The topic of ecosystem diversity involves biodiversity, natural selection, evolution, and ecosystem services.

Biodiversity

Biodiversity attempts to describe diversity at three levels: genetic, species, and ecosystem. Genetic diversity involves the range of all genetic traits, both expressed and recessive, that makes up the gene pool for a particular species. Species diversity is the number of different species that inhabit a specific area. For example, tropical rain forests have a higher species diversity than extreme deserts. Ecosystem diversity concerns the range of habitats that can be found in a defined area. Ecosystems are composed of both biotic and abiotic components.

Biodiversity

Diversity Increasers	Diversity Decreasers
Diverse habitats	Environmental stress
Disturbance in the habitat (fires, storms, etc.)	Extreme environments
Environmental conditions with low variation	Extreme limitations in the supply of a fundamental resource
Trophic levels with high diversity	Extreme amounts of disturbance
Middle states of succession	Introduction of species from other areas
Evolution	Geographic isolation

Most scientists believe the total number of species on Earth to be between 10 and 30 million. Tropical rain forests cover approximately 7% of the total dry surface of Earth but hold over half of all of the species. Scientists have given names to only about 1.5 million species.

Natural Selection

Natural selection is the mechanism of how organisms evolve. Natural selection works on the individual level by determining which organisms have adaptations that allow them to survive, reproduce, and be able to pass on those adaptive traits to their offspring. Natural selection occurs over successive generations. Evolution works on the species level by describing how the species attains the genetic adaptations that allow them to survive in a changing environment. Without a changing environment, neither evolution nor natural selection would exist.

The range of genetic variation within a species's gene pool determines whether or not the species, not the individual, has the capacity to adapt and survive to changes in the environment. The expression of that variation in phenotypic and behavioral expressions determines whether the individual survives and is commonly referred to as survival of the fittest. "Fittest" means the ability to reproduce and pass on genes to offspring. New genes enter the gene pool through mutation and when combined with a change in gene frequency, results in evolution at the species level. Natural selection operates in three ways: stabilizing, directional, and disruptive.

STABILIZING SELECTION

Stabilizing selection affects the extremes of a population and is the most common form of natural selection. The individuals that deviate too far from the average conditions are removed. The results are a decrease in diversity, maintenance of a stable gene pool, and no evolution. For example, human babies that are too low in weight or too high in weight have survival problems.

Natural Selection

Process of Natural Selection	Description
Competition	There is a struggle to survive, and competition exists for limited resources.
Disproportionate increase and persistence in phenotypic adaptation in successive populations	Variations that are advantageous to the individual in terms of survival allow more organisms possessing the trait(s) to survive, reproduce, and pass on the characteristic(s) to future generations.
Geometric (or exponential) increase in population	If all offspring survived, there would be astronomical numbers of individuals.
Individual variations	There is variation in offspring. For natural selection, the variations must be gene expressed and be capable of being inherited.
Limited resources	Earth has finite resources.

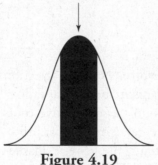

Figure 4.19
Stabilizing selection

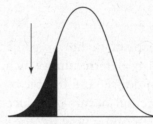

Figure 4.20
Directional selection

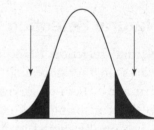

Figure 4.21
Disruptive selection

DIRECTIONAL SELECTION

Directional selection affects the extremes of a population. Individuals toward one end of the distribution may do especially well, resulting in a frequency distribution toward this advantage in future generations. An example is the industrial melanism of the peppered moth and the evolution of the horse. The early horse present during the early Eocene was a small-bodied creature that moved well through heavy brush and woodlands. Today's horse, with its long legs designed for speed in the open grassland and changes in its dental and toe structure, looks nothing like its ancestor.

DISRUPTIVE SELECTION

Disruptive selection acts against individuals that have the average condition and favors individuals at the extreme ends (bimodal). The population is essentially split into two. Both directional and disruptive selection affects the gene pool. In both cases, the population changes and evolution occurs.

More often than not, natural selection is based upon the cumulative effects of numerous genes, each responsible for slight changes. When genes at more than one locus contribute to the same trait, the result is called a polygenic effect.

Polyploidy occurs in plants when the entire set of chromosomes is multiplied. It is an example of sympatric speciation in which species arise within the same, overlapping geographic range. This can occur through the process of hybridization. In hybridization, chromosomes from two different species are artificially combined to form a new species (hybrid). In another type of hybridization, chromosomes naturally fail to segregate during meiosis, producing diploid gametes. If the hybrid has adaptive traits to survive in the new environment, a new species of plant has been produced. Although the plant may not be able to reproduce sexually, it may be able to reproduce through vegetative means. Examples of polyploidy include cotton, tobacco, sugarcane, bananas, potatoes, and wheat. More than half of all known species of plants today (260,000 species) may have originated through polyploidy.

Evolution

Change generally takes a very long time and is supported by the fossil record. Evolution is the change in the genetic composition of a population during successive generations as a result of natural selection acting on the genetic variation among individuals and results in the development of new species. The concept of a "common ancestor" states that similarities among species can be traced back through a phylogenetic tree.

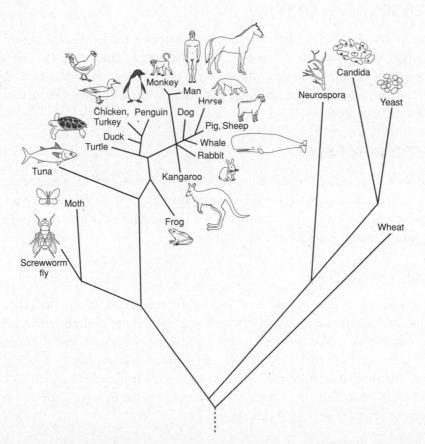

Figure 4.22 Phylogenetic tree traced through analysis of cytochrome C.

SPECIATION

Speciation results when segments of a population of one species become so isolated that gene flow stops. Adaptive radiation describes rapid speciation to fill ecological niches and is driven by either mutation and/or natural selection. Three basic types of adaptive radiation are: general adaptation, environmental change, and geographic isolation.

Adaptive Radiation

Type	Definition	Example
General	Species develop a radical new ability to reach new parts of its environment.	Bird flight
Environmental Change	Due to large changes in the environment, species branch into new species and occupy new niches.	Rapid spread and development of mammalian species after extinction of dinosaurs.
Geographic Isolation	Isolated ecosystems are colonized by species that undergo rapid divergent evolution.	Darwin finches in the Galapagos Islands.

CONVERGENT EVOLUTION

Convergent evolution describes the process whereby organisms not closely related to each other independently acquire similar (analogous) characteristics while evolving in separate and sometimes varying ecosystems. An example of convergent evolution is the development of wings in birds, insects, and bats even though the organisms are not closely related.

EVOLUTIONARY RELAY

Evolutionary relay occurs when independent species acquire similar characteristics through their evolution in similar ecosystems but not at the same time. An example of evolutionary relay is the development of a dorsal fin in both sharks and the prehistoric ichthyosaurs (extinct marine reptiles).

PARALLEL EVOLUTION

Parallel evolution occurs when two independent species evolve together at the same time and in the same ecosystem and acquire similar characteristics. A classic example of parallel evolution is the evolution of the placentals and the marsupials. Placental animals bear their young fully developed. Marsupials give birth prematurely and nurture their young in a pouch. In the plant kingdom, parallel evolution is seen in the similar forms of leaves, where similar patterns have appeared over and over again in separate genera and families.

GRADUALISM AND PUNCTUATED EQUILIBRIUM

Gradualism views evolution as a slow, stepwise development of a species over long periods of time (millions of years). Punctuated equilibrium proposes that some species arose suddenly in a short period of time (thousands of years) after long periods of stability. These bursts of rapid evolution are thought to have been triggered by changes in the physical or biological environment—perhaps a period of drought or the appearance of a new, more challenging predator. A classic example of punctuated equilibrium is the abrupt appearance of flowering plants without a fossil record.

Ecosystem Services

Ecosystem services are the processes by which the environment produces resources. Examples include clean water, timber, habitat for fisheries, and pollination of native and agricultural plants. Ecosystems provide the following services:

- Moderate weather extremes and their impacts
- Disperse seeds
- Mitigate droughts and floods
- Protect people from the sun's harmful ultraviolet rays
- Cycle and move nutrients
- Protect stream and river channels and coastal shores from erosion
- Detoxify and decompose wastes
- Control agricultural pests
- Maintain biodiversity
- Generate and preserve soils and renew their fertility
- Contribute to climate stability
- Purify the air and water
- Regulate disease-carrying organisms
- Pollinate crops and natural vegetation

Species Movements

Many of the different types of organisms that inhabit Earth have the ability to move. This movement can be accomplished either passively or actively. Examples of active movement include walking, running, flying, or swimming. In passive movement, the organism uses some external force to cause transit. For example, plants can use wind for seed dispersal while oyster larvae can passively travel great distances by sea currents.

One common factor of why organisms move is to disperse to new habitats to reduce intraspecific competition. By finding new suitable habitats, individuals can increase the range of their species. A larger range makes the species better off in terms of evolution.

The geographic distributions of plant and animal species are never fixed over time. Geographic ranges of organisms shift, expand, and contract. These changes are the result of two contrasting processes: (1) colonization and establishment, and (2) localized extinction. Colonization and establishment takes place when populations expand into new areas. A number of processes can initiate this process, including disturbance and abiotic environmental change. Localized extinction results in the elimination of populations from all or part of their former range. It can be caused by biotic interactions or abiotic environmental change.

Plants have developed a number of different mechanisms for dispersing their offspring:

1. The use of specialized structures to aid the transport of an individual by wind

2. The employment of particular structures to transport the individual by moving water

3. The production of fruit-encased seeds that other organisms consume and disperse

4. Adhesion mechanisms

5. The physical ejection of seeds

Once dispersed, an individual can colonize a new site only if it is devoid of other organisms and if the necessary abiotic requirements and conditions exist for its survival. Sites within ecosystems become devoid of organisms through disturbance. A disturbance can be caused by predation, climate variations, earthquakes, volcanoes, fire, animal burrowing, and even the impact of a raindrop. Often the struggle for survival does not end with colonization of an individual on a vacant site. Once colonized, an individual may not be able to establish itself over the long term because of abiotic and biotic influences. The death of the individual may occur through competitive interaction, predation, or an abiotic factor like fire.

Ecological Succession

Succession is the gradual and orderly process of ecosystem development brought about by changes in community composition and the production of a climax community characteristic of a particular geographic region. It describes the changes in an ecosystem through time and disturbance. Rates of succession are affected by several factors.

- Facilitation is when one species modifies an environment to the extent it meets the needs of another species.
- Inhibition is when one species modifies the environment to an extent that it is not suitable for another species.
- Tolerance is when species are not affected by the presence of other species.

Earlier successional species frequently called pioneer species are generalists. Pioneer plants have short reproductive times (annuals), and pioneer animals have low biomass and fast reproductive rates. Later successional species include larger perennial plants and animals with greater biomass, longer generational times, and higher parental care.

Types of Succession

Type	Description
Allogenic	Changes in the environmental conditions create conditions beneficial to new plant communities.
Primary	The colonization and establishment of pioneer plant species on bare ground. (Example: lichens on bare rocks)
Progressive	Communities become more complex over time by having a higher species diversity and greater biomass.
Retrogressive	The environment deteriorates and results in less biodiversity and less biomass.
Secondary	Begins in an area where the natural community has been disturbed but topsoil remains. (Example: forest fire)

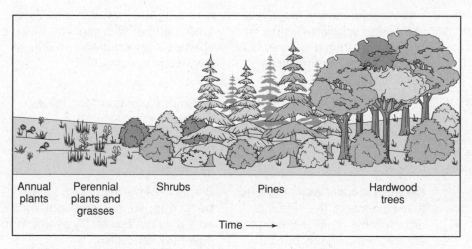

| Annual plants | Perennial plants and grasses | Shrubs | Pines | Hardwood trees |

Time ———→

Figure 4.23 Stages of succession in a temperate deciduous forest. The time span from annual plants to hardwood trees is over 100 years

The following table lists the characteristics of succession within plant communities:

Characteristics of Succession-Plant Communities

Characteristic	Early Successional Stage	Late Successional Stage
Biomass	Limited.	High in tropics and wetlands; limited in deserts.
Consumption of soil nutrients	Nutrients are quickly absorbed by simpler plants.	Since biomass is greater and more nutrients are contained within plant structures, nutrient cycling between the plant and soil tends to be slower.
Impact of macroenvironment	Early plants depend primarily on conditions created by macro-environmental changes (fires, floods, etc.).	These plant species appear only after pioneer plant communities have adequately prepared the soil.
Life span of seed	Long. Seeds may become dormant and able to withstand wide environmental fluctuations.	Short. Not able to withstand wide environmental fluctuations.
Life strategy	r-strategists: mature rapidly; short-lived species; number of organisms within a species is high; low biodiversity; niche generalists.	K-strategists: mature slowly; long-lived; number of organisms within a species is lower; greater biodiversity; niche specialists.
Location of nutrients	In the soil and in leaf litter.	Within the plant and top layers of soil.
Net primary productivity	High.	Low.
Nutrient cycling by decomposers	Limited.	Complex.
Nutrient cycling through biogeochemical cycles	Because nutrient sinks have not fully developed, the nutrients are available to cycle through established biogeochemical cycles fairly rapidly.	Because of nutrient sinks (carbon being trapped in vegetation), nutrients may not be readily available to flow through cycles.
Photosynthesis efficiency	Low.	High.
Plant structure complexity	Simple.	More complex.
Recovery rate of plants from environmental stress	Plants quickly and easily come back.	Recovery is slow.
Seed dispersal	Widespread	Limited in range.
Species diversity	Limited	High.
Stability of ecosystem	Since diversity is limited, ecosystem is subject to instability.	Due to high diversity, ecosystem can withstand stress.

QUICK REVIEW CHECKLIST

- ☐ **Biological Populations and Communities**
  - ☐ ecosystem characteristics
  - ☐ population dispersal patterns
  - ☐ ecological niches—generalists vs. specialists
  - ☐ biological interactions
    - ☐ amensalism
    - ☐ commensalism
    - ☐ competition
    - ☐ mutualism
    - ☐ parasitism
    - ☐ predation
    - ☐ saprotrophism
    - ☐ keystone species (examples)

- ☐ **Species Diversity (Adaptations)**
  - ☐ aquatic organisms
  - ☐ desert organisms
  - ☐ grassland organisms
  - ☐ forest organisms
  - ☐ temperate scrub forest land organisms
  - ☐ tundra organisms

- ☐ **Edge Effects**

- ☐ **Biomes**
  - ☐ Antarctic
  - ☐ benthos
  - ☐ coastal zones
  - ☐ coral reefs
  - ☐ deserts
  - ☐ freshwater wetlands
  - ☐ grasslands
  - ☐ hydrothermal vents
  - ☐ intertidal zones
  - ☐ ocean
  - ☐ savannas
  - ☐ taiga
  - ☐ temperate deciduous forests
  - ☐ temperate rain forests
  - ☐ temperate scrubland (chaparral)
  - ☐ temperate woodlands
  - ☐ tropical rain forests
  - ☐ tropical seasonal forests
  - ☐ tundra

- ☐ **Energy flow**
  - ☐ photosynthesis
  - ☐ cellular respiration

QUICK REVIEW CHECKLIST (continued)

☐ **Food Webs and Trophic Levels**
- ☐ desert food web
- ☐ oceanic food web
- ☐ coniferous forest food web
- ☐ deciduous forest food web
- ☐ tropical rain forest food web
- ☐ temperate rain forest food web
- ☐ tundra food web

☐ **Ecological Pyramid**
- ☐ aquatic ecosystem (inverted)
- ☐ desert biomass
- ☐ marine biomass
- ☐ deciduous forest biomass
- ☐ temperate rain forest biomass
- ☐ tundra biomass
- ☐ net primary productivity

☐ **Ecosystem Diversity**
- ☐ diversity increasers
- ☐ diversity decreasers

☐ **Natural Selection**
- ☐ stabilizing selection
- ☐ directional selection
- ☐ disruptive selection

☐ **Evolution**
- ☐ speciation
- ☐ convergent evolution
- ☐ evolutionary relay
- ☐ parallel evolution
- ☐ gradualism
- ☐ punctuated equilibrium

☐ **Ecosystem Services**

☐ **Species Movements**

☐ **Ecological Succession**
- ☐ types of succession
  - ☐ allogenic
  - ☐ primary
  - ☐ progressive
  - ☐ retrogressive
  - ☐ secondary
- ☐ stages of succession

☐ **Ecological Succession**
- ☐ characteristics of succession (plant communities)

MULTIPLE-CHOICE QUESTIONS

For questions 1–3, choose from the following items:

    (A) Tropical rain forest
    (B) Temperate deciduous forest
    (C) Savanna
    (D) Taiga
    (E) Tundra

1. Forests of cold climates of high latitudes and high altitudes.

2. Warm year-round; prolonged dry seasons; scattered trees.

3. Low biodiversity due to lots of shade, which limits food for herbivores. Major resource for timber.

4. The annual productivity of any ecosystem is greater than the annual increase in biomass of the herbivores in the ecosystem because

    (A) plants convert energy input into biomass more efficiently than animals
    (B) there are always more animals than plants in any ecosystem
    (C) plants have a greater longevity than animals
    (D) during each energy transformation, some energy is lost
    (E) animals convert energy input into biomass more efficiently than plants do

5. Net primary productivity per square meter is highest in which biome listed below?

    (A) Deserts
    (B) Grasslands
    (C) Boreal forests
    (D) Open oceans
    (E) Estuaries

6. All of the following are factors that increase population size EXCEPT

    (A) ability to adapt
    (B) specialized niche
    (C) few competitors
    (D) generalized niche
    (E) high birthrate

7. A specialist faces _____ competition for resources and has _____ ability to adapt to environmental changes. A generalist faces _____ competition for resources and has _____ ability to adapt to environmental changes.

    (A) less, greater, greater, less
    (B) greater, less, less, greater
    (C) less, less, greater, greater
    (D) greater, greater, less, less
    (E) None of the above

8. Whether a land area supports a deciduous forest or grassland depends primarily on

    (A) changes in temperature
    (B) latitude north or south of the equator
    (C) consistency of rainfall from year to year and the effect that it has on fires
    (D) changes in length of the growing season
    (E) None of the above

9. The main difference between primary and secondary succession is that

    (A) primary succession occurs in the year before secondary succession
    (B) primary succession occurs on barren, rocky areas and secondary succession does not
    (C) secondary succession ends in a climax species and primary succession ends in a pioneer species
    (D) secondary succession occurs on barren, rocky areas and primary succession does not
    (E) All of the above statements are true

10. The biggest threat to species is

    (A) low reproductive rates
    (B) disease
    (C) alien, invasive species
    (D) collecting, hunting, and poaching
    (E) loss of habitat

11. Darwin noted that the Patagonian hare was similar in appearance and had a niche similar to the European hare. However, the Patagonian hare is not a rabbit. It is a rodent related to the guinea pig. This example illustrates the principle known as

    (A) allopatric speciation
    (B) adaptive radiation
    (C) divergent evolution
    (D) coevolution
    (E) convergent evolution

For questions 12–14, choose from the following items:

    (A) Adaptive radiation
    (B) Isolation
    (C) Natural selection
    (D) Stable gene pool
    (E) Convergent evolution

12. Members of the same species of moths are prevented from interbreeding because they live on opposites sides of a mountain range.

13. Darwin's finches are a good example of this biological principle.

14. In the evolutionary history of the horse, the early horse (*Eohippus*) was replaced by the modern one-toed horse.

15. As one travels from the tropical rain forests near the equator to the frozen tundra, one passes through a variety of biomes including deserts, temperate and deciduous forests, etc. The primary reason for the change in vegetation is due to

    (A) evolution
    (B) succession
    (C) the hours of sunlight reaching a particular region
    (D) changes in temperature and amount of rainfall
    (E) the amount of relative cloud cover and ocean currents

16. Species that serve as early warnings of environmental damage are called

    (A) keystone species
    (B) native species
    (C) specialist species
    (D) indicator species
    (E) generalist species

17. Which one of the following statements is false?

    (A) When environmental conditions are changing rapidly, a generalist is usually better off than a specialist.
    (B) The fundamental niche of a species is the full range of physical, chemical, and biological factors it could use if there were no competition.
    (C) The competitive exclusion principle states that no two species with the same fundamental niche can indefinitely occupy the same habitat.
    (D) Interspecific competition is competition between two members of the same species.
    (E) Resource partitioning limits competition by two species using the same scarce resource at different times, in different ways, or in different places.

18. Which of the following best describes a nonanthropogenic secondary succession?

    (A) Plants and other vegetation die gradually due to drought
    (B) Wildflowers grow in an area that was previously destroyed by fire
    (C) A farmer removes weeds using a herbicide
    (D) Lichens and mosses secrete acids that allow other plants to grow
    (E) None of the above

19. The location of where an organism lives would be best described as its

    (A) niche
    (B) habitat
    (C) range
    (D) biome
    (E) ecosystem

For Question 20, refer to the locations marked by letters in the world map below

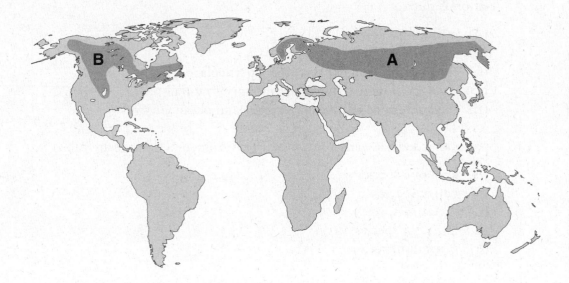

20. The locations shaded in the map above are known as

    (A) tropical rain forests
    (B) grasslands
    (C) taiga (boreal forest)
    (D) chaparral
    (E) tundra

FREE-RESPONSE QUESTION

There are over 25,000 species of bees in the world today. Africanized honey bees (AHB), *Apis mellifera L. scutellata* (Lepetelier), sometimes referred to as killer bees, are the same species as European honeybees (EHB) but a different subspecies. Only through careful examination under a microscope, or through DNA tests, can AHBs be distinguished from EHBs. AHBs are called "Africanized" honeybees as a result of interbreeding experiments. The thought was that AHBs might be better suited for the tropical climates of South America than EHBs. In the 1950s, AHBs were inadvertently released in Brazil. AHBs began migrating (about 300 miles per year), arriving in southern Texas in 1990, Arizona and New Mexico in 1993, and southern California in 1994. Since 1994, migration has appeared to slow down significantly. Although the two subspecies are virtually indistinguishable, it is the behavior of the two subspecies that sets them apart. AHBs defend their colonies much more vigorously than do EHBs. When AHBs sting, more of them participate in extremely aggressive behavior. AHBs nest where EHBs do not, such as small, confined spaces near the ground such as flowerpots, abandoned tires, and cracks in building foundations.

(a) Discuss the influx of AHBs in the southwest United States in terms of the processes of natural selection.

    (i)   In what areas do the processes of natural selection apply to the colonization of AHBs?

    (ii)   In what areas do the processes of natural selection NOT apply to the colonization of AHBs?

    (iii) What issues regarding the influx of AHBs warrant further study?

(b) Provide possible explanations of why the expansion of AHBs may have stopped.

(c) What adaptations have honeybees in general developed to be able to survive?

(d) Provide some possible methods to control AHBs.

(e) What are possible economic implications of AHBs to agriculture?

TIP

Be sure to explain technical terms when you use them in an answer. Dropping in terms like "bioaccumulation" without demonstrating an understanding of the term will not earn you any credit.

MULTIPLE-CHOICE ANSWERS AND EXPLANATIONS

1. **(D)** Refer to the biome descriptions (pages 170–173).
2. **(C)** Refer to the biome descriptions (pages 170–173).
3. **(B)** Refer to the biome descriptions (pages 170–173).
4. **(D)** Less energy is available at each trophic level because energy is lost by organisms through cellular respiration and incomplete digestion of food sources.
5. **(E)** Estuaries form a transition zone between river environments and ocean environments and are subject to both marine influences, such as tides, waves, and the influx of saline water; and riverine influences, such as flows of freshwater and sediment. The inflow of both seawater and freshwater provide high levels of nutrients in both the water column and sediment, making estuaries the most productive natural habitats in the world and also the most threatened.
6. **(B)** Specialized niches are more susceptible to environmental changes and have a direct effect on the stability of populations (population size).
7. **(C)** Specialist species are adapted to a narrow range of habitats and conditions, while generalist species are able to live in a variety of habitats.
8. **(C)** The question of determining whether it is a deciduous forest or grassland is dependent on yearly patterns of rainfall since both biomes can exist over similar temperature ranges. Frequent fires are an important factor in determining grasslands.
9. **(B)** Primary succession occurs on bare rocks and starts with lichens. Secondary succession occurs in areas where there is intact topsoil.
10. **(E)** Scientists warn that human activities may be bringing about the sixth mass extinction of species in the world's history.
11. **(E)** Convergent evolution describes the process whereby organisms not closely related independently acquire similar characteristics while evolving in separate and sometimes varying environments.
12. **(B)** "Species" is defined as a group of organisms that look similar, have the ability to interbreed, and produce fertile offspring. Two forms of isolation that prevent interbreeding are geographic isolation and reproductive isolation.
13. **(A)** Adaptive radiation is the development of many species that are derived from a single, ancestral population.
14. **(C)** Natural selection is the process by which only the organisms that are best adapted to their environment tend to survive and transmit their genes to successive generations.
15. **(D)** The distribution of biomes is primarily determined by global climatic zones as measured by average annual temperature and rainfall. An area's climate will determine what sort of biome can be sustained in that area. Other determining factors include sun angle, heat import, and the angle of the Earth's rotation. Regional climates, which also determine the distribution of biomes, are affected by rain shadows and cold or warm ocean currents.
16. **(D)** An indicator species is a species whose presence, absence, or relative well-being in a given environment is indicative of the health of the ecosystem as a whole.
17. **(D)** Interspecific competition is competition among members of different species.

18. **(B)** Ecosystems undergo secondary succession following some artificial or natural disturbance such as a forest fire or farming. The question included the word "nonanthropogenic" which would rule out choice (C) since anthropogenic refers to something caused by humans.

19. **(B)** A habitat is the location where a particular species lives and grows. A microhabitat is the immediate surroundings and other physical factors of an individual plant or animal within its habitat. Habitat destruction is a major factor in causing a species population to decrease, eventually leading to it being endangered or becoming extinct. A biome is the set of flora and fauna that lives in a habitat and occupy certain geography.

20. **(C)** Taiga, also known as boreal forest, is the world's largest land biome, and makes up almost a third of the world's forest cover with the largest areas located in Russia and Canada. The taiga has a subarctic climate with very large temperature range between seasons, but the long and cold winter is the dominant feature.

    Lakes and other water bodies are very common. The taiga experiences relatively low precipitation, 8"–30" (200–750 mm), throughout the year, primarily as rain during the summer months, but also as fog and snow. As evaporation is consequently low for most of the year, precipitation exceeds evaporation, and is sufficient to sustain the dense vegetation growth. Snow may remain on the ground for as long as nine months in the northernmost extensions of the taiga.

    Taiga soil tends to be young and poor in nutrients and lacks the deep, organically enriched profile present in temperate deciduous forests. The thinness of the soil is due largely to the cold, which hinders the development of soil and the ease with which plants can use its nutrients. Since the soil is acidic due to the falling pine needles, the forest floor has only lichens and some mosses growing on it. Diversity of soil organisms in the boreal forest is high, comparable to the tropical rain forest.

    Because the sun is low in the horizon for most of the year, it is difficult for plants to generate energy from photosynthesis. Pine, spruce, and fir do not lose their leaves seasonally and are able to photosynthesize with their older leaves in late winter and spring when light is good but temperatures are still too low for new growth to commence. The adaptation of evergreen needles limits the water lost due to transpiration and their dark green color increases their absorption of sunlight. Although precipitation is not a limiting factor, the ground freezes during the winter months and plant roots are unable to absorb water, so desiccation can be a severe problem in late winter.

    The taiga stores enormous quantities of carbon, more than the world's temperate and tropical forests combined, much of it in wetlands and peatland. Current estimates place boreal forests as storing twice as much carbon per unit area as tropical forests.

FREE-RESPONSE ANSWER

Ten points possible for full credit. Parts A, B, C, D, and E are worth 2 points maximum each.

(a) The range of genetic variation within a species's gene pool determines whether or not the species, not the individual, has the capacity to adapt and survive to changes in the environment. The colonization of the American southwest by AHBs differs from classical theories of natural selection in that AHBs and EHBs belong to the same species and therefore are presumably able to interbreed and produce fertile offspring. The two subspecies seem to differ only in behavioral patterns. As to whether the behavioral patterns of the two subspecies are so radically different as to cause a prezygotic reproductive barrier or not is worth investigation.

  Another issue that differs from classical natural selection theory is that, in this case, the environment, over such a short time span, is not changing. In effect, it is not a variable involved in survival rates. Both subspecies can and do survive in the environment. Additionally, the environment is not affecting disproportionate natality rates between the two subspecies based upon one subspecies having a survival advantage over the other.

  In terms of following classic natural selection theory, what exists in this scenario is competition between the two subspecies for limited resources. Both subspecies are competing for space and food supplies. As to whether or not the more aggressive behavior of protecting the nest allows more AHBs to survive disproportionately, thereby increasing the ratio of AHBs to EHBs is another factor that warrants further study.

(b) It appears that since 1994, the expansion of AHBs into the southwestern United States has slowed down. Possible reasons might include the following.

  (1) AHBs are adapted to tropical climates (Africa and Brazil). Their ability to survive may be restricted to climatic zones that have warm temperatures most of the year (similar to the American Southwest). Cold temperatures (common in the Midwest and northeast U.S.) may influence the viability of offspring and, consequently, their expansion into these areas.

  (2) Geographic barriers may be another reason for a decline in the expansion of AHBs. Mountain ranges, dry deserts with low humidity, and the lack of adequate food resources may also be slowing expansion. Finally, variation in seasonal photoperiod, timing of forage availability, and/or the existence of parasites such as mites, pathogenic bacteria, or viruses may be having an effect on the newly introduced subspecies.

(c) Three adaptations of honeybees have been essential to their evolution and biology. First is clustering behavior, which is working as a social unit with specific predetermined behavioral patterns and duties of each class in the hierarchy. The second is the ability to cool the hive through the process of evaporation of water collected outside the hive, ensuring a stable internal temperature. The ability to ensure temperature stability within the nest allows honeybees to colonize a wide variety of environments, as opposed to bees that lack this trait and are therefore restricted to thermally stable environments (tropics). The third is the ability to communicate information about food sources, direction, and distance through dance behavior. The ability to communicate such a wide

Notable points:

(a) Recognition of genetic variation as a determinant in the process of natural selection.

Recognition that subspecies are able to interbreed.

Recognition that behavior may be a reproductive barrier.

Recognition that the environment is remaining constant and not acting to select one subspecies over another.

Recognition that the two subspecies are competing for the same limited natural resources—a basic tenet of natural selection.

Recognition that more study is needed in the advantage of aggressive behavior and its effect on survivability of the colony.

(b) Specific adaptations of AHBs to tropical climates.

Temperature tolerances may be narrower in AHBs.

Geographic barriers.

Mountain ranges.

Low humidity in deserts.

Lack of sufficient food resources in deserts.

Variation in seasonal photoperiod.

Timing of forage availability (seasons) as opposed to no season in the tropics.

Parasites.

Pathogenic organisms.

(c) Clustering behavior.

Predetermined roles.

Temperature regulation of hive to create stable environment for young.

Internal stability of nest temperature allows colonization into more diverse areas.

Ability to communicate.

(d) Recognition that wide use of pesticides would not be effective.

EHBs are more suitable for agriculture.

Bacterial agents or viruses would affect both subspecies.

Only way to control AHBs would be to ensure competitive population numbers of EHBs.

Reduce carrying capacity of environment.

Interbreeding and its role to diminish aggressive behavior.

Introduction of sterile males.

(e) Bees are necessary for agriculture.

People are closer today to agricultural areas and are thereby more impacted.

Public fear and hysteria of AHBs.

Availability of suitable sites for EHBs.

Public pressure.

Liability issues (insurance).

Legislative mandates.

Fewer beekeepers.

Cost of maintaining EHBs would increase.

As cost of bees goes up, so would cost of agricultural products.

variety of complex information is unparalleled in the animal kingdom (except for humans).

(d) Since AHBs and EHBs are coexisting within the same areas, any attempt to control the AHBs through pesticides or other extermination techniques would probably have an effect on the EHBs as well. For the most part, EHBs and their less-aggressive behavior are more suited to coexist with humans for agricultural pollination purposes than are AHBs. Introduction of any bacterial agent, parasite, or other biological controlling mechanism would also probably affect the EHBs. One method that might work to control the domination of an area by AHBs would be to ensure that there are sufficient EHBs in the area. First, this would provide competition for resources between the two subspecies, thereby reducing the carrying capacity of the area. In turn, this would limit expansion of the AHB population. Second, ensuring sufficient EHBs in the area might allow interbreeding of an AHB queen with EHB males. The goal of interbreeding would be to possibly dampen the highly aggressive behavioral characteristics of the AHBs. Third, introducing large numbers of sterile, male AHBs into hives might result in smaller populations. Physically disrupting hives would only enhance recolonization.

(e) Agriculture as practiced today requires the use of honeybees for pollination. Honeybees are rented and placed into fields for this purpose by beekeepers. Urban expansion, human population increases, and sophisticated transportation systems have all made rural areas more accessible. AHBs strike terror in many people, and yet documented cases of deaths due to AHBs are rather low and exaggerated. Public pressure to eradicate "killer bees" will certainly involve economics. The cost of renting EHBs would certainly increase due to issues of suitability and availability of sites, public concerns, legislative mandates, and destroying EHBs in an attempt to control AHBs. As costs of renting EHBs increase, profit margins could decrease, forcing many beekeepers out of business. The result would be fewer bees for the needs of agriculture. This, in turn, could result in less agricultural production and, consequently, higher prices to the consumer.

YOUR TURN

- Let's practice the skills required for answering Free Response Questions. Read and review the essay above several times. Take very brief notes of important points and use them for now if you wish. When you are ready, set a timer for 23 minutes—the average amount of time you will have for each FRQ. Begin answering each question as best you can in your best writing style. STOP after 23 minutes. Take a breath! Then later, go back over the essay you wrote. REMEMBER: I had days to write my answer and you had 23 minutes. Your essay will be **much** shorter than mine but the thing you are looking for now is: (1) Did you answer the questions? (2) Do you feel comfortable with what you wrote? (3) Were you able to incorporate some of the material in my answers into your answers? (4) Do your answers flow and seem organized? (5) Were you able to back up or support your statements with facts, examples, etc.? (6) Does your essay look neat and legible? If any of the answers above were "no"—then try this same exercise again until you are finally satisfied. As you continue this method for each chapter, you are getting closer and closer to that 5!

Natural Biogeochemical Cycles

CARBON CYCLE

Carbon is the basic building block of life and the fundamental element found in carbohydrates, fats, proteins, and nucleic acids (DNA and RNA). Carbon is exchanged among the biosphere, geosphere, hydrosphere, and atmosphere. Although carbon is found in rocks, it is a minor component when compared with the mass of either oxygen or silicon atoms in rocks. Carbon is found in carbon dioxide (CO_2), which makes up less than 1% of the atmosphere.

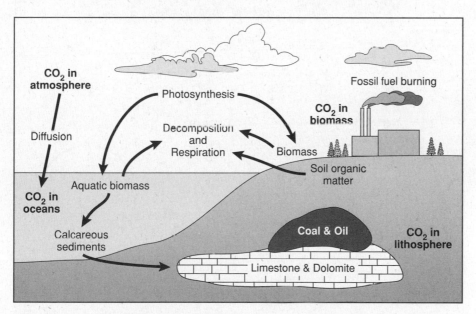

Figure 5.1 The carbon cycle viewed as chemical processes

The major reservoirs or "sinks" of carbon include:

- **Plant matter:** A portion of atmospheric carbon (15%) is removed through photosynthesis by which carbon is incorporated into plant structures and compounds:

$$6CO_2 + 6H_2O + \text{energy (sunlight)} \rightarrow C_6H_{12}O_6 + 6O_2$$

- **Terrestrial biosphere:** Forests store 86% of the planet's above-ground carbon and 73% of the planet's soil carbon. Carbon can be stored up to several hundreds of years in trees and up to thousands of years in soils.

- **Oceans:** Dissolved inorganic carbon in the form of CO_2 and living and nonliving marine biota (i.e., shells and skeletons) are included. The oceans are gaining approximately 2 gigatons (4×10^{12} kg) of carbon each year; however, most of it is not involved with rapid exchanges with the atmosphere. Removing carbon dioxide from water raises the pH, making the water more basic.
- **Sedimentary deposits:** Limestone CaCO₃) and carbon trapped in fossil fuels and coal. Limestone is the largest reservoir of carbon in the carbon cycle. The calcium comes from the weathering of calcium-silicate rocks, which causes the silicon in the rocks to combine with oxygen to form sand or quartz (silicon dioxide), leaving calcium ions available to form limestone.

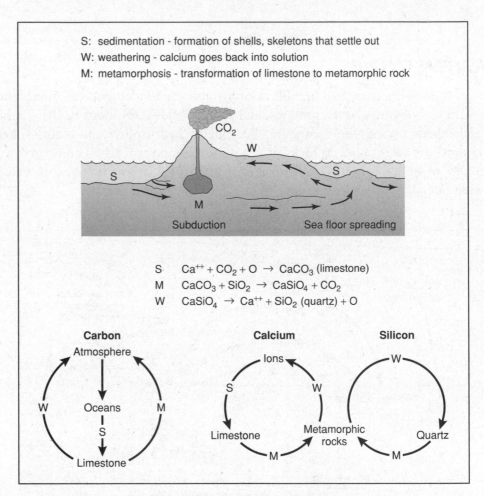

Figure 5.2 Inorganic carbon cycle

Carbon is released back into the atmosphere through:

1. Cellular respiration of plants and animals that break down glucose into carbon dioxide and water: $C_6H_{12}O_6 + 6O_2 \rightarrow 6CO_2 + 6H_2O + energy$. When oxygen is not present, anaerobic respiration occurs and releases carbon into the atmosphere in the form of methane (CH_4) (e.g., as marsh gas or flatulence).

2. Decay of organic material by the action of decomposers; if oxygen is present, the carbon is released in the form of carbon dioxide; if oxygen is absent, it is released in the form of methane (CH_4).

3. Burning fossil fuels, wood, coal, etc.

4. Weatherization of rocks and especially the erosion of limestone, marble, and chalk, which break down to carbon dioxide and carbonic acid (H_2CO_3).

5. Volcanic eruptions.

6. Release of carbon dioxide by warmer ocean waters.

Prior to the Industrial Revolution, transfer rates of carbon dioxide due to photosynthesis and respiration (including decay) were fairly balanced. However, since the Industrial Revolution, more carbon dioxide is being deposited in the Earth's atmosphere than is being removed. This increase is believed to be due to the burning of wood and fossil fuels and deforestation.

Carbon Sinks

Carbon Sink	Amount (Billions of Metric Tons)
Marine sediments and sedimentary rocks	~ 75,000,000
Ocean	~ 40,000
Fossil fuel deposits	~ 4000
Soil organic matter	~ 1500
Atmosphere	578 (in 1700 C.E.) to 766 (in 2000 C.E.)
Terrestrial plants	~ 580

NITROGEN CYCLE

Nitrogen makes up 78% of the atmosphere. Nitrogen is also an essential element needed to make amino acids, proteins, and nucleic acids. Other nitrogen stores include organic matter in the soil and the oceans (1 million times more nitrogen is found in the atmosphere than is contained in either land or ocean waters).

The natural cycling of nitrogen, in which atmospheric nitrogen is converted to nitrogen oxides by lightning and deposited in the soil by rain where it is assimilated by plants and either eaten by animals (and returned as feces) or decomposed back to elemental nitrogen by bacteria, includes the following processes:

- Nitrogen fixation
- Nitrification
- Assimilation
- Ammonification
- Denitrification

Nitrogen Fixation

Nitrogen fixation is the conversion of atmospheric nitrogen (N_2) to ammonia (NH_3) or nitrate (NO_3^-) ions. Nitrate is the product of high-energy fixation by lightning, cosmic radiation, or meteorite trails. In high-energy fixation, atmospheric nitrogen and oxygen combine to form nitrates, which are carried to Earth's surface in rainfall

TIP

Know *all* steps involved in the nitrogen cycle. You will find more questions about the nitrogen cycle than any other cycle on the APES exam.

as nitric acid (HNO_3). High-energy fixation accounts for about 10% of the nitrate entering the nitrogen cycle.

In contrast, biological fixation accounts for 90% of the fixed nitrogen in the cycle. In biological fixation, molecular nitrogen (N_2) is split into two free nitrogen atoms ($N_2 \rightarrow N + N$). The nitrogen atoms combine with hydrogen to yield ammonia (NH_3).

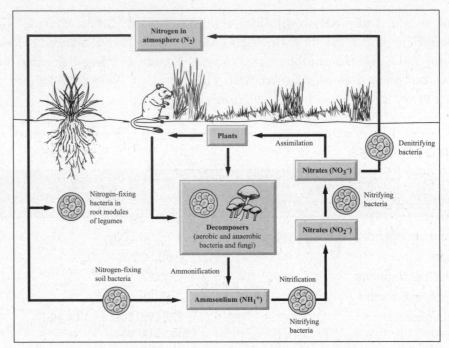

Figure 5.3 The nitrogen cycle

The fixation process is accomplished by a series of different microorganisms. The symbiotic bacteria *Rhizobium* is associated with the roots of legumes such as beans or clover. To a lesser extent, some non-leguminous plants also exhibit symbiotic relationships with bacteria.

Nitrification

Nitrification is the process in which ammonia is oxidized to nitrite (NO_2^-) and nitrate (NO_3^-), the forms most usable by plants. Two groups of microorganisms are involved in nitrification. *Nitrosomonas* oxidize ammonia to nitrite and water. *Nitrobacter* oxidize the nitrite to nitrate.

Assimilation

Nitrates are the form of nitrogen most commonly assimilated by plants through their root hairs. Nitrogen is used by plants to synthesize amino acids, oils, and nucleic acids. Rains and extensive irrigation can leach soluble nitrates and nitrites into groundwater, which, in high amounts, can interfere with blood-oxygen levels in human infants. Soluble nitrates that run off the land and enter aquatic and wetland habitats result in cultural eutrophication and the destruction of these habitats. Animals assimilate nitrogen-based compounds by consuming plants and other organisms that consume plants.

Ammonification

When a plant or animal dies, or an animal excretes, the initial form of nitrogen is found in amino acids and nucleic acids. Bacteria, or in some cases, fungi, convert this organic nitrogen within the remains back into ammonia (NH_3).

Denitrification

Denitrification is the process in which nitrates are reduced to gaseous nitrogen. This process is used by facultative anaerobes. These organisms flourish in an aerobic environment but are also capable of breaking down oxygen-containing compounds (NO_3^-) to obtain oxygen in anaerobic environments. Examples include fungi and the bacteria *Pseudomonas*.

Effects of Excess Nitrogen

Fossil fuel combustion has contributed to a sevenfold increase in nitrogen oxides (NO_x) to the atmosphere, particularly nitrogen dioxide (NO_2). NO_x is a precursor of tropospheric (lower atmosphere) ozone production and contributes to smog and acid rain and increases nitrogen inputs to ecosystems.

Ammonia (NH_3) in the atmosphere has tripled as the result of human activities since the Industrial Revolution. Ammonia acts as an aerosol and decreases air quality.

Nitrous oxide (N_2O) is a significant greenhouse gas and has deleterious effects in the stratosphere, where it breaks down and acts as a catalyst in the destruction of atmospheric ozone. N_2O is in a large part emitted during nitrification (conversion of ammonium to nitrate and nitrite) and denitrification (converting oxides back to nitrogen gas or nitrous oxides for energy generation) processes that take place in the soil. The largest N_2O emissions are observed where nitrogen-containing fertilizer is applied in agriculture.

Human activity has more than doubled the annual transfer of nitrogen into biological available forms through:

- Extensive cultivation of legumes (particularly soy, alfalfa, and clover)
- The extensive use of chemical fertilizers and pollution emitted by vehicles and industrial plants (NO_x)
- Biomass burning
- Cattle and feedlots
- Industrial processes

PHOSPHORUS CYCLE

Phosphorus is essential for the production of nucleotides, production of ATP, fats in cell membranes, bones, teeth, and shells. Phosphorus is not found in the atmosphere but, rather, the primary sink for phosphorus is in sedimentary rocks. Generally, phosphorus is found in the form of the phosphate ion (PO_4^{3-}) or the hydrogen phosphate ion (HPO_4^{2-}). Phosphorus is slowly released from terrestrial rocks by weathering and the action of acid rain. It then dissolves into the soil and is taken up by plants. It is often a limiting factor for soils due to its low concentration and solubility. Phosphorus is a key element in fertilizer. A fertilizer labeled 6-24-26 contains 6% nitrogen, 24% phosphorus, and 26% potassium.

Humans have impacted the phosphorus cycle in several ways: First, humans have mined large quantities of rocks containing phosphorus for inorganic fertilizers and detergents. Second, clear-cutting tropical habitats for agriculture decreases the amount of available phosphorus as it is contained in the vegetation. Third, humans allow runoff from feedlots, from fertilizers, and from the discharge of municipal sewage plants. The runoff collects in lakes, streams, and ponds. It causes an increase in the growth of cyanobacteria (blue-green bacteria), green algae, and aquatic plants. In turn, this growth results in decreased oxygen content in the water, which then kills other aquatic organism in the food web. Fourth, humans apply phosphorus-rich guano and other phosphate-containing fertilizers to fields.

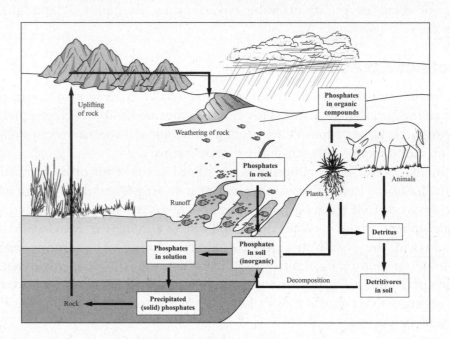

Figure 5.4 The phosphorus cycle

SULFUR CYCLE

Elemental or pure sulfur is commonly found in underground deposits or near natural hot springs or volcanoes. It's also found in the combined form in many minerals.

Sulfur in Lithosphere

Sulfur is the thirteenth most abundant element in Earth's crust and the ninth most abundant in sediments. Sulfur content of rocks varies considerably (e.g., sedimentary rocks have the most while igneous rocks have the least). Sulfur in lithosphere is mobilized by slow weathering of rock material. Dissolved in runoff, it moves with river water and is deposited in continental shield sediments in oceans. Eventually on geological timescale, this uplifts to surface again thus completing the geological part of the sulfur cycle.

Sulfur in Hydrosphere

Main storage of sulfur in the oceans is through dissolved sulfate. The most volatile sulfur compound in seawater is dimethyl sulfide (DMS; $(CH_3)_2S$), which is produced by algal and bacterial decay. Its highest concentrations are in coastal marshes and wetlands. Sulfur is the second most abundant compound in rivers with concentrations fluctuating highly with seasons and frequency of drought, flood, and normal flow. Rivers transport about 110 million tons (100 Tg) of sulfur per year to the oceans.

Sulfur in Soil and Biosphere

Sulfur is a major essential nutrient in the biosphere and is concentrated mainly in soil from where it enters the biosphere through plant uptake. Its main sources are deposition from the atmosphere, weathering of rocks, release from decay of organic matter and anthropogenic fertilizer, pesticides, and irrigation water. In soil, it is present mainly in the oxidized state (e.g., SO_4^{2-}). Rich organic soils may have up to 0.5% sulfur by dry weight. Sulfur in soil may be in bound or unbound form, as organic or inorganic compounds, organic sulfur being most prevalent. Plants take up sulfur from the soil mainly as sulfate and it is passed on with the food chain in the biosphere. It leaves the biosphere on death of living organisms when aerobic decay and decomposition brings back sulfate into the soil. Finally, anaerobic decomposition in soil releases part of organic sulfur as hydrogen sulfide (H_2S), dimethyl sulfide ($(CH_3)_2S$), and other organic compounds into the atmosphere. The release of sulfur is dependent upon warmer temperatures.

Sulfur in the Atmosphere

Six important sulfur compounds are released into the atmosphere due to interaction of processes between Earth's surface and the atmosphere.

1. **Carbonyl sulfide (COS):** The most abundant sulfur species in atmosphere and in nature is mainly produced by decomposition processes in soil, marshes and wetlands along ocean coasts, and areas of ocean upwelling that are rich in nutrients. Anthropogenic combustion processes produce less than 25% of COS. It has a lifetime of 44 years with the only sink being stratospheric photolysis and slow photochemical reactions in troposphere. Oceans may act both as source and sink. About 80% of total atmospheric sulfur is COS, but it is relatively inert and does not add much to the atmospheric sulfur pollution problem.

2. **Carbon disulfide (CS_2):** It is far more reactive than COS and has similar sources. It has a lifetime of 12 days and its major sink is photochemical reactions. Its concentration decreases rapidly with altitude. The most important source of the compound is microbial processes in warm tropical soils. Major secondary sources are marshes and wetlands along sea coasts. Small anthropogenic inputs are from fossil fuel combustion.

3. **Dimethyl sulfide ($(CH_3)_2S$):** DMS is released from oceans in much greater amounts than COS or CS_2 and is rapidly oxidized to sulfur dioxide or is redeposited to the oceans. In the sulfur cycle, most of the natural gas released from oceans is DMS. Its concentrations are highest during the night.

4. **Hydrogen sulfide (H_2S):** Mainly produced in nature during anaerobic decay in soils, wetlands, salt marshes, and other areas of stagnant water with maximum concentrations occurring over tropical forests. This highly

reactive compound is removed by reaction with hydroxyl radical (OH⁻) and COS. Its highest concentrations occur at night and in the early morning when photochemical activity is at a minimum.

5. **Sulfur dioxide (SO_2):** Its natural source is oxidation of H_2S and its major anthropogenic source is combustion of fossil fuels. Its atmospheric concentrations are most influenced by anthropogenic emissions. In some industrialized areas such as eastern North America, over 90% of SO_2 is from anthropogenic sources while half of global SO_2 originates from natural sources. The lifetime of SO_2 is 2–4 days indicating losses due to photochemical conversion. The rest of the gas is removed from the atmosphere by wet and dry deposition.

6. **Sulfate aerosol (SO_4^{2-}):** The largest natural source of sulfate aerosol particles originate from sea spray. About 3.3 million tons (3 Tg) per year of sulfate is added to the atmosphere from anthropogenic sources directly but much greater amounts are formed through secondary reactions from various sulfur species in the atmosphere. Most of the salt spray sulfate falls back to oceans but some is carried over the continents to be included in deposition processes there.

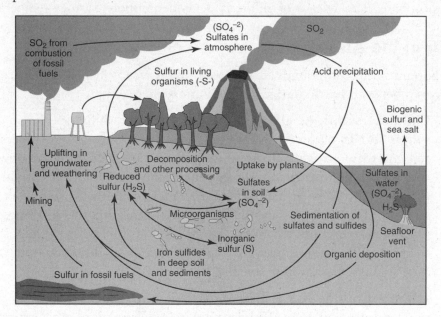

Figure 5.5 The sulfur cycle

WATER CYCLE

The water cycle is powered by energy from the sun. Solar energy evaporates water from oceans, lakes, rivers, streams, soil, and vegetation. The oceans hold 97% of all water on the planet and are the source of 78% of all global precipitation. Oceans are also the source of 86% of all global evaporation, with evaporation from the sea surface keeping Earth from overheating. If there were no oceans, surface temperatures on land would rise to an average of 153°F (67°C).

The water cycle is in a state of dynamic equilibrium by which the rate of evaporation equals the rate of precipitation. Warm air holds more water vapor than cold air. Processes involved in the water cycle include evaporation, evapotranspiration, condensation, infiltration, runoff, and precipitation.

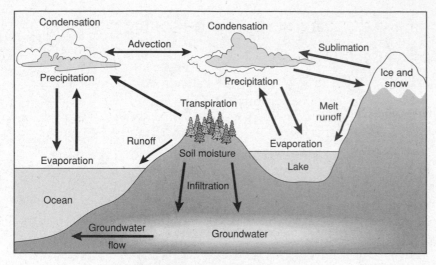

Figure 5.6 The water cycle

Human Impact on Water Cycle

Human Activity	Impact on Water Cycle
Withdrawing water from lakes, aquifers, and rivers	Groundwater depletion and saltwater intrusion
Clearing of land for agriculture and urbanization	Increased runoff Decreased infiltration Increased flood risks Accelerated soil erosion
Agriculture	Runoff contains nitrates, phosphates, ammonia, etc.
Destruction of wetlands	Disturbing natural processes that purify water
Pollution of water sources	Increased occurrences of infectious agents such as cholera, dysentery, etc.
Sewage runoff, feedlot runoff	Cultural eutrophication
Building power plants	Increased thermal pollution

CONSERVATION OF MATTER AND ENERGY

The law of conservation of matter states that during an ordinary chemical change, there is no detectable increase or decrease in the quantity of matter. The total quantity of matter and energy available in the universe is a fixed amount and is never any more or less. The law of conservation of energy states that energy cannot be created or destroyed but can change its form. Conversion of one type of matter into another is always accompanied by the conversion of one form of energy into another. Usually heat is either given off or absorbed, but sometimes the conversion involves light or electrical energy in addition to heat.

QUICK REVIEW CHECKLIST

☐ **Carbon Cycle**
    ☐ carbon sink

☐ **Nitrogen Cycle**
    ☐ nitrogen fixation
    ☐ nitrification
    ☐ assimilation
    ☐ ammonification
    ☐ denitrification

☐ **Phosphorus Cycle**

☐ **Sulfur Cycle**

☐ **Energy Flow**
    ☐ photosynthesis
    ☐ cellular respiration

☐ **Water Cycle**
    ☐ human impact on water cycle

☐ **Conservation of Matter and Energy**

MULTIPLE-CHOICE QUESTIONS

1. Which of the following is NOT a primary depository for the element listed?

    (A) Carbon—coal
    (B) Nitrogen—nitrogen gas in the atmosphere
    (C) Phosphorus—marble and limestone
    (D) Sulfur—deep ocean deposits
    (E) All are correct

2. Burning fossil fuels coupled with deforestation increases the amount of _____ in the atmosphere.

    (A) NO_2
    (B) CO_2
    (C) SO_2
    (D) O_3
    (E) All of the above are correct

3. In the nitrogen fixation cycle, cyanobacteria in the soil and water and *Rhizobium* bacteria in root systems are responsible for converting

   (A) organic material to ammonia and ammonium ions
   (B) ammonia, ammonium ions, and nitrate ions to DNA, amino acids, and proteins
   (C) ammonia and nitrite ions to nitrate ions
   (D) nitrogen and hydrogen gas to ammonia
   (E) ammonia to nitrite and nitrate ions

4. The cycle listed that has the most immediate effect on acid precipitation would be the

   (A) carbon cycle
   (B) sulfur cycle
   (C) water cycle
   (D) phosphorous cycle
   (E) rock cycle

5. Nitrogen is assimilated in plants in what form?

   (A) Nitrite, NO_2^-
   (B) Ammonia, NH_3
   (C) Ammonium, NH_4^+
   (D) Nitrate, NO_3^-
   (E) Choices B, C, and D

6. Plants assimilate sulfur primarily in what form?

   (A) Sulfates, SO_4^{2-}
   (B) Sulfites, SO_3^{2-}
   (C) Hydrogen sulfide, H_2S
   (D) Sulfur dioxide, SO_2
   (E) Elemental sulfur, S

7. Humans increase sulfur in the atmosphere and thereby increase acid deposition by all of the following activities EXCEPT

   (A) industrial processing
   (B) processing (smelting) ores to produce metals
   (C) burning coal
   (D) refining petroleum
   (E) clear-cutting

8. Phosphorus is being added to the environment by all of the following activities EXCEPT

   (A) runoff from feedlots
   (B) slashing and burning in tropical areas
   (C) stream runoff
   (D) burning coal and fossil fuels
   (E) mining to produce inorganic fertilizer

9. Carbon dioxide is a reactant in

   (A) photosynthesis
   (B) cellular respiration
   (C) the Haber-Bosch process
   (D) nitrogen fixation
   (E) None of the above

10. Human activity adds significant amounts of carbon dioxide to the atmosphere by all of the following EXCEPT

   (A) brush clearing
   (B) burning wood
   (C) burning fossil fuels
   (D) clear-cutting
   (E) agricultural runoff

11. Clearing of land for either habitation or agriculture does all of the following EXCEPT

   (A) increases runoff
   (B) increases flood risks
   (C) increases potential for landslides
   (D) increases infiltration
   (E) accelerates soil erosion

12. All of the following have an impact on the nitrogen cycle EXCEPT

   (A) the application of inorganic fertilizers applied to the soil
   (B) the action of aerobic bacteria acting on livestock wastes
   (C) the overplanting of nitrogen-rich crops
   (D) the discharge of municipal sewage
   (E) the burning of fossil fuels

13. Which of the following is a macronutrient essential for the formation of proteins?

    (A) Sulfur
    (B) Nitrogen
    (C) Iron
    (D) Cobalt
    (E) Molybdenum

14. Which of the following bacteria are able to convert soil nitrites to nitrates?
    (A) *Nitrosomonas*
    (B) *Nitrobacter*
    (C) *Rhizobium*
    (D) *Penicillium*
    (E) *Clostridium*

15. Which of the following cycles would be considered sedimentary?

    (A) Carbon cycle
    (B) Water cycle
    (C) Phosphorus cycle
    (D) Nitrogen cycle
    (E) Sulfur cycle

16. The cycle that is most common to all other cycles is the

    (A) nitrogen cycle
    (B) carbon cycle
    (C) hydrologic cycle
    (D) life cycle
    (E) the rock cycle

17. The energy that drives the hydrologic cycle comes primarily from

    (A) trade winds
    (B) solar energy and gravity
    (C) Earth's rotation on its axis
    (D) ocean currents and wind patterns
    (E) solar radiation

18. Which one of the following processes is working against gravity?

    (A) Precipitation
    (B) Percolation
    (C) Runoff
    (D) Transpiration
    (E) Infiltration

19. Ammonium ions (NH_4^+) are converted to nitrate (NO_3^-) and nitrite ions (NO_2^-) through which process?

    (A) Assimilation
    (B) Denitrification
    (C) Nitrogen fixation
    (D) Nitrification
    (E) Ammonification

20. Which form of nitrogen is most usable by plants?

    (A) Nitrate
    (B) Nitrite
    (C) Nitrogen gas
    (D) Ammonia
    (E) Atomic nitrogen

FREE-RESPONSE QUESTION

By: Sarah E. Utley
College Board APES Reader; Princeton, New Jersey
Environmental Content Specialist
Center for Digital Innovation
AP Environmental Science Division
University of California at Los Angeles

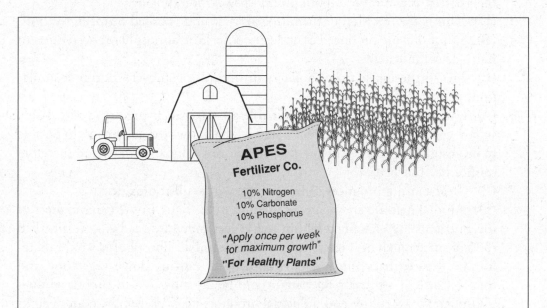

The APES Fertilizer Company manufactures a soil additive that contains 10% (w/v) of nitrogen, carbonate, and phosphorus. Use this information to answer the following questions.

(a) Choose either carbon or nitrogen and clearly explain the biogeochemical cycle of that nutrient. Include all of the important steps in the cycle and indicate movement from the atmosphere, water, and soil where necessary.

(b) Agricultural runoff from overapplication of fertilizers is a major source of water pollution. Choose either nitrogen or phosphorus and fully explain the cause and effect of this excess on the aquatic ecosystem.

(c) Rather than rely on the application of fertilizers, current agricultural research is encouraging the practice of more conservation-minded farming. Name and explain one such environmental practice that can decrease dependence on human-applied nutrients without causing a significant decrease in crop yield.

MULTIPLE-CHOICE ANSWERS AND EXPLANATIONS

1. **(C)** The primary sinks for phosphorus are ocean sediments and certain islands rich in guano.

2. **(B)** Burning fossil fuels releases sulfur oxides (SO_x), carbon oxides (carbon dioxide on complete combustion, or carbon monoxide on incomplete combustion) and nitrogen oxides (NO_x). Ozone is not produced by burning fossil fuels. Deforestation releases carbon dioxide. Since the question said "coupled," the gas that is common to both processes is carbon dioxide.

3. **(D)** This is the first step in the nitrogen cycle and is called nitrogen fixation.

4. **(B)** Sulfur dioxide produced by industry enters the atmosphere and returns to Earth as sulfuric acid.

5. **(E)** The nitrite ion is toxic to plants. In the nitrogen cycle, during assimilation, plant roots absorb nitrate.

6. **(A)** Hydrogen sulfide and sulfur dioxide are toxic to living organisms. Some sulfate compounds are soluble in water, which allows the sulfate ion to be able to be absorbed by plants. Elemental sulfur is not soluble in water and therefore cannot be absorbed.

7. **(E)** Clear-cutting produces carbon dioxide, not sulfur dioxide.

8. **(D)** Animal manure and guano are rich in phosphate. In the tropics, most of the nutrients are contained within the vegetation. Little is being retained in the soil since much of it leaches due to high rainfall. Phosphorus would therefore be released back into the environment by cutting down vegetation and then burning it—thereby releasing it and being subjected to runoff. Mining phosphates for fertilizer and industrial products takes phosphorus out of sinks and puts it into the environment. Burning coal and petroleum does not add appreciable amounts of phosphorus to the environment.

9. **(A)** The reaction for photosynthesis is $6CO_2 + 6H_2O + sunlight \rightarrow C_6H_{12}O_6 + 6O_2$. Cellular respiration is the reverse of photosynthesis.

10. **(E)** Agricultural runoff, primarily from fertilizers and feedlots, adds nitrates and phosphates to streams and results in cultural eutrophication. All other choices involve combustion, which produces carbon dioxide.

11. **(D)** Infiltration is the movement of water into the soil. Removing vegetation decreases infiltration by not allowing water to percolate through the soil slowly. Since runoff is increased, the potential for flooding increases. As the soil structure loses its integrity, the chances of landslides increase. Runoff also carries with it topsoil and nutrients, thus accelerating soil erosion.

12. **(B)** The bacteria that normally work to decompose livestock wastes are anaerobic, operate only in environments with little or no oxygen, and produce nitrous oxide (N_2O). Digesters can be constructed to reduce livestock wastes to methane gas (CH_4), which can then be burned as a fuel.

13. **(B)** The growth of all organisms depends on the availability of macro- and micronutrients, and none is more important than nitrogen, which is required in large amounts as an essential component of proteins, nucleic acids, and other cellular constituents.

14. **(B)** *Nitrosomonas* bacteria oxidize ammonia to nitrite: $NH_3 \rightarrow NO_2^-$. *Nitrobacter* then oxidize nitrite to nitrate: $NO_2^- \rightarrow NO_3^-$. Denitrifying bacteria anaerobically reduce nitrate to nitrogen gas: $NO_3^- \rightarrow N_2$.

15. **(C)** The largest reservoir of phosphorus is sedimentary rock. The phosphorus cycle originates with the introduction of phosphate (PO_4^{3-}) into soils from the weathering of rocks. Phosphate enters living ecosystems when plants take up phosphate ions from the soil.

16. **(C)** Water plays a part somewhere in every cycle.

17. **(B)** Solar energy allows water to change phase (water to gas, liquid to ice, etc.) and gravity causes rain to fall and rivers to flow.

18. **(D)** Transpiration is the process by which water moves upward (against gravity) from the soil through the roots and out through the leaves of plants.

19. **(D)** Nitrification is the biological oxidation of ammonia into nitrite ions followed by the oxidation of nitrites into nitrates. The oxidation of ammonia into nitrites is done by bacteria belonging to the genera *Nitrosomonas* and *Nitrosococcus*. The oxidation of nitrite ions into nitrate ions is done by bacteria belonging to the genus *Nitrobacter*.

20. **(A)** Nitrite and ammonia are more toxic to plants than nitrate. Plants cannot use nitrogen from the atmosphere directly.

FREE-RESPONSE ANSWER

Part (a): Maximum 5 points
Part (b): Maximum 3 points
Part (c): Maximum 2 points

Total: maximum 10 points

(a) Earth is a closed system. A finite amount of matter (such as life-supporting nutrients) is found in the biosphere. These nutrients cycle through Earth's varied support systems in different forms. Nitrogen, an essential nutrient for life, is found primarily in the atmosphere (making up 78% by volume). However, it also cycles through the lithosphere and the biomass of organisms. Unfortunately, the producers cannot absorb nitrogen in its most commonly found form, N_2. In order for this gaseous element to be used by living organisms, it has to be converted in a series of steps called the nitrogen cycle.

> *Notable points:*
>
> *Movement through Earth's life support systems (atmosphere, lithosphere, biomass).*
>
> *Introduction to nitrogen cycle.*

The first step in the cycle is called nitrogen fixation. This is the process by which atmospheric nitrogen is converted to ammonia (NH_3). There are some natural sources of nitrogen fixing such as lightning and volcanoes. However, most of the biological fixing is done by nitrogen-fixing bacteria that live in the root nodules of certain plants called legumes, such as peas and clover. The nitrogen (N_2) is converted to ammonia (NH_3) using the enzyme nitrogenase.

> *Definition of nitrogen fixation.*
>
> *Sources of fixation.*
>
> *Next step—nitrification.*

After the nitrogen is fixed into ammonia, it goes through a process called nitrification. This is a two-step process where special soil bacteria first convert the ammonia to nitrite (NO_2^-) and then to nitrate (NO_3^-), the form that plants can most easily absorb through their roots. Plants can also take up a limited amount of ammonia. The process by which the roots absorb the nitrate and ammonia is called assimilation. During this step, the plants incorporate nitrogen into their tissues. This absorbed nitrogen is used by plants to form nitrogen-containing molecules such as proteins and nucleic acids.

> *Definition of nitrification (step 1).*
>
> *Definition of nitrification (step 2).*
>
> *Next step in cycle—assimilation.*
>
> *Next step in cycle—ammonification.*
>
> *Next step in cycle—denitrification.*

Since animals cannot directly absorb nitrogen, they assimilate it into their body tissues when they eat plants or other animals. After animals absorb the nitrogen into their tissues, it is excreted as urea or uric acid. Uric acid is the end product of nitrogen metabolism in birds and reptiles. In humans and most other higher animals, the main product of nitrogen detoxification is urea. In fish, bacteria, and protozoa, it is ammonia. In those animals that do produce uric acid in high quantities, it is excreted in the feces as a dry mass. Although this compound is produced through a complex and energetically costly metabolic pathway (in comparison with the production of other nitrogenous wastes), its elimination minimizes water loss and is therefore commonly found in the excretions of animals that live in desert habitats.

In addition, when the organism dies, the nitrogen in the tissues is released back into the cycle. There nitrogen-containing substances are decomposed into ammonia in a process called ammonification. Then during denitrification, other bacteria convert the ammonia found in the soil into nitrate and nitrite. The nitrite and nitrate are further converted into gaseous nitrogen, which is then available to be refixed and used by other organisms.

(b) When fertilizers such as phosphorus are overapplied, rain or irrigation causes the excess fertilizer to run off into local waterways. Since phosphorus is often a limiting factor in aquatic ecosystems, this excess amount results in proliferation of algae and is known as an algal bloom. When these algal populations die, they are decomposed by large amounts of bacteria that consume much of the dissolved oxygen in the water. This, in turn, causes aquatic organisms to die from lack of oxygen. The result is a sterile lake or stream that lacks the nutrients necessary to support life.

(c) Conservation-minded agriculture can lead to a decreased dependence on human-applied nutrients. One of the most common, environmentally conscious agricultural practices is conservation or no-till farming. In this practice, the fields are not plowed under at the end of the growing season. The old stalks and vegetation are left on the fields over the winter. Then, in the spring, special machinery is used to plant the new crop on the untilled field. By leaving the previous season's biomass on the untilled land, the natural decomposition process can occur. The nutrients that would normally have been removed and lost are allowed to decompose back into the soil. This leads to a decreased need for artificial fertilizers without a loss of crop yield.

UNIT III: POPULATION
(10–15%)

Areas on Which You Will Be Tested

A. Population Biology Concepts—population ecology, carrying capacity, reproductive strategies, and survivorship.

B. Human Population—
1. Human population dynamics—historical population sizes, distribution, fertility rates, growth rates and doubling times, demographic transition, and age-structure diagrams.
2. Population size—strategies for sustainability, case studies, and national policies.
3. Impacts of population growth—hunger, disease, economic effects, resource use, habitat destruction.

Populations

CHAPTER 6

POPULATION BIOLOGY CONCEPTS

Several topics must be considered when discussing population biology:

- Population ecology
- Carrying capacity (K)
- Reproductive strategies
- Survivorship

Population Ecology

Population ecology studies the dynamics of species' populations and how these populations interact with the environment. Most organisms live in groups (flocks, schools, nests, etc.) and living in groups provides several advantages: increased protection from predators, increased chances for mating, and division of labor.

Population ecology plays an important role in the development of the field of conservation biology, especially in predicting the long-term probability of a species persisting in a given habitat. The following table lists those factors that affect population viability.

Carrying Capacity (K)

Carrying capacity refers to the number of organisms that can be supported in a given area sustainably. It varies from species to species and is subject to changes over time. As an environment degrades, the carrying capacity decreases. Factors that keep population sizes in balance with the carrying capacity are called regulating factors. They include food availability, space, oxygen content in aquatic ecosystems, nutrient levels in soil profiles, and the amount of sunlight. Below the carrying capacity, populations tend to increase in size. Population size cannot be sustained above the carrying capacity; eventually the population will crash.

A population's biotic potential is the maximum rate at which a population can grow. Its biotic potential, given optimal conditions (food, water, and space), occurs when resources are unlimited. Factors that influence biotic potential include age at reproduction, frequency of reproduction, number of offspring produced, reproductive life span, and average death rate under ideal conditions. If a population in a community is left unchecked, the maximum population growth rate can increase exponentially and takes on a J-shaped curve.

Factors That Affect Population Viability

Increase (+) Viability	Decrease (–) Viability
Favorable environmental conditions (light, temperature, and nutrients)	Unfavorable environmental conditions (insufficient light, temperature extremes, and/or poor supply of nutrients)
High birthrate	Low birthrate
Generalized niche	Specialized niche
Satisfactory habitat	Habitat not satisfactory or has been seriously impacted
Few competitors	Too many competitors
Suitable predatory defense mechanism(s)	Unsuitable predatory defense mechanism(s)
Adequate resistance to diseases and parasites	Little or no suitable defense mechanisms against diseases or parasites
Able to migrate	Unable to migrate
Flexible—able to adapt	Inflexible—unable to adapt
Sufficient food supply	Deficient food supply

Predation not only removes the very old, the very young, and the weaker members from the population, it may also reduce the population of the prey. If the predators do not keep the prey population in balance, the carrying capacity is exceeded and the prey may starve. Predator and prey populations are closely interdependent. Populations of algae, annual plants, and insects with short life spans are controlled by seasonal and nutritional environmental changes. A J-shaped growth curve may show a plunge in population as the reproductive potential declines because of environmental changes. The following year, there may be an exponential increase in the population.

Figure 6.1 J-shaped curve

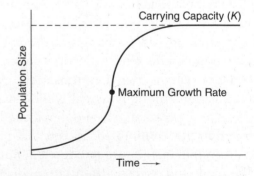

Figure 6.2 S-shaped curve (logistics model)

An S-shaped curve is used to describe the pattern of growth over extended periods of time when organisms move into an empty niche. Growth rates are density-dependent. They are characterized by maximum population growth rate and the carrying capacity. In an S-shaped curve, population size initially increases due to unlimited resources for a small population. However, as resources become limited, the population growth rate slows down and stabilizes around the limits of the carrying capacity at the point where the birthrate equals the death rate. The average American's ecological footprint (the demand of an individual with an average amount of resources) is about 12 acres.

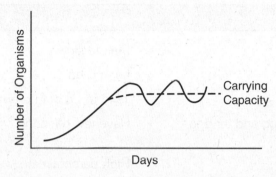

Figure 6.3 Fluctuations around the carrying capacity

Thomas Malthus

Thomas Malthus was a political economist who was concerned about what he saw as the decline of living conditions in 19th-century England. He blamed this decline on three elements: (1) the overproduction of young; (2) the inability of resources to keep up with the rising human population; and (3) the irresponsibility of the lower classes. To combat this, Malthus suggested the family size of the lower class ought to be regulated such that poor families do not produce more children than they can support. In his 1798 work, *Essay on Population*, Malthus hypothesized that unchecked population growth always exceeds the means of supporting a larger population. He argued that actual population growth is kept in line with food supply by "positive checks" such as starvation and disease, which elevate the death rate, and by "preventive checks" (e.g., postponement of marriage), which keep down the birthrate. Malthus's hypothesis implied that actual population growth always has the tendency to push above the available food supply. Because of this tendency, any attempt to correct the condition of the lower classes by increasing their living standards or improving agricultural productivity would not be possible, as any extra means of subsistence would be completely absorbed by an increase in the population. Charles Darwin incorporated some of Malthus's ideas into his 1859 book, *On the Origin of Species*, by stating that limited resources result in competition, with those organisms that survive being able to pass on those adaptations through their genes to their offspring.

Reproductive Strategies

Organisms have adapted either to maximize growth rates in environments that lack limits or to maintain population size at close to the carrying capacity in stable environments. Species that have high reproductive rates are known as *r*-strategists. Species that reproduce later in life and with fewer offspring are known as *K*-strategists.

Reproductive Strategies

r-Strategists	*K*-Strategists
Mature rapidly	Mature slowly
Short lived	Long lived
Tend to be prey	Tend to be both predator and prey
Have many offspring and tend to overproduce	Have few offspring
Low parental care	High parental care
Are generally not endangered	Most endangered species are *K*-strategists
Wide fluctuations in population density (booms and busts)	Population size stabilizes near the carrying capacity
Population size limited by density-independent limiting factors, including climate, weather, natural disasters, and requirements for growth	Density-dependent limiting factors to population growth stem from intra-specific competition and include competition, predation, parasitism, and migration
Tend to be small	Tend to be larger
Type III survivorship curve	Type I or II survivorship curve
Examples: most insects, algae, bacteria, rodents, and annual plants	Examples: humans, elephants, cacti, and sharks

Survivorship

Survivorship curves show age distribution characteristics of species, reproductive strategies, and life history. Reproductive success is measured by how many organisms are able to mature and reproduce. Each survivorship curve represents a balance between natural resource limitations and interspecific and intraspecific competition. For example, humans could not survive in a Type III survivorship mode, where human females would produce thousands of offspring. Likewise, ants could not survive in a Type I mode, where each queen ant would produce only a few eggs during her lifetime and where she would spent most of her time and energy raising offspring.

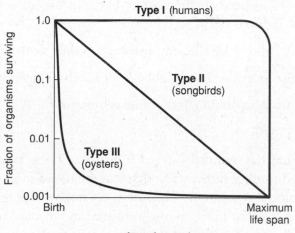

Figure 6.4 Survivorship curve

Survivorship Curves

Type	Description
I Late Loss	Reproduction occurs fairly early in life. Most deaths occur at the limit of biological life span. Low mortality at birth; high probability of surviving to advanced age. Death rates increase during old age. Advances in pre-natal care, nutrition, disease prevention, and cures including immunization have meant longer life spans for humans. Examples: humans, annual plants, sheep, and elephants.
II Constant Loss	Individuals in all age categories have fairly uniform death rates. Predation affecting all age categories is primary means of death. Typical of organisms that reach adult stages quickly. Examples: rodents, perennial plants, and songbirds.
III Early Loss	Typical of species that have great numbers of offspring and reproduce for most of their lifetime. Death is prevalent for younger members of the species (environmental loss and predation) and declines with age. Examples: sea turtles, trees, internal parasites, fish, and oysters.

HUMAN POPULATION DYNAMICS

Many different factors affect the human population: historical population sizes, population distribution, fertility rates, growth rates, doubling times, and demographic transition. Age-structure diagrams act as indicators of future population trends.

Historical Population Sizes

The rapid growth of the world's human population over the past 100 years has been due primarily to a decrease in death rates. In 1900, the overall death rate in the United States was 1.7%. In 2000, the death rate had dropped to 0.9% (almost half). Children in 1900 were 10 times more likely to die than children in 2000. Several factors have reduced death rates:

1. Increased food and more efficient distribution that result in better nutrition

2. Improvements in medical and public health technology

3. Improvements in sanitation and personal hygiene

4. Safer water supplies

Human population has had three surges in growth. These surges in population have been attributed to three factors: The first was the use of tools and fire. The second was the agricultural revolution, when humans stopped being hunter-gatherers and began to raise crops. The third was the industrial and medical revolutions within the last 200 years. Birthrate or crude birthrate is equal to the number of live births per 1,000 members of the population in one year. Death rate or crude death rate is equal to the number of deaths per 1,000 members of the population in one year. Immigration refers to the number of individuals that enter the population, while emigration refers to the number of individuals that leave the population. Population change can be calculated from the following formula:

Population change = (crude birthrate + immigration) − (crude death rate + emigration)

EXAMPLE

In 1950, the population of a small suburb in Los Angeles, California, was 20,000. The birthrate was measured at 25 per 1,000 population per year, while the death rate was measured at 7 per 1,000 population per year. Immigration was measured at 600 per year, while emigration was measured at 200 per year. By how much did the population increase (or decrease) in that year?

Answer:

Population change = (crude birthrate + immigration) − (crude death rate + emigration)

$$= \quad (25(20) + 600) \quad - \quad (7(20) + 200)$$

$$= +760$$

The population grew from 20,000 to 20,760 in one year.

The actual rate of population change can be determined by using the following formula:

$$\text{Actual growth rate (\%)} = \frac{\text{birthrate} - \text{death rate}}{10}$$

EXAMPLE

The United States had a birthrate of 14.6 live births per 1,000 population in one year, compared to India's birthrate of 22.2 in that same year. The death rate in that year for the United States was 8.3 deaths per 1,000 population, compared to India's rate of 6.4. Calculate the population growth rates (%) for both the United States and India for that year.

Answer:

$$\text{United States: } \frac{\text{birthrate} - \text{death rate}}{10} = \frac{14.6 - 8.3}{10} = 0.6$$

$$\text{India: } \frac{\text{birthrate} - \text{death rate}}{10} = \frac{22.2 - 6.4}{10} = 1.6$$

A current world growth rate of approximately 1.3% when applied to the about 7 billion people on Earth yields an annual increase of about 85 million people. Because of the large and increasing population size, the number of people added to the global population will remain high for several decades, even if growth rates decline.

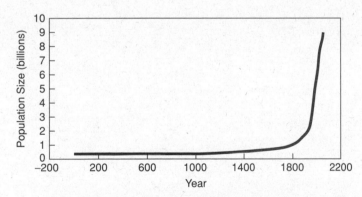

Figure 6.5 Estimated human population growth from 200 B.C.E. to 2200 C.E.

Distribution

In 1800, the vast majority of the world's population (65%) resided in Asia and Europe. By 1900, 25% of the human population lived in Europe largely due to the Industrial Revolution. Between 2000 and 2030, most of the growth will occur in the less-developed countries in Africa, Asia, and Latin America whose growth rates are much higher than those in more-developed countries. The more-developed countries in Europe and North America will have growth rates less than 1%. Some countries such as Russia, Germany, Italy, and Japan will even experience negative growth rates.

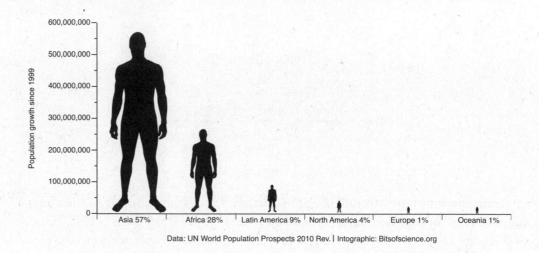

Data: UN World Population Prospects 2010 Rev. | Intographic: Bitsofscience.org

Figure 6.6 Distribution of the last billion people on Earth

Human Population Growth

Time Period	Description	Practicing Worldview
Before Agricultural Revolution	~ 1 million to 3 million humans. Hunter-gatherer lifestyle.	**Earth Wisdom**—Natural cycles that can serve as a model for human behavior.
8000 B.C.E. to 5000 B.C.E.	~ 50 million humans. Increases due to advances in agriculture, domestication of animals, and the end of a nomadic lifestyle.	
5000 B.C.E. to 1 B.C.E.	~ 200 million humans. Rate of population growth during this period was about 0.03 to 0.05%, compared with today's growth rate of 1.3%.	**Frontier Worldview**—Viewed undeveloped land as a hostile wilderness to be cleared and planted, then exploited for its resources as quickly as possible.
0 C.E. to 1300 C.E.	~ 500 million humans. Population rate increased during the Middle Ages because new habitats were discovered. Factors that reduced population growth rate during this time were famines, wars, and disease (density-dependent factors).	
1300 C.E. to 1650 C.E.	~ 600 million humans. Plagues reduced population growth rate. Up to 25% mortality rates are attributed to the plagues that reached their peak in the mid-1600s.	
1650 C.E. to present	Currently ~ 6 billion humans. In 1650 C.E. growth rate was ~ 0.1%. Today it is ~ 1.3%. Health care, health insurance, vaccines, medical cures, preventative care, advanced drugs and antibiotics, improvements in hygiene and sanitation, advances in agriculture and distribution, and education are factors that have increased the growth rate.	**Planetary Management**—Beliefs that as the planet's most important species, we are in charge of Earth; we will not run out of resources because of our ability to develop and find new ones; the potential for economic growth is essentially unlimited; and our success depends on how well we manage Earth's life-support systems mostly for our own benefit.
Present to 2050 C.E.	Estimates are as high as ~ 15 billion.	**Earth Wisdom**—Beliefs that nature exists for all Earth's species and we are not in charge of the Earth; resources are limited and should not be wasted. We should encourage Earth-sustaining forms of economic growth and discourage Earth-degrading forms of economic growth; and our success depends on learning how Earth sustains itself and integrating such lessons from nature into the ways we think and act.

Fertility Rates

Replacement level fertility (RLF) is the level of fertility at which a couple has only enough children to replace themselves, or about 2 children per couple. It takes a RLF of 2.1 to replace each generation since some children will die before they grow up to have their own two children. RLF rates are lower in moderately developed countries (MDC) and higher in less-developed countries (LDC) due to higher infant mortality rates in LDCs. Infant mortality rates are good indicators of comparative standards of living. The total fertility rate (TFR) is the average number of children that each woman will have during her lifetime. The country of Niger in Africa leads the world's TFR at 7.68.

Worldwide Total Fertility Rate

Country	TFR
Niger	7.68
India	2.65
Israel	2.72
Mexico	2.31
United States	2.06
China	1.54
Russia	1.54
Japan	1.21
World average	**2.59**

Despite half of the world's population having subreplacement fertility rates, the world's population is still growing quickly. This growth is due to nations with above-replacement TFRs and a population momentum caused by large numbers of younger females who have as yet not had children.

Declines in fertility rates can be attributed to several factors. Urbanization results in a higher cost of living. Urbanization reduces the need for extra children to work on farms. There is a greater personal acceptance and government encouragement of contraception and abortion. The numbers of females in the workforce and female educational opportunities are increasing. More individuals desire to increase their standard of living by having less children. Many people are postponing marriage until their careers are established.

The two main effects of TFRs less than 2.1 without additions through immigration are population decline and population aging. In the United States the TFR has been inconsistent. The greatest TFR occurred during the post-World War II years (baby boomers). New immigrants and their descendants are projected to contribute 66% of the expected growth by 2050.

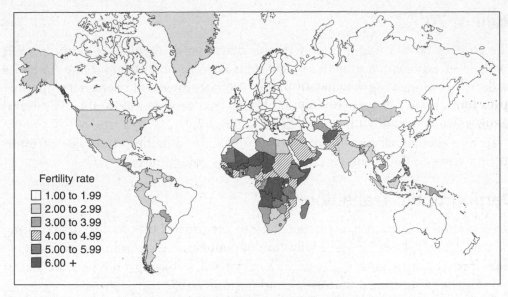

Figure 6.7 Worldwide total fertility rates

Growth Rates and Doubling Times

The 20th century saw the largest increase in the world's population in human history. The following chart shows the doubling times of the world's population:

950 C.E. to 1600	650 years
1600 to 1800	200 years
1800 to 1925	125 years
1925 to 1975	50 years
1975 to 2025	50 years

Note that the doubling times from 1925 to 1975 and projected from 1975 to 2025 remain constant. This does NOT mean that the world is not increasing in population—it means that the growth *rate* has decreased as shown in the following figure.

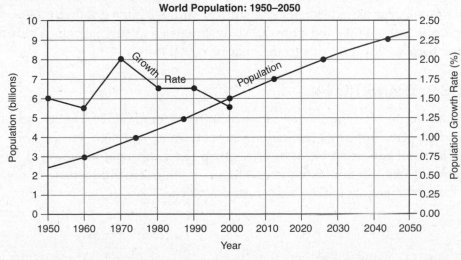

Source: U.S. Census Bureau. International Data Base, December 2010 Update

Figure 6.8 Yearly growth rate of human population (1950–2050)

Rule of 70

The Rule of 70 is a useful rule of thumb that roughly explains the time periods involved in exponential growth at a constant rate. To find the doubling time of a quantity growing at a given annual percentage rate, divide the percentage number into 70 to obtain the approximate number of years required to double. For example, at a 2% annual growth rate, doubling time is 70 / 2 = 35 years.

To find the annual growth rate, divide 70 by the doubling time. For example, 70 / 35 years doubling time = 2, or a 2% annual growth rate.

Demographic Transition

Demographic transition is the name given to the process that has occurred during the past century. It leads to a stabilization of population growth in the more highly developed countries and is generally characterized as having four separate stages: preindustrial, transitional, industrial, and postindustrial.

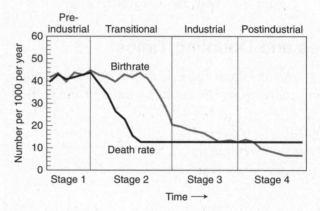

Figure 6.9 Demographic transitions occurring in human populations.

STAGE 1: PRE-INDUSTRIAL

Living conditions are severe, medical care is poor or nonexistent, and the food supply is limited due to poor agricultural techniques, preservation, and pestilence. Birthrates are high to replace individuals lost through high mortality rates. The net result is little population growth. Many countries in sub-Saharan Africa have reverted back to this stage due to the increase of AIDS.

STAGE 2: TRANSITIONAL

This stage occurs after the start of industrialization. Standards of hygiene, advances in medical care, improved sanitation, cleaner water supplies, vaccination, and higher levels of education begin to drive down the death rate, leading to a significant upward trend in population size. The net result is a rapid increase in population. Examples include India, Pakistan, and Mexico.

STAGE 3: INDUSTRIAL

Urbanization decreases the economic incentives for large families. The cost of supporting an urban family grows, and parents are more actively discouraged from having large families. Educational and work opportunities for women decrease birthrates. Obtaining food is not a major focus of the day. Leisure time is available. Retirement safety nets are in place, reducing the need for extra children to support parents. In response to these economic pressures, the birthrate starts to drop, ultimately coming close to the death rate. An example is China.

STAGE 4: POST-INDUSTRIAL

Birthrates equal mortality rates, and zero population growth is achieved. Birth and death rates are both relatively low, and the standard of living is much higher than during the earlier periods. In some countries, birthrates may actually fall below mortality rates and result in net losses in population. Examples of declining populations include Russia, Japan, and many European countries. The developed world currently remains in this fourth stage of demographic transition.

STAGE 5: SUB-REPLACEMENT FERTILITY

The original demographic transition model has just four stages; however, some theorists consider that a fifth stage is needed to represent countries that have sub-replacement fertility (that is, below 2.1 children per woman). Most European and many East Asian countries now have higher death rates than birthrates. In this stage, population aging and population decline will eventually occur to some extent, presuming that sustained mass immigration does not occur.

Age-Structure Diagrams

A good indicator of future trends in population growth is furnished by age-structure diagrams. Age-structure diagrams are determined by birthrate, generation time, death rate, and sex ratios. There are three major age groups in a population: pre-reproductive, reproductive, and post-reproductive. A pyramid-shaped age structure diagram (i.e., Nigeria) indicates the population has high birthrates and the majority of the population is in the reproductive age group (generally late teens to mid-40s). A bell-shape indicates that pre-reproductive and reproductive age groups are more nearly equal, with the post-reproductive group being smallest due to mortality; this is characteristic of stable populations (i.e., United States). An urn-shaped diagram indicates the post-reproductive group is largest and the pre-reproductive group is smallest, a result of the birthrate falling below the death rate and is characteristic of declining populations (i.e., Germany).

The age-structure diagrams in the figure below show age distributions of less-developed countries compared with more developed countries. When the base is large (greater number of younger individuals in the population), there is a potential for an increase in the population as these younger individuals mature and have children of their own (population momentum). When the top of the pyramid is larger, it indicates a large segment of the population is past their reproductive years (post-reproductive) and indicates a future slowdown in population growth. Age-structure diagrams reflect demographic transitions.

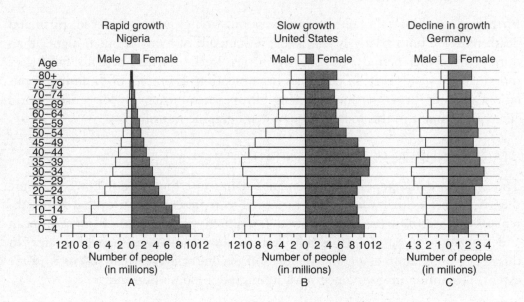

Figure 6.10 Age structure diagrams for countries with rapid, slow, and declining birthrates

In Mexico, large family size is due to the necessity for farm labor, the need to support parents when they no longer work, a need to increase family income, and cultural and religious beliefs. The death rate has declined due to social and medical programs. However, the birthrate continues to remain high.

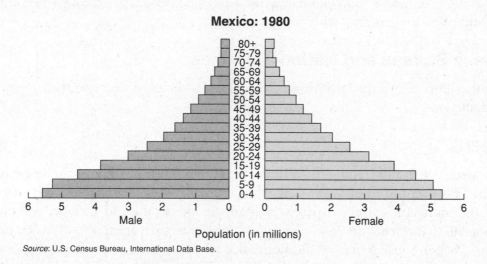

Source: U.S. Census Bureau, International Data Base.

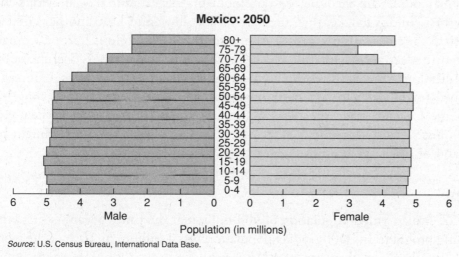

Source: U.S. Census Bureau, International Data Base.

Figure 6.11 Age structure in Mexico, 1980 and 2050 (estimated)

POPULATION SIZE

This section discusses several strategies to sustain population size. It then provides a few case studies.

Strategies for Sustainability

- Provide economic incentives for having fewer children.
- Empower and educate women. The low status of women is the number one problem.
- Education usually leads to higher incomes. Higher incomes decrease the need for having extra children to take care of older parents.
- Higher education usually results in having children later in life.

- Provide government family planning services. These practices have ranged from education and birth control to sterilization and abortion.
- Improve prenatal and infant health care. Women would not need more children if the ones they had survived.
- Increase economic development in less-developed countries through free trade and private investment with tax incentives.

Case Studies and National Policies

Both China and India have instituted national policies to decrease their growing populations.

CHINA

Between 1972 and 2000, China dramatically reduced its crude birthrate by half (the TFR dropped from 5.7 to 1.8). The family planning program in China is one of the most efficient and strictest programs in the world and includes a mobile program that reaches the rural population. Incentives such as extra food, larger pensions, better housing, free medical care, free school tuition, and salary bonuses for parents who limit their number of children are responsible for the success of this program. Couples are encouraged to postpone marriage and to have only one child. (If the first child is female, then the parents are allowed to have another child and still retain the financial incentives.) After a mother's first child is born, a woman is required to wear an intrauterine device. Removal of the device is a crime subject to sterilization. Physicians receive bonuses for sterilization procedures. Couples are punished for refusing to terminate unapproved pregnancies and for giving birth under the legal marriage age. Penalties include fines (up to 50% of their yearly salary), loss of land, less food, a decrease in farm supplies, loss of government benefits, and/or discharge from the Communist Party.

INDIA

In 1952, India (with a population of 400 million at the time) began its first family planning program. In 2000, India's population was 1 billion, or 16% of the world's population. Each day there are 50,000 live births in India. One-third of the population of India earns less than 40 cents per day, and cropland has decreased 50% per capita since 1960. In the 1970s, India instituted a mandatory sterilization program involving vasectomies. Some of the reasons for India's failures were poor planning, low status of women, favoring male children, and insensitivity to cultures and religion. Tubal ligation is the preferred method of family planning in India today. Condoms are free from the Indian government but have less than 10% use. Other birth control methods are usually accepted by only the upper/educated class.

IMPACTS OF POPULATION GROWTH

When left unchecked, population growth often results in hunger. Diseases and economics also affect population growth.

Hunger

A large portion of the world's population—25%—is malnourished. Areas of greatest malnutrition are Africa, Asia, and parts of Latin America. Consequently, these are areas with the highest TFRs. Several factors contribute to malnutrition.

1. Poverty

2. Droughts, which will only increase as the impact of global warming becomes more severe

3. Populations that have surpassed their carrying capacity

4. Political instability and wars, which cause mass migrations

5. Pestilence

6. Foreign investors who own large landholdings and whose sole motivation is profit (selling the food to the highest bidder, which often means exporting it)

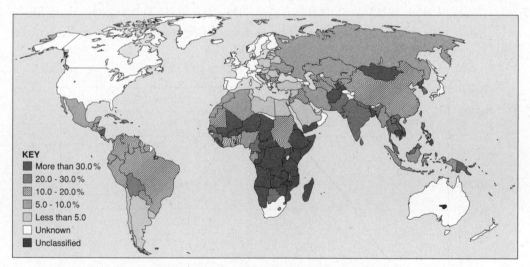

Figure 6.12 Worldwide malnutrition

Advances made during the first and second green revolutions (which focused on food production) are not ending world famine. In India, Mexico, and the Philippines, where large advances were made in crop production, famine is still common. Exports from these and other poor countries caused by distributors receiving higher profits from richer countries is actually making famine worse. Profit motives result in removing food from countries that grow it and sending the food to other countries that are able to pay higher prices.

Large-scale agribusiness or corporate ownership of farmland will not solve world hunger. Land reform in Japan, Zimbabwe, Taiwan, and Brazil has redistributed land into smaller holdings. It has also raised agricultural output an average of 80%.

The issue of malnutrition is not that the world does not produce enough food. The issue is that too many people cannot afford it or that it is not distributed efficiently. Enough wheat, rice, and grain are produced on Earth each day to provide each human with 3,500 calories per day (2,500 calories is the minimum daily requirement for men, and 2,000 calories is the minimum daily requirement for women). This does not even include vegetables, beans, nuts, meats, and fish processed each day. If all food grown and raised on Earth was distributed equally, it would result in 4.3 pounds (2 kg) per person per day.

Disease and Economic Impact

This section discusses how disease and economic realities affect population growth.

AIDS

In the United States, more than half a million people have died from complications arising from AIDS since 1981. An estimated 15,000 currently die annually according to the Federal Centers for Disease Control and Prevention. More than 1 million people in the United States are living with the virus, and 40,000 become infected each year. African-Americans, who make up approximately 13% of the U.S. population, account for half of new U.S. infections and a third of all AIDS-related deaths. African-American males are 7 times more likely as Caucasian males to be infected with HIV, while African-American females are 20 times more likely to be infected as Caucasian females. The fastest growing population in the United States for new infections are 15 to 24 years old.

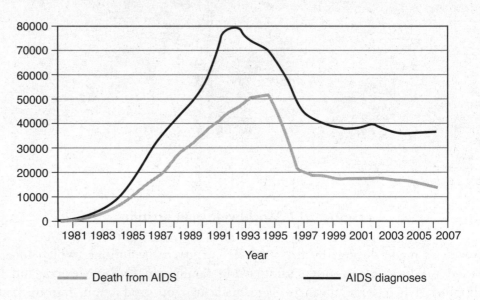

Figure 6.13 AIDS diagnoses and death rates in the United States between 1981–2007

AIDS—WORLDWIDE

In 2009, 33 million people around the world were living with HIV/AIDS. More than 60 million people have been infected with HIV and 25 million people have died of AIDS since the pandemic began in the early 1980s with currently over

7,000 new HIV infections occurring each day. AIDS is the leading cause of death in sub-Saharan Africa and the fourth leading cause of death globally with 5,000 people worldwide dying each day due to AIDS. About half of all new adult HIV infections occur among 15–24-year-olds. Approximately 97% of people living with HIV/AIDS live in low- and middle-income countries. Sub-Saharan Africa is the hardest-hit region and is home to two-thirds of all people living with HIV worldwide. Parts of Asia and Latin America are experiencing severe epidemics at the national or local level. Eastern Europe and central Asia is the region with the fastest growing HIV/AIDS epidemic in the world.

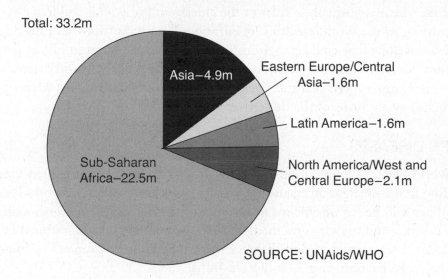

Figure 6.14 People living with HIV (by region)

AIDS—ECONOMIC IMPACT

For many less-developed countries with higher incidences of HIV-AIDS, the disease affects economic growth by reducing human capital. Those infected require significant medical care straining emerging economies. Higher mortality rates also creates a smaller skilled labor force, which results in a smaller tax base, which in turn reduces the resources available for public expenditures such as education, other health care services, etc. On the level of the household, AIDS results in both the loss of income and increased costs for medical care.

Worldwide, at least 25 million people have died from AIDS, and 2.8 million died in 2006, according to the World Health Organization. AIDS could kill 31 million people in India and 18 million in China by 2025, according to projections by U.N. population researchers. The toll from AIDS over the next 25 years will go far beyond the 34 million people thought to have died from the bubonic plague in the 14th century or the 20 to 40 million people who perished in the 1918 Spanish flu pandemic. AIDS is the leading cause of death in Africa, which has accounted for nearly half of all global AIDS deaths.

The epidemic is still growing. Its peak could be a decade or more away. In at least seven countries, the U.N. estimates that AIDS has reduced life expectancy to 40 years or less. Botswana has the highest HIV prevalence in the world; 36% of the country's 1.6 million people are HIV positive. AIDS experts predict that in the next few years, more than 50% of Botswana's children will be AIDS orphans and the average life expectancy will have fallen from 47 years to 27 years.

PANDEMICS

The Spanish influenza pandemic of 1918 to 1919 killed somewhere between 20 and 40 million people worldwide, with 28% of all Americans infected. It has been cited as the most devastating pandemic in recorded world history. More people died of Spanish influenza in a single year than in the four years of the plague from 1347 to 1351. The flu was most deadly for people ages 20 to 40. This was unusual for influenza, which is normally a killer of the elderly and of young children.

According to the World Health Organization, if a mutant form of bird flu virus (H5N1) develops that could pass from person to person, it could kill as many as 7 million people worldwide and infect nearly one-third of the world's population. Infectious disease experts fear the virus may mutate within a pig that harbors both human and avian forms of the flu virus.

OTHER DISEASES

The World Health Organization estimates that by 2020, tobacco-related illnesses, including heart disease, cancer, and respiratory disorders, will be the world's leading killer. They will be responsible for more deaths than AIDS, tuberculosis, road accidents, murder, and suicide combined. In 2005, estimated losses in national income from heart disease, strokes, and diabetes was $18 billion in China, $11 billion in the Russian Federation, and $9 billion in India.

Tuberculosis is the leading cause of death in many poorer countries. Its economic impact on global economies is estimated to be $12 billion.

Each year, approximately 500 million malaria infections lead to over 1 million deaths, 75% of which occur among children living in Africa. Mortality rates due to malaria are rising among young children, and drug therapies that were once fairly effective in treating malaria are becoming less effective as newer strains become more drug-resistant.

Resource Use and Habitat Destruction

Three methods are used to estimate the effects of humans on patterns of resource utilization: measure net primary productivity, estimate how much impact humans have had on Earth, and examine finite resources and from that draw conclusions on increasing productivity. Net primary productivity (NPP) is the total amount of solar energy converted into biochemical energy through photosynthesis minus the energy needed by those plants for their own metabolic requirements. It is a quantifiable measure of resources available on Earth. The NPP without human activity has been estimated to be 150 billion tons of organic matter per year. Human activity has caused a 12% decline in the NPP due to deforestation. Humans utilize about 27% of the NPP for their own purposes (food, building material, energy, and so on) or by converting productive land to nonproductive purposes. When added up, humans utilize about 40% of the NPP and leave all other life on Earth with 60%. The difference grows with the ever-increasing human population growth.

Of the NPP of the oceans, about 8% is utilized for human purposes. In nutrient-rich upwelling areas, humans utilize about 25%. In temperate continental shelf waters, humans utilize approximately 35% of the NPP.

If humans utilized 100% of the NPP (leaving nothing for all other life-forms on Earth), the theoretical maximum sustainable human population at 100% of the carrying capacity would be 15 billion people—which could be achieved during the 21st century. In this scenario, every square centimeter of Earth would be utilized for human needs. The following table lists factors that affect resource utilization.

Factors That Affect Resource Utilization

Factor	Description
Carrying capacity	See the section "Carrying Capacity (K)" in this chapter.
Energy resources	One average American consumes as much energy as 500 Ethiopians or 35 people from India.
Environmental degradation	Due to increased population size, erosion, desertification, pollution, impact on the ozone layer, and gases that contribute to global warming all increase.
Exploitation of natural resources as a function of gross domestic product	The richest 20% of the world's population contribute to resource depletion of energy and raw materials through overconsumption. This, in turn, leads to disproportionate amounts of pollution. The poorest 20% of the world's population are forced to deplete resources by being forced to cut down forests, clear land, and farm marginal land.
Extinction of animal and plant species	Close to 50% of all species of animals and plants on Earth could be on a path toward extinction within 100 years. 11% of all bird species and 13% of all plant species are at risk for extinction.
Famine	See the section "Hunger" in this chapter.
Political unrest	Affects employment, food distribution and standard of living that, in turn, affect utilization of resources.
Population density	Density, more than population size, has a greater effect on the amount of pollution and use of energy.
Population size	Large numbers of people lead to high rates of habitat loss and natural resource depletion.
Poverty	20% of the world's richest countries control 80% of the world's wealth. The poorest 20% of the world's population controls just 1.5% of the world's economic resources.
Technological development	More-developed countries consume more resources than less-developed countries. The United States represents about 5% of the world's population but consumes 25% of the world's resources and generates 25% of the world's waste.

QUICK REVIEW CHECKLIST

☐ **Population Biology Concepts**
- ☐ population ecology
- ☐ factors that affect population viability
- ☐ carrying capacity
- ☐ J-curves
- ☐ S-curves
- ☐ reproductive strategies
  - ☐ *r*-strategists
  - ☐ *K*-strategists
- ☐ survivorship
  - ☐ survivorship curves
    - ☐ Type I
    - ☐ Type II
    - ☐ Type III

☐ **Human Population Dynamics**
- ☐ historical population sizes
- ☐ calculating population change
  - ☐ crude birthrate
  - ☐ crude death rate
  - ☐ immigration
  - ☐ emigration
- ☐ distribution

☐ **Fertility Rates**
- ☐ replacement fertility rate (RLF)
- ☐ total fertility rate
- ☐ growth rates
- ☐ doubling time

☐ **Demographic Transition**
- ☐ stage 1
- ☐ stage 2
- ☐ stage 3
- ☐ stage 4

☐ **Age Structure Diagrams**

☐ **Population Size**
- ☐ strategies for sustainability

☐ **Case Studies**
- ☐ China
- ☐ India

QUICK REVIEW CHECKLIST (continued)

☐ **Impacts of Population Growth**
- ☐ hunger
- ☐ disease and economic impact
  - ☐ AIDS
  - ☐ pandemics
  - ☐ other diseases

☐ **Resource Use and Habitat Destruction**
- ☐ factors that affect resource utilization

MULTIPLE-CHOICE QUESTIONS

1. Which would be least likely to be affected by a density-dependent limiting factor?

   (A) A small, scattered population
   (B) A population with a high birthrate
   (C) A large, dense population
   (D) A population with a high immigration rate
   (E) None of the above

2. A population showing a growth rate of 20, 40, 60, 80 . . . would be characteristic of
   (A) logarithmic growth
   (B) exponential growth
   (C) static growth
   (D) arithmetic (linear) growth
   (E) power curve growth

3. If a population doubles in about 70 years, it is showing a ___ % growth rate.

   (A) 1
   (B) 5
   (C) 35
   (D) 140
   (E) 200

4. An island off Costa Rica includes 500 birds of a particular species. Population biologists determined that this bird population was isolated with no immigration or emigration. After one year, the scientists were able to count 60 births and 10 deaths. The net growth for this population was

    (A) 0.5
    (B) 0.9
    (C) 1.0
    (D) 1.1
    (E) 1.5

5. Afghanistan has a current growth rate of 4.8%, representing a doubling time of approximately

    (A) 4.8 years
    (B) 9.6 years
    (C) 14.5 years
    (D) 35 years
    (E) 70 years

6. Biotic potential refers to

    (A) an estimate of the maximum capacity of living things to survive and reproduce under optimal environmental conditions
    (B) the proportion of the population of each sex at each age category
    (C) the ratio of total live births to total population
    (D) a factor that influences population growth and that increases in magnitude with an increase in the size or density of the population
    (E) events and phenomena of nature that act to keep population sizes stable

7. The number of children an average woman would have, assuming that she livers her full reproductive lifetime, is known as the
    (A) birthrate
    (B) crude birthrate
    (C) TFR
    (D) RLF
    (E) zero population growth rate

8. The average American's ecological footprint is approximately

    (A) the size of a shoe
    (B) 0.5 acres
    (C) 3 acres
    (D) 6 acres
    (E) 12 acres

9. Which of the following statements is FALSE?

   (A) The United States, while having only 5% of the world's population, consumes 25% of the world's resources.

   (B) Up to 50% of all plants and animals could become extinct within the next 100 years.

   (C) In 1990, 20% of the world's population controlled 80% of the world's wealth.

   (D) Every second, five people are born and two people die, a net gain of three people.

   (E) They are all true statements.

10. Pronatalists (people in favor of having many children) would include all of the following EXCEPT

    (A) many children die early due to health and environmental conditions

    (B) children are expensive and time intensive

    (C) children provide extra income for families

    (D) children provide security for parents when the parents reach old age

    (E) status of the family is often determined by the number of children

11. The most successful method of controlling a country's population has been

    (A) required sterilization

    (B) government quotas on children produced

    (C) birth control

    (D) financial incentives

    (E) All of the above

12. The following age-structure diagram would be typical of

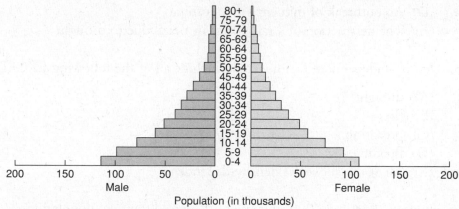

Source: U.S. Census Bureau, International Data Base.

    (A) Russia

    (B) China

    (C) the United States

    (D) Niger or Peru

    (E) Canada

13. Examine the following age-structure diagram.

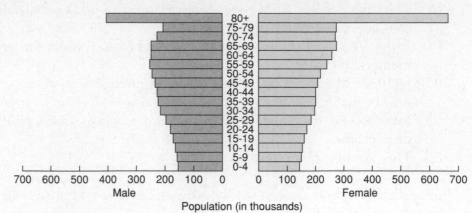

Source: U.S. Census Bureau, International Data Base.

This population will be

(A) declining rapidly in the future
(B) growing slowly in the future
(C) remaining stable
(D) growing rapidly in the future
(E) declining slowly in the future

14. Which of the following examples does NOT demonstrate a density-dependent factor affecting the population size?

(A) The "black plagues" that occurred in Europe during the mid 14th century and which are estimated to have killed up to ⅓ of the European population
(B) The number of lions inhabiting a grassland in Africa
(C) Tropical plants located near the ground in a tropical rain forest
(D) An outbreak of influenza in a hospital
(E) The destruction of a rain forest in Brazil due to drought

15. Density-independent factors would include all of the following EXCEPT

(A) drought
(B) fires
(C) predation
(D) flooding
(E) All of the above are density-dependent

16. Between 1963 and 2000, the rate of the world's annual population change _____ but the human population size _____ .

(A) dropped 40%; rose 90%
(B) rose 90%; dropped 40%
(C) dropped 90%; rose 40%
(D) remained stable; tripled
(E) rose 40%; dropped 2.9%

17. RLF for a couple is

   (A) 1.0
   (B) 2.0
   (C) 2.1
   (D) 3.0
   (E) varies depending on country

18. Which of the following factors is associated with the highest potential for population growth?

   (A) High percentage of people under age 18
   (B) High percentage of people in their 30s
   (C) High percentage of people in their 50s
   (D) High percentage of people in high-income groups
   (E) High percentage of people in low-income groups

19. Using the demographic transition model, what stage would be characteristic of death rates falling while birthrates remain high?

   (A) Preindustrial
   (B) Industrial
   (C) Postindustrial
   (D) Transitional
   (E) None of the above

20. All of the following factors tend to cause women to have fewer children EXCEPT

   (A) higher education
   (B) high infant mortality
   (C) better prenatal care
   (D) birth control education
   (E) human rights are protected

FREE-RESPONSE QUESTION

By: William Aghassi
Brooklyn Technical High School, Brooklyn, NY
B.S. (Mechanical Engineering- Polytechnic University)
M.S. (Environmental Engineering- New Jersey Inst. of Technology)

The 1990 U.S. census revealed that, for the first time, a majority of Americans live in suburbs of major cities or in suburban-like communities.

(a) List three factors why this trend may not be sustainable from a resource point of view.

(b) Explain why each of the three reasons you have listed is not sustainable and the possible consequence of each.

(c) What national public policy decision(s) made the United States into a suburban country as opposed to Europe, which is largely urbanized?

(d) Many environmentalists think of the 21st century as the "Century of the City." State whether or not you agree or disagree with this statement and include your reasons.

MULTIPLE-CHOICE ANSWERS AND EXPLANATIONS

1. **(A)** Increasing population size reduces available resources, thus limiting population growth. In restricting population growth, a density-dependent factor intensifies as the population size increases, affecting each individual more strongly. Population growth declines because of death rate increase, birthrate decrease, or both. There is a reduction in the food supply, which restricts reproduction resulting in less offspring. The competition for space to establish territories is a behavioral mechanism that may restrict population growth. Predators concentrate in areas where there is a high concentration of prey. As long as the natural resources are available in sufficient quantity, the population will remain constant. As the population decreases, so do predators. The accumulation of toxic wastes may also limit the size of a population. Intrinsic factors may play a role in limiting a population size. High densities may cause stress syndromes, resulting in hormonal changes that may delay the onset of reproduction. It has also been reported that immune disorders are related to stress in high densely populated areas. Density-independent factors include weather, climate, and natural disasters such as freezes, seasonal changes, hurricanes, and fires. These factors affect all individuals in the population, regardless of population size.

2. **(D)** Arithmetic or linear growth is characterized by a constant increase per unit of time. In this case, the constant is an increase of 20.

3. **(A)** A 1% growth rate would cause a population to double in 70 years. Hint: divide 70 by the annual percentage growth rate to get the doubling time in years.

4. **(D)** Population size = original size (500) + births (60) − deaths (10) + immigration (0) − emigration (0) = 550. Net growth *rate* = 550/500 = 1.1.

5. **(C)** Using the Rule of 70, 70/4.8 = 14.5 years.
6. **(A)** The maximum reproductive rate is called the biotic potential.
7. **(C)** Global TFR is approximately 2.6.
8. **(E)** An ecological footprint is a metaphor used to depict the amount of land a person would hypothetically need to provide the resources required to support himself.
9. **(E)** No explanation needed.
10. **(B)** Pronatalists urge people to have many children. Choice (B) is the only argument provided that does not promote having children.
11. **(E)** All of the choices listed have had some measure of success.
12. **(D)** Age-structure diagrams with a wide base are populations that have a high proportion of young, which results in a powerful, built-in momentum to increase population size, assuming death rates do not unexpectedly increase.
13. **(A)** This graph is the reverse of the diagram in Question 12. In this case, the majority of the population is beyond reproductive years and the death rate exceeds the birthrate. This projected age structure diagram is for Hong Kong in 2050.
14. **(E)** Choices A–D are examples of density-dependent factors, wherein large, dense populations are more strongly affected than small, less crowded ones. Drought is a naturally occurring event and does not depend on the density of any organism(s) occurring in the area. However, had the destruction of the rain forest been due to ranching, timber harvesting, and/or agriculture (as is the most common cause), then an increase in the number of humans in the rain forest would increase the amount of rain forest destruction in which case the destruction would be due to density-dependent elements.
15. **(C)** Density-independent factors influence population growth and do not depend on the size or density of the population. Predation rates are affected by population size.
16. **(A)** Population change is an increase or decrease in the size of a population. It is equal to (births + immigration) – (deaths + emigration). Population size is the number of individuals in the population.
17. **(C)** A TFR of 2.1 is considered the replacement rate. Once the TFR of a population reaches 2.1, the population will remain stable, assuming no immigration or emigration takes place.
18. **(A)** Population momentum is the tendency for changes in population growth rates to lag behind changes in childbearing behavior and mortality conditions. Momentum operates through the population age distribution. A population that has been growing rapidly for a long time acquires a "young" age distribution that will result in positive population growth rates for many decades into the future.
19. **(D)** In the preindustrial stage, living conditions are harsh, birth and death rates are high, and there is little increase in population size. In the transitional stage, living conditions improve, the death rate drops, and birthrates remain high. In the industrial stage, growth slows. In the postindustrial stage, zero population growth is reached and the birthrate falls below the death rate.
20. **(B)** Countries with high infant mortality rates generally have high TFRs.

FREE-RESPONSE ANSWER

(a) Three factors may limit the trend to American suburbanization:
1. Limit to the amount of open space available near cities
2. Non-point-source pollution (runoff) that could jeopardize the integrity of the region's watershed
3. Traffic congestion
4. Diversity of plant and animal ecosystems could be jeopardized
5. Degradation of air quality
6. Loss of agricultural lands
7. Scarcity of available and inexpensive energy sources

(b) 1. The two features that distinguish land use in suburban areas are low-density housing and that over 60% of available land is devoted to the needs of the automobile. As suburbanization increases, these two factors tend to diminish the availability of open space.
2. Suburbanization near cities tends to encroach upon the watersheds of these regions. Suburban lawns, parking lots, and roads are responsible for a much greater nutrient and chemical loading than undeveloped land and greatly affect the water quality of the region's drinking water.
3. Suburban communities are low density and widely dispersed. Since suburban residents do not commute from or to a central location, they are usually ill served by mass transit and carpooling is sometimes not practical. Commuting times therefore lengthen to intolerable levels, resulting in high air pollution levels.
4. Suburban sprawl can destroy many unique and/or productive ecosystems such as wetlands and forests, thus limiting regional diversity.
5. Since suburbs are decentralized, residents must rely on the automobile for every errand, and all of their goods must be transported by truck. This often results in traffic congestion and air quality degradation that leads to respiratory distress in humans, damage to plants, and damage to buildings.
6. Land goes toward the highest value use. As suburbs encroach on agricultural areas, farmers may be forced to sell to developers due to rezoning and a resulting increase in real estate taxes or simply a higher rate of return than that of farming.
7. Low-density single-family housing and exclusive reliance on the automobile to get around are energy inefficient. As imported oil supplies become expensive or hard to obtain, this could limit suburban growth.

(c) After World War II, the decision to invest infrastructure dollars on roads and disinvest in mass transit made suburbanization possible.

 The public policy decision to keep energy prices low through relatively low taxation compared with Europe, which has high energy taxation and high energy prices, made suburban sprawl possible.

(d) Agree: Cities are more energy efficient and, if well planned, are more convenient and offer a better sense of community than many suburbs. This will attract more Americans to move back to cities.

 Disagree: Americans are reluctant to forgo the American dream of the single-family home and the automobile to ever return to cities in large numbers.

UNIT IV: LAND AND WATER USE (10–15%)

Areas on Which You Will Be Tested

A. Agriculture
1. Feeding a growing population—human nutritional requirements, types of agriculture, green revolution, genetic engineering, crop production, deforestation, irrigation, and sustainable agriculture.
2. Controlling pests—types of pesticides, costs and benefits of pesticide use, integrated pest management, and relevant laws.

B. Forestry—tree plantations, old-growth forests, forest fires, forest management, and national forests.

C. Rangelands—overgrazing, deforestation, desertification, rangeland management, and federal rangelands.

D. Other Land Use
1. Urban land development—planned development, suburban sprawl, and urbanization.
2. Transportation infrastructure—federal highway system, canals and channels, roadless areas, and ecosystem impacts.
3. Public and federal lands—management, wilderness areas, national parks, wildlife refuges, forests, and wetlands.
4. Land conservation options—preservation, remediation, mitigation, and restoration.
5. Sustainable land use strategies.

E. Mining—mineral formation, extraction, global reserves, and relevant laws and treaties.

F. Fishing—fishing techniques, overfishing, aquaculture, and relevant laws and treaties.

G. Global Economics—globalization, World Bank, "Tragedy of the Commons," and relevant laws and treaties.

Land and Water Use

CHAPTER 7

FEEDING A GROWING POPULATION

In order to feed a population adequately, several factors must be taken into account. The following sections discuss these in detail.

Human Nutritional Requirements

A healthy diet generally requires 2,500 calories per day for the average male and 2,000 calories per day for the average female. Proper nutrition also requires a balanced intake of protein, carbohydrates, and fat. Protein produces 4 calories of energy per gram and should make up about 30% of all calories. Carbohydrates also produce 4 calories of energy per gram and should make up approximately 60% of the daily diet. Fats produce 9 calories of energy per gram and should not make up more than 10% of the total daily caloric intake.

Only about 100 species of plants of the 350,000 known are commercially grown to meet human nutritional needs. Of these, wheat and rice supply over half the human caloric intake. Just 8 species of animal protein supply over 90% of the world's needs. It takes about 16 pounds (7 kg) of grain to produce 1 pound (0.5 kg) of edible meat, and 20% of the world's richest countries consume 80% of the world's meat production. About 90% of the grain grown in the United States is grown for animal feed. By consuming the grain directly instead of consuming the animals that feed upon it, there would be a 20-fold increase in the amount of calories available and an 8-fold increase in the amount of protein available. The benefit of consuming meat products is that they are concentrated sources of protein that are broken down through digestion into amino acids.

Malnutrition

In terms of famine and malnutrition, about 11 million children die each year from starvation, with 850 million people (13% of the world population) considered malnourished. Chronic undernourishment and vitamin or mineral deficiencies result in stunted growth, weakness, and increased susceptibility to illness.

Most of the undernourished are in developing countries. Marasmus and kwashiorkor are protein deficiency diseases in which victims may become so emaciated that they may be less than 80% of their normal weight for their height.

Types of Agriculture

AGROFORESTRY

A system of land use in which harvestable trees or shrubs are grown among or around crops or on pastureland as a means of preserving or enhancing the productivity of the land.

ALLEY CROPPING

A method of planting crops in strips with rows of trees or shrubs on each side. Alley cropping increases biodiversity, reduces surface water runoff and erosion, improves the utilization of nutrients, reduces wind erosion, modifies the microclimate for improved crop production, and improves wildlife habitat.

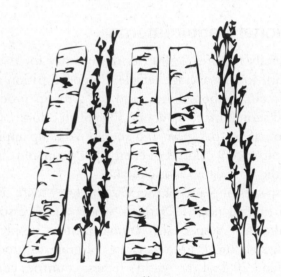

Figure 7.1 Alley cropping

CROP ROTATION

Planting a field with different crops from year to year to reduce soil nutrient depletion. Example: Rotating corn or cotton, which removes large amounts of nitrogen from the soil, with soybeans that then add nitrogen to the soil.

Crop Rotation Schedule			
	Field 1	**Field 2**	**Field 3**
1st Year	Wheat and rye	Oats and barley	Fallow
2nd Year	Fallow	Wheat and rye	Peas and beans
3rd Year	Peas and beans	Fallow	Wheat and rye

HIGH-INPUT AGRICULTURE

Includes the use of mechanized equipment, chemical fertilizers, and pesticides.

INDUSTRIAL AGRICULTURE OR CORPORATE FARMING

A system characterized by mechanization, monocultures, and the use of synthetic inputs such as chemical fertilizers and pesticides, with an emphasis on maximizing productivity and profitability.

INTERCROPPING

To grow more than one crop in the same field, especially in alternating rows or sections.

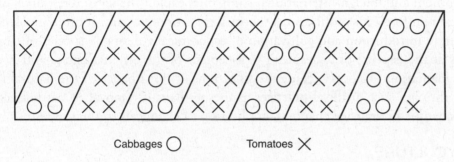

Cabbages ◯   Tomatoes ✕

Figure 7.2 Example of intercropping

INTERPLANTING

Growing two different crops in an area at the same time. To interplant successfully, plants should have similar nutrient and moisture requirements.

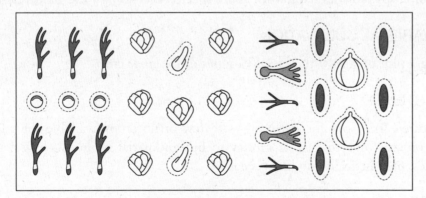

Figure 7.3 Example of interplanting

LOW INPUT

Depends on hand tools and natural fertilizers; lacks large-scale irrigation.

LOW-TILL, NO-TILL, OR CONSERVATION-TILL AGRICULTURE

Soil is disturbed little or not at all to reduce soil erosion. Has lower labor costs, reduces the need for fertilizer, and saves energy.

MONOCULTURE

The cultivation of a single crop.

ORGANIC FARMING

A form of agriculture that relies on crop rotation, green manure, compost, biological pest control, and mechanical cultivation to maintain soil productivity and control pests. This practice excludes or strictly limits the use of synthetic fertilizers, synthetic pesticides, plant growth regulators, livestock feed additives, and genetically modified organisms.

PLANTATION

A commercial tropical agriculture system that is essentially export oriented. The local government and foreign/international companies exploit the natural resources of the tropical rain forest for profit, usually short-term economic gain. It often involves the deliberate introduction and cultivation of economically desirable species of tropical plants at the expense of widespread replacement of the original native and natural flora. Plantation practices include modifications or disturbance of the natural landscape through such artificial practices as the permanent removal of natural vegetation, changes in drainage channels, application of chemicals to the soil, etc.

POLYCULTURE

Polyculture uses *different* crops in the same space, in imitation of the diversity of natural ecosystems, and avoids large stands of a single crop (monoculture). It includes crop rotation, multicropping, intercropping, and alley cropping. Polyculture, though it often requires more labor, has several advantages over monoculture. The diversity of crops avoids the susceptibility of monocultures to disease. The greater variety of crops provides habitat for more species, increasing local biodiversity.

POLYVARIETAL CULTIVATION

Planting a plot of land with several varieties of the *same* crop.

SUBSISTENCE

Agriculture carried out for survival—with few or no crops available for sale. It is usually organic, due to lack of money to buy industrial inputs such as fertilizer, pesticides, or genetically modified seeds.

TILLAGE

Conventional method in which the surface is plowed which then breaks up and exposes the soil. This is then followed by smoothing the surface and planting. This method exposes the land to water and wind erosion.

Green Revolution

The first green revolution occurred between 1950 and 1970. It involved planting monocultures, using high applications of inorganic fertilizers and pesticides, and the widespread use of artificial irrigation systems. Before the first green revolution, crop production was correlated with increases in acreage under cultivation. After the first green revolution, crop acreage increased about 25%, but crop yield increased 200%. Crop yield then reached a plateau since it was easier and more economical to increase crop production through various agricultural techniques than to buy and clear new land.

The second green revolution began during the 1970s and is continuing today. It involves growing genetically engineered crops that produce the most yields per acre. It is in contrast with past agricultural practices in which farmers planted a variety of locally adapted strains. For example, of all wheat grown in the United States today, 50% comes from 9 different genotypes.

CRITICISMS OF THE GREEN REVOLUTION

Critics cite the following as problems and/or failures of the Green Revolution:

- The Green Revolution is unsustainable.
- Increasing food production is not synonymous with increasing food security, i.e., famines are not caused by decreases in food supply but by socioeconomic dynamics and a failure of public action.
- Green Revolution agriculture produces monocultures of cereal grains, while traditional agriculture usually incorporates polycultures.
- There has been a drop in the productivity of land that has been intensively farmed for the past 30 years due to desertification and other forms of land degradation.
- The necessary purchase of inputs led to the widespread establishment of rural credit institutions, which then caused many smaller farmers to go into debt and in many cases resulted in a loss of their farmland.
- Green Revolution agriculture increases the use of pesticides, which are necessary to limit the high levels of pest damage that inevitably occur in monoculture.
- Salinization, water logging, and lowering of water levels in certain areas increased as consequences of increased irrigation.
- The Green Revolution reduced agricultural biodiversity, as it relied upon only a few high-yield varieties of each crop. This led to the susceptibility of the food supply to pathogens that cannot be controlled by agrochemicals, as well as the permanent loss of many valuable genetic traits bred into traditional varieties over thousands of years.

Genetic Engineering and Crop Production

Genetic engineering involves moving genes from one species to another or designing gene sequences with desirable characteristics. These include pest, drought, mold, and saline resistance, higher protein yields, and higher vitamin content. About 75% of all crops grown derive from genetically engineered or transgenic crop species.

In 2006, 10 million farmers in 22 countries planted 252 million acres of transgenic crops. The majority of these crops were herbicide- and insect-resistant soybeans, corn, cotton, canola, and alfalfa. Other examples included sweet potatoes resistant to a virus that could seriously reduce most of the African harvest, rice with increased iron and vitamins that may alleviate chronic malnutrition (golden rice), and a variety of plants able to survive weather extremes. In the near future, genetically modified bananas may produce human vaccines to prevent infectious diseases such as hepatitis B. Genetic modification helps produce fish that mature more quickly, provides cows with a resistance to bovine spongiform encephalopathy (mad cow disease), and produces fruit and nut trees that yield years earlier. The future will see exponential growth in developing genetically modified products.

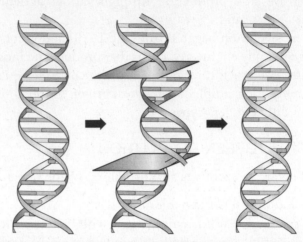

Figure 7.4 Example of gene splicing, a process by which a segment of DNA from one source is attached to or inserted into a strand of DNA from another source.

CASE STUDY

Golden rice is produced by splicing three foreign genes, two from the daffodil and one from a bacterium, into a variety of rice that supplies vitamin A to populations that frequently suffer from vitamin A deficiency.

Genetically Engineered Crops

Pros	Cons
May require less water and fertilizer	Unknown ecological effects
Higher crop yields	Less biodiversity
Less spoilage	May harm beneficial insects
Faster growth which may mean greater productivity, resulting in lower operating costs	May pose allergen risk
More resistant to disease, drought, frost, and insects	May result in mutations with unknown consequences
May be able to grow in saltier soils	May cause pesticide-resistant strains

Irrigation

Three-quarters of all freshwater used on Earth is used for agriculture. Worldwide, approximately 40% of all crop yields come from 16% of all cropland that is irrigated. The use of irrigation depends on the climate and the degree of industrialization. For example, Canada irrigates about 10% of its crops, whereas India requires 90% of its crops to be irrigated. With inefficiencies such as seepage, leakage, and evaporation, up to 70% of all irrigation water can be lost. A drip irrigation system, which solves many of these problems but is more expensive to install, is used on approximately 1% of crops worldwide.

Increases in human population growth are outpacing the rate of land that is being irrigated. Sustainable irrigation is limited as a result of increases in costs, depletion of current sources of water, competition for water by urban areas, restoration of wetlands and fisheries, waterlogging, and salinization. Future water capacity will increase through increases in efficiency.

Sustainable Agriculture

Sustainable agriculture involves a variety of approaches. It integrates three main goals: environmental health, economic profitability, and social and economic equity. Specific strategies must take into account topography, soil characteristics, climate, pests, local availability of inputs, and the individual grower's goals. Despite the site-specific and individual nature of sustainable agriculture, several general principles can be applied to help growers select appropriate management practices and are described below.

Changes in agriculture have had many positive effects and reduced many risks in farming. However, there have also been significant costs. Prominent among these are topsoil depletion, groundwater contamination, the decline of family farms, continued neglect of the living and working conditions for farm laborers, increasing costs of production, and the disintegration of economic and social conditions in rural communities.

EFFICIENT USE OF INPUTS

Sustainable farmers maximize reliance on natural, renewable farm inputs with the goal to develop efficient, biological systems that do not need high levels of material inputs. Sustainable approaches are those that are the least toxic and least energy intensive and yet maintain productivity and profitability. Preventive strategies and other alternatives (such as integrated pest management) should be employed before using chemical inputs from any source.

SELECTION OF SITE, SPECIES, AND VARIETY

Preventive strategies, when adopted early, can reduce inputs and help establish a sustainable production system. When possible, pest-resistant crops should be selected that are tolerant of existing soil or site conditions. When site selection is an option, factors such as soil type and depth, previous crop history, and location (climate and topography) should be taken into account before planting.

SOIL MANAGEMENT

Proper soil, water, and nutrient management can help prevent some pest problems brought on by crop stress or nutrient imbalance. Furthermore, crop management systems that impair soil quality often result in greater inputs of water, nutrients, pesticides, and/or energy for tillage to maintain yields. In sustainable systems, the soil is viewed as a fragile and living medium that must be protected and nurtured to ensure its long-term productivity and stability. Methods to protect and enhance the productivity of the soil include using cover crops, compost, and/or manures, reducing tillage, and maintaining soil cover with plants and/or mulches. Regular additions of organic matter or the use of cover crops can increase soil aggregate stability, soil tilth, and the diversity of soil microbial life.

SPECIES DIVERSITY

By growing a variety of crops, farmers spread out the economic risk and are less susceptible to the radical price fluctuations associated with changes in supply and demand. For example, in annual cropping systems, crop rotation can be used to suppress weeds, pathogens, and insect pests. Also, cover crops can have stabilizing effects on the agroecosystem by holding soil and nutrients in place, conserving soil moisture with dead mulches, and increasing the water infiltration rate and soil water-holding capacity. Optimum diversity may be obtained by integrating both crops and livestock in the same farming operation. Growing row crops on more level land and pasture or forages on steeper slopes will reduce soil erosion. Planting pasture and forage crops in rotation enhances soil quality and reduces erosion. Livestock manure, in turn, contributes to soil fertility. Livestock can buffer the negative impacts of low rainfall periods by consuming crop residue that in plant-only systems would have been considered crop failures. Feeding and marketing are flexible in animal production systems. This can help cushion farmers against trade and price fluctuations and, in conjunction with cropping operations, make more efficient use of farm labor and resources.

CONTROLLING PESTS

Pesticides can be used to control pests, but their use has drawbacks. Integrated pest management is another strategy to control pests.

Types of Pesticides

Pesticides differ in several ways. Their chemistry, how long they remain effective in the environment (environmental persistence), and their effect on the food web (bio-accumulation and biomagnification) are just a few concerns. Others include what type of organisms are affected, how the pesticides work (nervous system, reproductive cycles, blood chemistry), how fast they work, and their application.

BIOLOGICAL

Living organisms are used to control pests. Examples include bacteria, Bt (*Bacillus thuringiensis*), ladybugs, milky spore disease, parasitic wasps, and certain viruses.

CASE STUDY

Bacillus thuringiensis (Bt) is a soil-dwelling bacterium that also occurs naturally in the gut of caterpillars of various types of moths and butterflies, as well as on the dark surface of plants. Proteins produced by Bt are used as specific insecticides. It works by secreting one or more toxins after being ingested by an insect. The toxins are often specific to a family of insects, and because of their specificity, these pesticides are regarded as environmentally friendly. Advantages of using Bt include:

(i) The level of toxin can be very high, thus delivering sufficient dosage to the pest.

(ii) It is contained within the plant system; therefore only those insects that feed on the crop perish.

(iii) It replaces the use of synthetic pesticides in the environment.

A possible drawback to Bt may be that constant exposure to a toxin creates evolutionary pressure for pests resistant to that toxin.

CARBAMATES

Carbamates, also known as urethanes, affect the nervous system of pests. 100 grams of a carbamate has the same effect as 2,000 grams of a chlorinated hydrocarbon such as DDT. Carbamates are more water soluble than chlorinated hydrocarbons, which brings a greater risk of them being dissolved in surface water and percolating into groundwater.

CASE STUDY

On December 2, 1984 at the Union Carbide pesticide plant in Bhopal, India, a leak of chemicals from a plant manufacturing a carbamate pesticide leaked, which resulted in the exposure of hundreds of thousands of people. Roughly 8,000 people have since died from gas-related diseases with 200,000 people having permanent injuries.

CHLORINATED HYDROCARBONS AND OTHER PERSISTENT ORGANIC COMPOUNDS (POPS)

Chlorinated hydrocarbons, such as the pesticide DDT, are synthetic organic compounds belonging to a group of chemicals known as persistent organic pollutants or POPs. Persistent organic pollutants (POPs) are organic compounds that are resistant to environmental degradation through chemical, biological, and photolytic processes. Because of this, they have been observed to persist in the environment, to be capable of long-range transport, bioaccumulate in human and animal tissue, biomagnify in food chains, and to have potential significant impacts on human health and the environment. Many POPs are currently or were in the past used as pesticides. Others are used in industrial processes and in the production of a range of goods such as solvents, polyvinyl chloride, and pharmaceuticals. Some of the chemical characteristics of POPs include low water solubility, high lipid solubility, semi-volatility, and high molecular masses. POPs are frequently halogenated, usually with chlorine. The more chlorine or other halogen groups a POP has, the more resistant it is to being broken down over time. One important factor of their chemical properties, such as lipid solubility, results in the ability to pass through biological phospholipid membranes and bioaccumulate in the fatty tissues of living organisms. The chemicals' semi-volatility allows them to travel long distances through the atmosphere before being deposited. Thus POPs can be found all over the world, including in areas where they have never been used and remote regions such as the middle of oceans and Antarctica. The chemicals' semi-volatility also means that they tend to volatilize in hot regions and accumulate in cold regions, where they tend to condense and persist. As pesticides, they affect the nervous system of pests. In the 1950s, DDT was linked with the thinning of eggshells in certain species of birds (e.g., bald eagle).

FUMIGANTS

Used to sterilize soil and prevent pest infestation of stored grain.

INORGANIC

Broad-based pesticides. Includes arsenic, copper, lead, and mercury. Highly toxic and accumulate in the environment.

ORGANIC OR NATURAL

Natural poisons derived from plants such as tobacco or chrysanthemum.

ORGANOPHOSPHATES

Extremely toxic but remain in the environment for only a brief time. Examples include malathion to control mosquitoes and West Nile Virus, and parathion.

Costs and Benefits of Pesticide Use

Despite the use of pesticides, many pests worldwide have increased in numbers. This is due to genetic resistance, reduced crop rotation, increased mobility of pests due to increased world trade, and reduction in crop diversity. The following table shows the pros and cons of pesticide use.

The Pros and Cons of Pesticides

Pros	Cons
Kill unwanted pests that carry disease	Accumulate in food chains
Increase food supplies	Pests develop resistance and create a pesticide treadmill
More food means food is less expensive	Estimates range from $5 to $10 in damage done to the environment for every $1 spent on pesticides; pesticides are expensive to purchase and apply
Newer pesticides are safer and more specific	Pesticide runoff and its effect on aquatic environments through biomagnification
Reduces labor costs	Inefficiency—only 5% of a pesticide reaches a pest
Agriculture is more profitable	Threatens endangered species and pollinators; also affects human health

Integrated Pest Management

Integrated pest management (IPM) is an ecological pest control strategy that uses a variety of methods. When used in combination, these methods working together can reduce or eliminate the use of traditional pesticides. The aim of IPM is not to eradicate pests but to control their numbers to acceptable levels. In integrated pest management, chemical pesticides are the last resort. Methods employed in IPM include:

- Polyculture
- Intercropping
- Planting pest-repellent crops
- Using mulch to control weeds
- Using pyrethroids or naturally occurring microorganisms (such as Bt) instead of toxic pesticides
- Natural insect predators
- Rotating crops often to disrupt insect cycles
- Using pheromones or hormone interrupters
- Releasing sterilized insects
- Developing genetically modified crops that are more insect resistant
- Regular monitoring through visual inspection and traps followed by record keeping
- Construction of mechanical controls such as traps, tillage, insect barriers, or agricultural vacuums equipped with lights

> ### RELEVANT LAWS
>
> **Federal Insecticide, Fungicide and Rodenticide Control Act (FIFRA) (1947):** Regulates the manufacture and use of pesticides. Pesticides must be registered and approved. Labels require directions for use and disposal.
>
> **Federal Environmental Pesticides Control Act (1972):** Requires registration of all pesticides in U.S. commerce.
>
> **Food Quality Protection Act (FQPA) (1996):** Emphasizes the protection of infants and children in reference to pesticide residue in food.

FORESTRY

Forestry involves the management of forests. Sometimes this involves planting new trees, and sometimes it involves fires.

Ecological Services

Forests are an important global reserve. The ecological services of forests include:

1. Providing wildlife habitats
2. Carbon sinks
3. Affecting local climate patterns
4. Purifying air and water
5. Reducing soil erosion as they serve as a watershed, absorbing and releasing controlled amounts water
6. Providing energy and nutrient cycling

Tree Plantations

Tree plantations are large, managed commercial or government-owned farms with uniformly aged trees of one species (monoculture). Trees may not be native to the area and may be hybrids (genetically modified). The primary use of plantation trees is for pulp and lumber. Pine, spruce, and eucalyptus are widely used due to their fast growth rates and can be used for paper and timber. Trees are harvested by clear-cutting. Short rotation cycles of 25–30 years or 6–10 years in the tropics are economically important factors. Just 5% of the world's forests are tree plantations, but they account for 20% of the current world wood production. In comparison, 63% of the world's forests are secondary-growth forests, and 22% are old-growth forests.

Annually, tropical tree plantations yield much more wood (25 m^3/hectare) than traditional forests (1–3 m^3/hectare). Some of the natural closed forests—7%—are being lost in the tropics due to land conversion to tree plantations. Tree plantations do not support food webs found in old-growth forests, and they contain little biodiversity. Decaying wood is absent, which provides a vital link in an old-growth forest. Conversion to tree plantations may result in draining wetlands and replacing traditional hardwoods. Newer techniques allow leaving blocks of native species within the plantation or retaining corridors of natural forest. The Kyoto Protocol encourages use of tree plantations to reduce carbon dioxide levels although carbon dioxide may eventually re-enter the atmosphere after harvesting.

Old-Growth Forests

Old-growth forests are forests that have not been seriously impacted by human activities for hundreds of years. Old-growth forests are rich in biodiversity. Old-growth forests are characterized by:

TIP

"in situ" means "in its original place."

- Older and mixed-aged trees
- Minimal signs of human activity
- Multilayered canopy openings due to tree falls
- Pit-and-mound topography due to trees falling and creating new microenvironments by recycling carbon-rich organic material directly to the soil and providing a substrate for mosses, fungi (necessary for *in situ* recycling), and seedlings
- Decaying wood and ground layer that provides a rich carbon sink
- Dead trees (snags) that are necessary nesting sites for woodpeckers and spotted owls
- Healthy soil profiles
- Indicator species
- Little vegetation on the forest floor due to light being a limiting factor.

Depletion of old-growth forests increases the risk of climatic change. Many old-growth forests contain species of trees that have high economic value but that require a long time to mature (mahogany, oak, etc.).

Forest Fires

Forests are unique in terms of their ecological significance and the number and frequency of forest fires. Current wildfire frequency in the United States is about 4 times the average of 1970–1986. The total area burned by current fires is about 7 times its previous level. The U.S. Forest Service has lengthened the wildfire season by 78 days. The change in wildfire frequency appears to be linked to annual spring and summer temperatures. Longer, warmer summers are documented with an increase in the number of forest fires. A correlation exists between early arrivals of the spring snowmelt in the mountainous regions and the incidence of large forest fires. An earlier snowmelt can lead to an earlier and longer dry season, which provides greater opportunities for larger fires. As forests burn, they release their stored carbon as carbon dioxide into the atmosphere, further compounding the problems of global warming. Another reason for the increase in forest fires is a change in fire management philosophy. Any naturally started fire on federal land that is not threatening resources (homes or commercial structures) is allowed to burn.

CROWN FIRES

Occur in forests that have not had surface fires for a long time. Extremely hot. Burn entire trees and leap from treetop to treetop. Kills wildlife, increases soil erosion, and destroys structures.

GROUND FIRES

Fires that occur underground and burn partially decayed leaves. Common in peat bogs. Difficult to detect and extinguish.

SURFACE FIRES

Burns undergrowth and leaf litter. Kills seedlings and small trees. Spares older trees, and allows many wild animals to escape. Advantages: burns away flammable ground litter, reducing larger fires later; releases minerals back into soil profile; stimulates germination for some species with cones that require heat to open up and release the seeds, such as giant sequoia and jack pine; helps to keep pathogens and insects in check; and allows vegetation to grow in clearings that provides food for deer, moose, elk, muskrat, and quail.

METHODS TO CONTROL FIRES

Two methods are used to control forest fires: prevention and prescribed burning. Prevention involves burning permits, closing parts of the forest during times of the year when the number of visitors is high and during periods of drought, and educating the public. Prescribed burning involves purposely setting controlled surface fires and setting small, prescribed fires to thin out underbrush in high-risk areas. It requires careful planning and monitoring. Other strategies include allowing fires to burn themselves out and creating large, clear areas around structures.

Deforestation

Deforestation is the conversion of forested areas to nonforested areas. They are then used as grasslands for livestock grazing, grain fields, mining, petroleum extraction, fuel wood cutting, commercial logging, tree plantations, and urban sprawl. Natural deforestation can be caused by tsunamis, forest fires, volcanic eruptions, glaciation, and desertification. Deforestation results in a degraded environment with reduced biodiversity and reduced ecological services. Deforestation threatens the extinction of species with specialized niches, reduces the available habits for migratory species of birds and butterflies, decreases soil fertility brought about by erosion, and allows runoff into aquatic ecosystems. It also causes changes in local climate patterns and increases the amount of carbon dioxide released into the air from burning and tree decay. In addition to the direct effects brought about by deforestation, indirect effects caused by edge effects and habitat fragmentation can also occur.

Methods that are currently employed to manage and harvest trees include:

- **Even-age management**—Essentially the practice of tree plantations.
- **Uneven-age management**—Maintain a stand with trees of all ages from seedling to mature.
- **Selective cutting**—Specific trees in an area are chosen and cut.
- **High grading**—Cutting and removing only the largest and best trees.
- **Shelterwood cutting**—Removes all mature trees in an area within a limited time.
- **Seed tree cutting**—Majority of trees are removed except for scattered, seed-producing trees used to regenerate a new stand.
- **Clear-cutting**—All of the trees in an area are cut at the same time. This technique is sometimes used to cultivate shade-intolerant tree species.
- **Strip cutting**—Clear-cutting a strip of trees that follows the land contour. The corridor is allowed to regenerate.

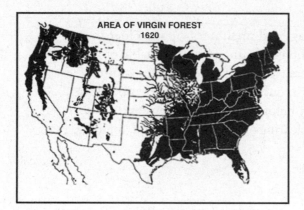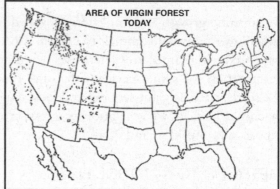

Figure 7.5 Deforestation in the United States (1620–present)

Deforestation alters the hydrologic cycle, potentially increasing or decreasing the amount of water in the soil and groundwater. This then affects the recharge of aquifers and the moisture in the atmosphere. Shrinking forest cover lessens the landscape's capacity to intercept, retain, and transport precipitation. Instead of trapping precipitation, which then percolates to groundwater systems, deforested areas become sources of surface water runoff, which moves much faster than subsurface flows. The faster transport of surface water can translate into flash flooding and more extreme floods than would occur with the forest cover.

Tree Plantations

Pros	Cons
Practical method for trees that require full or moderate sunlight in order to grow.	Reduces recreational value of land.
Efficient and economical method.	If done on steeply sloped areas, will often cause soil erosion, water pollution, and flooding.
Genetically improved species of trees that resist disease and grow faster can be grown.	Causes habitat fragmentation.
Increases economic returns on investments.	Reduces biodiversity.
Produces a high yield of timber at the lowest cost, and provides jobs.	Promotes monoculture and tree plantations that are prone to disease or infestation through the lack of diversity.

Deforestation also contributes to decreased evapotranspiration. This lessens atmospheric moisture and precipitation levels. It also affects precipitation levels downwind from the deforested area as water is not recycled to downwind forests but, instead, is lost in runoff and returns directly to the oceans. Forests are also important carbon sinks. Forests can extract carbon dioxide and pollutants from the air, thus contributing to biosphere stability and reducing the greenhouse effect. Forests are also valued for their aesthetic beauty and as a cultural resource and tourist attraction.

Three schools of thought exist with regard to the causes of deforestation—the impoverished school, the neoclassical school, and the political-ecology school. The impoverished school believes that the major cause of deforestation is the growing number of poor. The neoclassical school believes that the major cause is "open-access property rights." The political-ecology school believes that the major cause of deforestation is due to entrepreneurs.

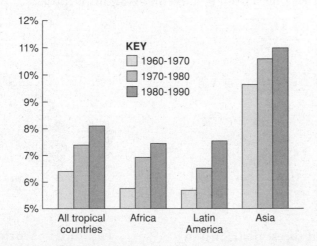

Figure 7.6 Rate of deforestation in tropical rain forests

Forest Management

Forests cover about one-third of all land surface worldwide. Most—80%—of these forests are closed canopies (tree crowns covering more than 20% of the ground), and 20% are classified as open canopy (tree crowns covering less than 20% of the ground surface). Most forests (70%) are located in North America, the Russian Federation, and South America. In the United States, the largest area of timbering is in the Pacific Northwest, employing 150,000 people and representing a $7 billion per year industry.

Forests account for about a third of the land in the United States, the largest of any land use category. Out of the 747 million acres of U.S. forest land, two-thirds (500 million acres) are nonfederal.

The Forest Service was established in 1905 as an agency of the U.S. Department of Agriculture and consists of 155 national forests and 22 grasslands. The Forest Service manages public lands in national forests and grasslands and encompasses 193 million acres (approximately the size of Texas). These resources are used for logging, farming, recreation, hunting, fishing, oil and gas extraction, watersheds, mining, livestock grazing, farming, and conservation purposes.

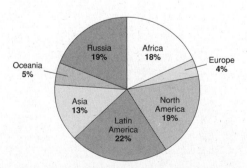

Figure 7.7 World forest distribution

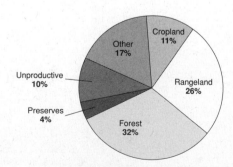

Figure 7.8 World land use

The Forest Service protects and manages natural resources on National Forest System lands. It sponsors research on all aspects of forestry, rangeland management, and forest resource utilization. It provides community assistance and cooperation with state and local governments, forest industries and private landowners to help protect and manage nonfederal forests, associated ranges, and watersheds to improve conditions in rural areas. The Forest Service also provides international assistance in formulating policy and coordinating U.S. support for the protection and sound management of the world's forest resources.

RELEVANT LAWS

Wilderness Act (1964): Created the legal definition of wilderness in the United States. Currently four agencies (National Park Service, U.S. Forest Service, U.S. Fish and Wildlife Service, and the Bureau of Land Management) are in charge of more than 106 million acres (429,000 km^2) of federal wilderness.

Wild and Scenic Rivers Act (1968): Preserves and protects certain rivers with outstanding natural, cultural, and recreational values in a free-flowing condition for the enjoyment of present and future generations. Classified rivers as wild, scenic, or recreational.

RANGELANDS

Rangelands are being compromised by overgrazing and desertification. The federal government is trying to manage and sustain the rangelanads.

Overgrazing

The consequences of overgrazing and its effect on sustainability, which was the theme of the essay "The Tragedy of the Commons," occurs when plants are exposed to grazing for too long without sufficient recovery periods. When a plant is grazed severely, it uses energy stored in its roots to support regrowth. As this energy is used, the roots die back. The degree to which the roots die back depends on the severity of the grazing. Root dieback does add organic matter to the soil, which increases soil porosity, the infiltration rate of water, and the soil's moisture-holding capacity. If sufficient time has passed, enough leaves will regrow and the roots will regrow as well. A plant is considered overgrazed when it is regrazed before the roots recover. Overgrazing can reduce root growth by up to 90%.

Consequences to overgrazing include pastures becoming less productive, soils having less organic matter and becoming less fertile, and a decrease in soil porosity. The infiltration rate and moisture-holding capacity of the soil drops and susceptibility to soil compaction increases. Additionally, desirable plants become stressed, while weedier species thrive in these harsher conditions. Overgrazing causes biodiversity to decrease by reducing native vegetation, which leads to erosion. Riparian areas are affected by cattle destroying banks and streambeds, thereby increasing silting. Eutrophication increases due to cattle wastes. Therefore, aquatic environments are negatively impacted. Predator-prey relationships and the balance achieved through predator control programs are affected. Overgrazing increases the incidence of disease in native plant species. Finally, land is affected to the point where sustainability is threatened.

Desertification

Desertification is the conversion of marginal rangeland or cropland to a more desert-like land type. It is often caused by overgrazing, soil erosion, prolonged drought, or climate changes as well as the overuse of available resources such as nutrients and water. Desertification proceeds with the following steps: First, overgrazing results in animals eating all available plant life. Next, rain washes away the trampled soil, since nothing holds water anymore. Wells, springs, and other sources of water dry up. What vegetation is left dies from drought or is taken for firewood. Then weeds that are unsuitable for grazing may begin to take over. The ground becomes unsuitable for seed germination. Finally, wind and dry heat blow away the topsoil.

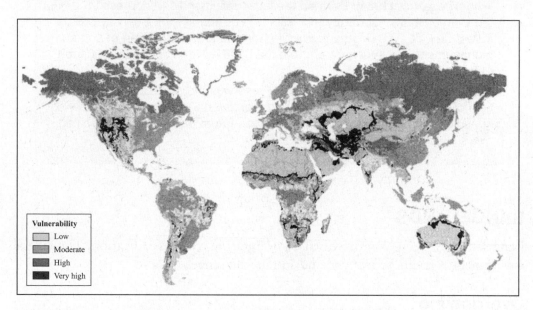

Figure 7.9 Desertification vulnerability

Federal Rangeland Management

Rangelands comprise about 40% of the landmass of the United States and are the dominant type of land in arid and semiarid regions. Nearly 80% of the lands of the western United States are classified as rangelands, whereas only 7% of some areas near the East Coast are classified as rangelands. Rangelands provide valuable grazing lands for livestock and wildlife. Rangelands serve as a source of high-quality water, clean air, and open spaces. They benefit people as a setting for recreation and as an economic means for agriculture, mining, and living communities. Rangelands serve multiple purposes:

- A habitat for a wide array of game and nongame animal species
- A habitat for a diverse and wide array of native plant species
- A source of high-quality water, clean air, and open spaces
- A setting for recreational hiking, camping, fishing, hunting, and nature experiences
- The foundation for low-input, fully renewable food production systems for the cattle industry

Jurisdiction of public grazing rangelands is coordinated through the Forest Service and the Bureau of Land Management (BLM). Before 1995, grazing policies were determined by rancher advisory boards composed of permit holders. After 1995, resource advisory councils were formed made up of diverse groups representing different viewpoints and interests. Forty percent of all federal grazing permits are owned by 3% (or approximately 2000) of all livestock operators. Federal grazing permits average about 5 cents per day per animal through federal subsidies. The true cost of doing business would make this fee closer to $10 to $20 per animal per day.

Methods of rangeland management include:

1. Controlling the number and distribution of livestock so that the carrying capacity is not exceeded

2. Restoring degraded rangeland

3. Moving livestock from one area to another to allow the rangeland to recover

4. Fencing off riparian (stream) areas to reduce damage to these sensitive areas

5. Suppressing the growth of invasive plant species

6. Replanting barren rangeland with native grass seed to reduce soil erosion

7. Providing supplemental feed at selected sites

8. Locating water holes, water tanks and salt blocks at strategic points that do not degrade the environment

Land administered by the BLM is inhabited by 219 endangered species of wildlife. Livestock grazing is the fifth-rated threat to endangered plant species, the fourth leading threat for all endangered wildlife, and the number one threat to all endangered species in arid regions of the United States.

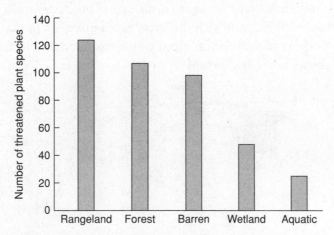

Figure 7.10 Endangered plant species as a function of biome

URBAN LAND DEVELOPMENT

The following sections discuss another form of land use—urban land development.

Planned Development

There are more than 76 million residential buildings and approximately 5 million commercial buildings in the United States alone. Together, these buildings use one-third of all the energy and two-thirds of all the electricity consumed in the United States. Energy needs of buildings account for almost half of the sulfur dioxide emissions, one-fourth of the nitrous oxide emissions, and one-third of the carbon dioxide emissions.

Green building and city characteristics focus on whole-system approaches. They include:

- Energy conservation through government and private industry rebates and tax incentives for solar and other less-polluting forms of energy
- Resource-efficient building techniques and materials
- Indoor air quality
- Water conservation through the use of xeriscaping
- Designs that minimize waste while utilizing recycled materials
- Placing buildings whenever possible near public transportation hubs that use a multitude of venues such as light rail, subways, and park and rides
- Creating environments that are pedestrian friendly by incorporating parks, green-belts, and shopping areas in accessible areas
- Preserving historical and cultural aspects of the community while at the same time blending into the natural feeling and aesthetics of a community

Suburban Sprawl and Urbanization

Urbanization refers to the movement of people from rural areas to cities and the changes that accompany it. Areas that are experiencing the greatest growth in urbanization are countries in Asia and Africa. Asia alone has close to half of the world's urban inhabitants even though 60% of its population still lives in rural areas. Africa, which is generally considered overwhelmingly rural, now has a larger urban population than North America. Reasons for this include access to jobs, higher standards of living, easier access to health care, mechanization of agriculture, and access to education. Nations with the most rapid increases in their urbanization rates are generally those with the most rapid economic growth. From 1950 to 1990, the world's economy increased fivefold.

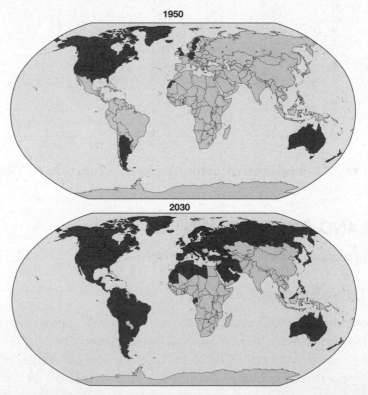

Figure 7.11 Projected worldwide urban growth from 1950–2030

Urbanization

Pros	Cons
Uses less land—less impact on the environment.	Impact on land is more concentrated and more pronounced. Examples include water runoff and flooding.
Better educational delivery system.	Overcrowded schools.
Mass transit systems decrease reliance on fossil fuels—commuting distances are shorter.	Commuting times are longer because the infrastructure cannot keep up with growth.
Better sanitation systems.	Sanitation systems have greater volumes of wastes to deal with.
Recycling systems are more efficient.	Solid-waste buildup is more pronounced. Landfill space becomes scarce and costly.
Large numbers of people generate high tax revenues.	Large numbers of poor people place strains on social services. This results in wealthier people moving away from urban areas into suburbs and decreasing the tax base.
Urban areas attract industry due to availability of raw materials, distribution networks, customers, and labor pool.	Higher population densities increase crime rates. Population increase may be higher than job growth.
Much of the pollution comes from point sources, enabling focused remediation techniques.	Since population densities are high, pollution levels are also high (urban heat islands, ozone levels, and water and soil pollution).

TRANSPORTATION INFRASTRUCTURE

Transportation can be via roadways or water channels. Areas without transportation infrastructure suffer an ecosystem impact.

Federal Highway System

The federal highway system contains approximately 160,000 miles (256,000 km) of roadway important to the nation's economy, defense, and mobility. Although interstate highways usually receive substantial federal funding and comply with federal standards, they are owned, built, and operated by the states in which they are located. Current federal highway taxes include an 18-cents per gallon tax on gasoline, a 25-cent per gallon tax on diesel, and a tax on heavy vehicles.

The system serves all major U.S. cities. Unlike counterparts in most industrialized countries, interstates go through downtown areas and facilitate urban sprawl. The distribution of virtually all goods and services involve interstate highways at some point. Residents of American cities commonly use urban interstates to travel to their employment.

An efficient and well-maintained federal highway system can have the following impacts on the environment:

LESS POLLUTANTS

Vehicles caught in stop-and-go traffic emit far more pollutants, particularly carbon monoxide, nitrogen oxides, and smog-forming volatile organic compounds, than they do without frequent braking and acceleration.

REDUCE GREENHOUSE GASES

Improving traffic flow and reducing congestion will decrease atmospheric carbon dioxide.

IMPROVE FUEL ECONOMY AND REDUCE FOREIGN OIL DEPENDENCE

For vehicles stuck in traffic, not only do tailpipe emissions go up but fuel economy goes down. Modest improvements to the nation's worst traffic bottlenecks would save 1 billion gallons of fuel each year.

IMPROVE THE ECONOMY

Interstates return $6 in economic productivity for every $1 invested.

IMPROVE THE QUALITY OF LIFE

An interstate highway system allows products (both perishable and nonperishable) to be distributed throughout the country in a short period of time.

Canals and Channels

The term channel is another word for strait, which is defined as a relatively narrow body of water that connects two larger bodies of water. Channels can occur naturally or be constructed. Repeated dredging of canals and channels is often necessary because of silting.

In the United States, channels frequented by ships are generally maintained by the United States Department of the Interior and monitored and policed by the United States Coast Guard. Smaller channels are maintained by the various states or local governments.

The two largest canals in the world of major economic value are the Panama Canal and the Suez Canal. The 48-mile (77 km) Panama Canal connects the Pacific Ocean with the Atlantic. It allows water transport without having to circumnavigate South America. The 163-mile (190 km) Suez Canal connects the Red Sea with the Mediterranean. It allows water transport between Europe and Asia without traveling around Africa, and 8% of the world's shipping runs through the Suez Canal.

> ### CASE STUDY
>
> Gatún Lake is an artificial lake created for the operation of the locks in the Panama Canal. A significant environmental problem that occurs with Gatún Lake is the drainage of water (up to 52 million gallons [197 million L] of freshwater leaves the lake every time a ship passes through the canal). Although there is sufficient annual rainfall to replenish the water used by the canal, the seasonal nature of the rainfall requires storage of water from one season to the next. The rainforest surrounding the lake traditionally played a major role by absorbing rainwater and releasing it slowly. However, due to recent deforestation of the land around the lake, rain flows rapidly down through the deforested slopes, carrying silt into the lake. The excess water is spilled out into the ocean. This results in a shortfall during the dry season and silt buildup in the lake, which must be periodically dredged.

Roadless Areas and Ecosystem Impacts

Roadless areas are places where no roads have been built and where, as a result, no logging or other development can occur. Roadless areas are havens for fish and wildlife whose habitats in many other forest areas has been fragmented or entirely destroyed. They provide habitats for more than 1,600 threatened, endangered, or sensitive plant and animal species and include watersheds that supply clean drinking water. The roadless rule protects 60 million acres, or 31%, of National Forest System lands—about 2% of the total land base of the United States.

PUBLIC AND FEDERAL LANDS

The federal government manages public lands. It sets aside areas as national parks, wildlife refuges, and wetlands.

Management

The Bureau of Land Management (BLM) is responsible for managing 262 million acres (105 million ha) of land, about one-eighth of the land in the United States. The BLM also manages about 300 million additional acres (120 million ha) of subsurface mineral resources. The bureau is also responsible for wildfire management and preservation on 400 million acres (162 million ha). Most of the lands the BLM manages are located in the western United States, including Alaska. They are dominated by extensive grasslands, forests, high mountains, arctic tundra, and deserts. The BLM manages a wide variety of resources and uses including energy and minerals, timber, forage, wild horse and burro populations, fish and wildlife habitats, wilderness areas, and archaeological, paleontological, and historical sites.

National Parks

There are over 1,100 national parks in the world today. However, many of them do not receive proper protection from poachers, loggers, miners, or farmers due to the costs involved.

The United States National Park System encompasses approximately 84 million acres (34 million ha), of which more than 4 million acres (1.6 million ha) remain in private ownership. The largest area is in Alaska and is more than 16% of the entire system. U.S. National parks are threatened by high demand by large numbers of visitors, which leads to congestion, eroded trails, noise that disrupts wildlife, and pollution from autos and visitors. Other threats include off-road vehicles, introduction of nonnative species that impact biodiversity, and commercial activities such as mining, logging, livestock grazing, and urban development.

There are several solutions to these problems:

* Reducing the amount of private land within national parks through incentives to current owners
* Providing education programs to the public
* Setting quotas on attendance through advanced reservation systems
* Adopting a fee system that covers external costs
* Banning off-road vehicles
* Banning autos and instead provide shuttle buses to control traffic
* Providing tax incentives for property owners near national parks to use land grants
* Conducting periodic and detailed wildlife and plant inventories

---

RELEVANT LAWS

Wilderness Act (1964): Wilderness was defined by its lack of noticeable human modification or presence. Federal officials are required to manage wilderness areas in a manner conducive to retention of their wilderness character.

Wild and Scenic Rivers Act (1968): Established a system of areas distinct from the traditional park concept to ensure the protection of each river's unique environment. It also preserves certain selected rivers that possess outstanding scenic, recreational, geological, cultural, or historic values and maintains their free-flowing condition.

Food Security Act (1985): Also known as "Swampbuster," this act contains provisions designed to discourage the conversion of wetlands into non-wetland areas. The act also created a system for farmers to regain lost federal benefits if they restored converted wetlands.

Wildlife Refuges

President Theodore Roosevelt designated 4-acre (1.6 ha) Pelican Island, off Florida, in 1903 as the first national wildlife refuge, designed to protect breeding birds. Roosevelt designated another 52 wildlife refuges before he left office in 1909. The early refuges were established primarily to protect wildlife such as the overhunted bison and birds killed by market hunters, such as egrets and waterfowl. During the drought years of the Great Depression, refuges were created to protect waterfowl. The system developed piecemeal largely in response to such wildlife crises. The National Wildlife Refuge System, consisting today of 547 refuges encompassing more than 93 million acres (37 million ha), is managed by the U.S. Fish and Wildlife Service.

Wetlands

Wetlands are areas that are covered by water and support plants that can grow in water-saturated soil. High plant productivity supports a rich diversity of animal life. Countries with the most wetlands are Canada (14% of land area), the Russian Federation (including Siberia), and Brazil. Wetlands were once about 10% of the land area in the United States but have been reduced to about 5%, with most of them in Louisiana and Florida. Most wetland habitat loss—90%—is due to conversion of the land to agriculture with the rest of the loss due to urbanization. Wetlands are home to a wide variety of species. One-third of all endangered species in the United States spend part of their life span in wetlands. Wetlands serve as natural water purification systems removing sediments, nutrients, and toxins from flowing water. Wetlands along lakes and oceans stabilize shorelines and reduce damage caused by storm surges, reduce the risks of flooding, and reduce saltwater intrusion.

Fens are wetlands characterized by continuous sources of groundwater rich in magnesium and calcium, which makes a fen very alkaline. This groundwater comes from glaciers that have melted, depositing their water in layers of gravel and sand. The water sits upon layers of soil (glacial drift) that are not permeable, thus keeping the water from sinking beneath the surface. The water is then forced to flow sideways along the surface, where it picks up minerals in its path.

A bog is a type of wetland that accumulates acidic peat, a deposit of dead plant material that can be dried and burned for fuel. Bogs are located in cold, temperate climates. They are usually in boreal biomes such as western Siberia, parts of Russia, Ireland, Canada, and the states of Minnesota and Michigan. Bogs are generally low in nutrients and highly acidic. Carnivorous plants have adapted to these conditions by using insects as their nutrient source.

Wilderness Areas

Wilderness areas are wild or primitive portions of national forests, parks, and wildlife refuges where timbering, most commercial activity, motor vehicles, and human-made structures are prohibited. The Wilderness Act (1964) created the National Wilderness Preservation System, the system that collectively unites all individual wilderness areas. This system encompasses a wide variety of ecosystems throughout the country including swamps in the Southeast, tundra in Alaska, snowcapped peaks in the Rocky Mountains, hardwoods forests in the Northeast, and deserts in the Southwest.

LAND CONSERVATION OPTIONS

Preservation, remediation, mitigation, restoration and sustainable land use strategies are land conservation options. Several principles can be employed in land conservation using a land-use ethic model:

1. Protect biodiversity, wildlife habitats, and the ecological functioning of public land ecosystems through careful monitoring and enforcement.

2. Adopt a user pay approach for extracting resources from public lands. Eliminate government subsidies and tax breaks to corporations that extract publicly owned resources.

3. Institute fair compensation for resources extracted from public land. Instead of the government subsidizing the extraction of resources, the corporations should be paying the government fair market value for natural resources.

4. Require responsibility for any user who damages or alters public land.

5. Adopt uneven-aged management forestry practices that foster maintaining a variety of tree species at various ages and sizes. Uneven-aged management fosters biological diversity, long-term sustainable production of high-quality timber, selective cutting, and the principle of multiple use of the forests for recreation, watershed protection, wildlife, and timber.

6. Include ecological services of trees in estimating valuation.

7. Reduce road building into uncut forest areas. Require restoration plans for those roads that are currently in place, and require such plans for any future roads.

8. Coordinate with the Forest Service on leaving fallen timber and standing dead trees in place to promote nutrient cycling and providing wildlife habitats.

9. Grow timber on longer rotations.

10. Reduce or eliminate clear-cutting, shelter wood cutting, or seed tree cutting on sloped land.

11. Rely on more sustainable tree-cutting methods such as selective and strip cutting.

12. Reduce fragmentation of remaining large forests.

13. Require certification of lumber that is cut according to sustainable forest practices.

14. Use sustainable techniques for tropical forests. These include: educating settlers about sustainable forest practices and their advantages, monitoring and enforcing cutting based on sound ecological principles, and reducing subsidies that encourage tropical deforestation. Other techniques include instituting debt-for-nature and conservation easements, creating subsidies for sustainable practices, and rehabilitating areas that have already been degraded.

15. Solutions to urban land use problems include zoning. Various parcels of land are designated for certain uses but can be influenced by developers. Local governments can limit permits, require environmental impact analyses, require developer fees for services, tax farmland on the basis of its actual (not potential) use, promote compact development and preservation of open space, and establish green spaces, urban boundaries, and land trusts.

The following items describe some land conservation options:

PRESERVATION OR SUSTAINABLE

To keep or maintain intact. Example: Land trusts are private, nonprofit organizations that actively work to conserve land by undertaking or assisting in land or conservation acquisition, or by stewardship of such land.

REMEDIATION

The act or process of correcting a fault or deficiency. Example: Cleanup from the 1989 *Exxon Valdez* Alaskan oil spill or from the 2010 BP Deepwater Horizon oil spill affecting the Gulf of Mexico and the gulf coast of the United States.

MITIGATION

To moderate or alleviate in force or intensity. Example: Numerous vehicle-deer collisions in Georgia led officials to implement a system of wildlife warning reflectors along local roads. The county installed several thousand reflectors designed to make deer freeze in place before approaching the roadway. The reflectors reduced wildlife mortality and saved motorists thousands of dollars in collision costs.

RESTORATION

To restore to its former good condition. Ecosystem restoration involves management actions designed to facilitate the recovery or reestablishment of native ecosystems. A central premise of ecological restoration is that restoration of natural systems to conditions consistent with their evolutionary environments will prevent their further degradation while simultaneously conserving their native plants and animals. Example: Removing a dam, which will allow the return of native fish to a river or the removal of harmful, invasive exotic species from a riparian environment.

MINING

The following table provides an overview of mining.

Overview of Mining

Steps	Descriptions	Environmental Effects and Issues
Mining	Removing mineral resource from the ground. Can involve underground mines, drilling, room-and-pillar mining, long-wall mining, open pit, dredging, contour strip mining, and mountaintop removal.	Mine wastes—acids and toxins. Displacement of native species. Reclamation of land and recycling.
Processing	Removing ore from gangue. Involves transportation, processing, purification, smelting, and manufacturing.	Pollution (air, water, soil, and noise). Human health concerns, risks, and hazards.
Use	Involves distribution to end user.	

Extraction

Before mining begins, economic decisions are made to determine whether a site will be profitable. Factors that enter into the decision include current and projected price, amount of ore at the site, concentration, type of mining required, cost of transporting the ore to a processing facility, and cost of reclamation. After all factors are analyzed, several steps are employed.

SITE DEVELOPMENT

Samples are taken from an area to determine the quality and quantity of minerals in a location. Roads and equipment are brought in.

EXTRACTION

Three main methods of extraction exist: surface mining, underground mining, and *in situ* leaching. Surface mining is a type of mining in which the soil and rock overlying the mineral deposit (known as overburden) is removed and stored (spoilbank). Surface mining is used where deposits of commercially useful minerals are found near the surface, and where the overburden is relatively thin or the area is unsuitable for tunneling. Surface mines are typically enlarged until either the mineral deposit is exhausted, or the cost of removing larger volumes of overburden makes further mining impractical.

There are five main forms of surface mining: (1) Strip mining, the most commonly used method to mine coal or tar sand, is the practice of mining a seam of mineral by first removing a long strip of overburden. Area stripping is used on fairly flat terrain to extract deposits over a large area. As each long strip is excavated, the overburden is placed in the excavation produced by the previous strip. Contour stripping involves removing the overburden above the mineral seam near the outcrop in hilly terrain, where the mineral outcrop usually follows the contour of the land. This method commonly leaves behind terraces in mountainsides. (2) Open-pit mining refers to a method of extracting rock or minerals from the earth by their removal from an open pit. (3) Mountaintop removal mining is used where a coal seam outcrops all the way around a mountaintop. All the rock and soil above the coal seam are removed, and the soil is placed in adjacent lows such as hollows or ravines. Mountaintop removal replaces previously steep topography with a relatively level surface. (4) Dredging is a method often used to bring up underwater mineral deposits. Although dredging is usually employed to clear or enlarge waterways for boats, it can also recover significant amounts of underwater minerals relatively efficiently and cheaply. (5) Highwall mining uses a continuous mining machine driven under remote control into the seam exposed by previous open-cut operations. A continuous haulage system carries the coal from the mine to an open-air installation for stockpiling and transport.

In underground mining, large shafts are dug into the earth. There is less surface destruction and waste rock produced than in surface mining, but it is unsafe. Subsurface mining often occurs below the water table, so water must be constantly pumped out of the mine in order to prevent flooding. When a mine is abandoned, the pumping ceases and water floods the mine. This introduction of water often results in acid rock drainage, which is caused by certain bacteria accelerating the decomposition of metal sulfide ions that have been exposed to air and water.

Finally, with *in situ* leaching, small holes are drilled into the site. Water-based chemical solvents are used by miners to extract the resources. Advantages of *in situ* mining include: it is a less-expensive method since rocks do not have to be broken up or removed; there are shorter lead times to production; and it requires less sur-face ground disturbance and less mediation. On the negative side, fluids injected into the earth are toxic and enter the groundwater supply.

PROCESSING

Processing involves intensive chemical processing during smelting. This is the method by which a metal is obtained from its ore, either as an element or as a simple compound. It is usually accomplished by heating beyond the melting point, ordinarily in the presence of reducing agents such as coke or oxidizing agents such as air. A metal whose ore is an oxygen compound (iron, zinc, or lead oxides) is heated (reduction smelting) in a blast furnace to a high temperature. The oxide combines with the carbon in the coke, escaping as carbon monoxide or carbon dioxide. Other impurities are removed by adding flux, with which they combine to form slag. If the ore is a sulfide mineral (copper, nickel, lead, or cobalt sulfides) air or oxygen is introduced to oxidize the sulfide to sulfur dioxide and any iron to slag, leaving the metal.

In cyanide heap leaching, gold ore is heaped into a large pile. Cyanide solution is then sprayed on top of the pile. As the cyanide percolates downward, the gold leaches out of the ore and collects in pools at the bottom. The gold extracted may be only 0.01% of the total ore processed. Liquid wastes containing cyanide and other toxins are kept in tailing ponds, which eventually leak and enter groundwater supplies.

Tailings are the materials left over after the process of separating the valuable frac-tion from the ore. Tailings represent an external cost of mining. In coal and oil sands mining, the word "tailings" refers specifically to fine waste suspended in water.

TIP

coke—a solid fuel made by heating coal in the absence of air so that the volatile components are driven off.

flux—a mineral added to the metals in a furnace to promote fusing or to prevent the formation of oxides.

slag—waste matter separated from metals during the smelting or refining of ore.

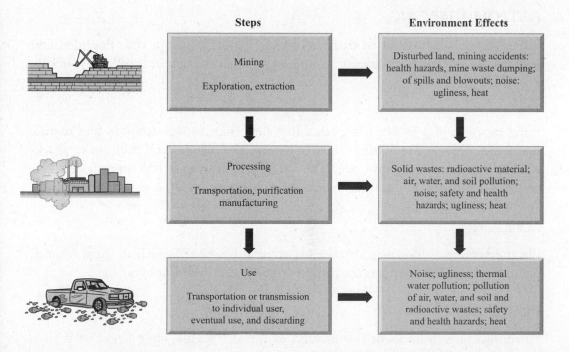

Figure 7.12 Steps required for manufacturing products and their environmental consequences

Global Reserves

Two billion tons (1.8 billion m.t.) of minerals are extracted and used each year in the United States (about 10 tons [9.1 m.t.] for every American). At the same time, the United States imports more than 50% of its most needed minerals. As mineral reserves become depleted, lower grades of ore are mined, which causes more processing and consequently more pollution.

The United States, Germany, and Russia represent 8% of the world's population, yet they consume 75% of the most widely used metals, with the United States consuming 20%.

RELEVANT LAWS

General Mining Law (1872): Grants free access to individuals and corporations to prospect for minerals in public domain lands and allows them, upon making a discovery, to stake a claim on that deposit.

Mineral Leasing Act (1920): Authorizes and governs leasing of public lands for developing deposits of coal, petroleum, natural gas and other hydrocarbons, phosphates, and sodium in the United States. Prior to the act, these materials were subject to mining claims under the General Mining Act of 1872.

Surface Mining Control and Reclamation Act (1977): Established a program for regulating surface coal mining and reclamation activities.

FISHING

Different techniques are used to fish the planet's waters:

BOTTOM TRAWLING

Uses a funnel-shaped net to drag the ocean bottom. Shrimp, cod, flounder, and scallops. Analogous to clear-cutting forests. Species not wanted is called bycatch.

DRIFT NET

Long expanses of nets that hang down in water. Traps turtles, seabirds, and marine mammals. During the 1980s, 10,000 dolphins and whales and millions of sharks were killed each year by drift nets. The 1992 U.N. voluntary ban on drift nets longer than 1.5 miles (2.4 km) has made some progress.

LONGLINE

Placing very long lines with thousands of baited hooks. Swordfish, tuna, sharks, halibut, and cod. Endangers sea turtles, pilot whales, and dolphins.

PURSE SEINE

Surrounds large schools of fish spotted by aircraft with a large net. Net is then drawn tight. Tuna, mackerel, anchovies, and herring.

Overfishing

The oceans have been looked on as unlimited resources. Ocean productivity is generally low and results from spatial separation of required plant nutrients. Light is restricted to surface waters.

The oceans supply 1% of all human food and represent 10% of the world's protein source. China is responsible for about one-third of all fish harvesting from the oceans. About one-third of the total catch of fish is used for purposes other than human consumption, such as fish oil, fish meal, and animal feed. Another one-third of global catches consists of bycatch. These are marine mammals, sea turtles, birds, noncommercial fish, and shellfish that are ensnared in fishing nets or dredged up by trawling and discarded.

Maximum sustained yield is the largest amount of marine organisms that can be continually harvested without causing the population to crash. This yield generally occurs when a population is maintained at half the carrying capacity.

Methods to manage fisheries in a sustainable manner include:

1. Regulate locations and numbers of fish farms and monitor their pollution output

2. Encourage the production of herbivorous fish species

3. Require and enforce labeling of fish products that were raised or caught according to sustainable methods

4. Set catch limits far below maximum sustainable yields

5. Eliminate government subsidies for commercial fishing

6. Prevent importation of fish from foreign countries that do not adhere to sustainable-harvesting methods

7. Place trading sanctions on foreign countries that do not respect the marine habitat, including countries that hunt whales

8. Assess fees for harvesting fish and shellfish from public waters

9. Increase the number of marine sanctuaries and no-fishing areas

10. Increase penalties for fishing techniques that do not allow escape of bycatch, including unwanted fish species, marine mammals, sea birds, and sea turtles

11. Ban the throwing back of bycatch

12. Monitor and destroy invasive species transported through ship ballast

Several methods can restore habitats suitable for freshwater fish. These include: planting native vegetation on stream banks, rehabilitating in-stream habitats, controlling erosion, and controlling invasive species. Other restsorative methods include restoring fish passages around human-made impediments, monitoring, regulating, and enforcing recreational and commercial fishing; and protecting coastal estuaries and wetlands.

Aquaculture

Aquaculture, commonly known as mariculture or fish farming, includes the commercial growing of aquatic organisms for food. It involves stocking, feeding, protection from predators, and harvesting. Aquaculture is growing about 6% annually and provides 5% of the total food production worldwide, most of it coming from less-developed countries. Currently, the most popular products being produced through aquaculture include seaweeds, mussels, oysters, shrimp, and certain species of fish (primarily salmon, trout, and catfish). Kelp makes up about 17% of aquaculture output and is used as a food product and as a source of various products used in the food industry. Aquaculture is used to raise 80% of all mollusks, 40% of all shrimp, and 75% of all kelp.

Aquaculture offers several advantages over raising livestock in that cold-blooded organisms convert more feed to usable protein. For example, for every 1 million calories of feed required, a trout raised on a farm produces about 35 grams of protein whereas a chicken produces 15 grams of protein and cattle produce 2 grams of protein. For every hectare of ocean, intense oyster farming can produce 58,000 kg of protein while natural harvesting of oysters produces 10 kg of protein.

For aquaculture to be profitable, the species must be marketable, inexpensive to raise, trophically efficient, at marketable size within 1 to 2 years, and disease resistant. Aquaculture creates dense monocultures that reduce biodiversity within habitats and requires large levels of nutrients in the water.

Aquaculture offers possibilities for sustainable protein-rich food production and for economic development to local communities. However, aquaculture on an industrial scale may pose several threats to marine and coastal biological diversity. It creates wide-scale destruction and degradation of natural habitats and leaves nutrients and antibiotics in aquaculture wastes. Accidental releases of alien or modified organisms into native waters, transmission of diseases to wild stocks, and displacement of local and indigenous human communities are also side effects.

CASE STUDY

Polychlorinated biphenyls (PCBs) were banned in the U.S. in the late 1970s and are slated for global phase-out under the United Nations treaty on persistent organic pollutants. PCBs are highly persistent and have been linked to cancer and impaired fetal brain development. Salmon farming has made salmon the third most popular fish in America and comprises 22 percent of all retail seafood counter sales. Many consumers eat more salmon today to avoid over-consumption of beef and poultry, and to benefit from anti-cancer and anti-heart disease properties of oily fish. However, analysis of U.S. government data found that farmed salmon are likely the most PCB-contaminated protein source in the current U.S. food supply. Approximately 800,000 U.S. adults have an increased cancer risk by eating PCB-contaminated salmon. Farmed salmon are fattened with ground fishmeal and fish oils that are high in PCBs. As a result, salmon farming operations that produce inexpensive fish unnaturally concentrate PCBs. Furthermore, farmed salmon contains 52 percent more fat than wild salmon, according to the U.S. Department of Agriculture.

> ### RELEVANT LAWS
>
> **Anadromous Fish Conservation Act (1965):** Authorizes the Secretary of the Interior to enter into agreements with states and other non-federal interests to conserve, develop, and enhance the anadromous fish (fish that migrate from the sea to fresh water to spawn) resources of the United States.
>
> **Magnuson Fishery Conservation and Management Act (1976):** Governs marine fisheries management in United States federal waters. Aids in the development of the domestic fishing industry by phasing out foreign fishing. To manage the fisheries and promote conservation, the Act created eight regional fishery management councils. The 1996 amendments focused on rebuilding overfished fisheries, protecting essential fish habitat, and reducing bycatch.
>
> **United Nations Treaty on the Law of the Sea (1982):** Defines the rights and responsibilities of nations in their use of the world's oceans, establishing guidelines for businesses, the environment, and the management of marine natural resources.

GLOBAL ECONOMICS

The economy and the environment are intrinsically linked such that both are simultaneously causes and effects and are inputs and outputs of each other. The environment contains all the resources that can be used in the economy. The use of resources for economic purposes continuously creates new environmental situations. For example, while some resources are depleted and transformed from usable to unusable states, economic resources are used to expand additional resources. This occurs through increasing the available supply of materials, opening land to agricultural production, transporting resources from locations where they are in surplus to areas of shortage, and so on.

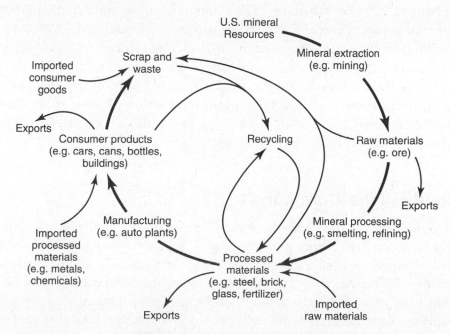

Figure 7.13 Resources cycle

Increased levels of economic activity and improvements in living standards have occurred since the end of World War II. People in the wealthiest countries constitute 15% of the global population and enjoy average levels of income that are about 20 times greater than the 85% of the global population that live in poorer countries.

If the income of the poorest 85% were raised to only one-third that of the richest countries, the level of total world production and consumption would have to double, with a similar increase in the use of Earth's resources. The conclusion is that continued increases in living standards in poorer countries will increase pressure on the carrying capacity of the planet.

Until recently, developments in local economies and local environments were dispersed and, to a great extent, isolated. They did not typically result in cumulative processes that had widespread or global impact. However, as with the economy, over the past century or so, the growth of human populations throughout the world (and the greater vigor with which these populations have assaulted the environment in their pursuit of higher production and consumption levels) has led to a significant increase in global environmental disruption. The effects of these disruptions have become increasingly interlinked. A global environment—a set of interrelated processes, causes and effects, not simply a group of unrelated ecological events—has come into being.

World Bank

The World Bank is a source of financial and technical assistance to developing countries around the world. The World Bank (owned by 184 member countries) provides low-interest loans, interest-free credit, and grants to developing countries for education, health, infrastructure, communications, and environmental issues. In 2001, the World Bank Board of Directors endorsed an environment strategy to guide the bank's actions in environmental areas. The strategy emphasizes three objectives: improving the quality of life, improving the quality of growth, and protecting the quality of the regional and global commons through the "greening" of investments in agriculture, water sanitation, and other environmental projects. The World Bank in 2005 has distributed $13.8 billion in public and private funds in the areas of biodiversity, conservation, climate change, and international waters; funded $740 million in projects to phase out ozone-depleting substances; and funded $1.6 billion into projects that reduce greenhouse gas emissions.

The World Bank is also the greatest single source of funds for large dam projects, having provided more than $50 billion for construction of more than 500 large dams in 92 countries. Since 1948, theses large dam projects have forcibly displaced approximately 10 million people from their land.

"Tragedy of the Commons"

Garrett Hardin wrote the "Tragedy of the Commons" in 1968, and it appeared in the journal *Science*. The story parallels what is happening worldwide in regard to resource depletion and pollution. The seas, air, water, animals, and minerals are all the commons. They are there for humans to use. Those who exploit them become rich. The price of depleting the resources of the commons is an external cost paid by all people on Earth. Examples of current environmental problems that parallel the issues of sustainability brought up in the "Tragedy of the Commons" include:

- Uncontrolled human population growth leading to overpopulation
- Air pollution
- Overextraction of groundwater and wasting water due to excessive irrigation
- Frontier logging of old growth forest and slash and burn
- Burning of fossil fuels and consequential global warming
- Habitat destruction and poaching
- Overfishing

Limits to the "Tragedy of the Commons" include the following:

- Economic decisions are generally short term, based on reactions in the world market. Environmental decisions are long term.
- Land that is privately owned is subject to market pressure. For example, if privately owned timberland is increasing in value at an annual rate of 3% but interest rates on loans to purchase the land are 7%, this could result in the land being sold or the timber being harvested for short-term profits.
- Some commons are easier to control than others. Land, lakes, rangeland, deserts, and forests are geographically defined and easier to control than air or the open oceans that do not belong to any one group.
- Incorporating discount rates into the valuation of resources would be an incentive for investors to bear a short-term cost for a long-term gain.
- Breaking a commons into smaller, privately owned parcels fragments the policies of governing the entire commons. Different standards and practices used on one parcel may or may not affect all other parcels.

NAMES TO KNOW

Rachel Carson: Wrote *Silent Spring*, which spurred a reversal in national pesticide policy and led to a nationwide ban on DDT and other pesticides, and which inspired a grassroots environmental movement that led to the creation of the Environmental Protection Agency.

Aldo Leopold: Best known for his book *A Sand County Almanac*. Influential in the development of modern environmental ethics and in wilderness conservation emphasizing biodiversity and ecology. Developed the science of wildlife management.

John Muir: Helped to save the Yosemite Valley, Sequoia National Park, and other wilderness areas. The Sierra Club, which he founded, is now one of the most important environmental conservation organizations in the United States.

Theodore Roosevelt: As the twenty-sixth president of the United States, he used his position to pave the way for environmentalists of the future. He is known for setting aside land for national forests, establishing wildlife refuges, developing the farmlands of the American West, and advocating protection of natural resources. Roosevelt set aside 150 million acres for forest reserves, created 50 wildlife refuges, turned much of the arid land of the southwestern United States into farmland, and initiated sixteen major reclamation projects in the southwest.

Henry David Thoreau: Author of *Walden*, which viewed unity and community as important aspects of nature, and wrote that all disturbances in these links are caused by human beings and that modern materialism would lead to the destruction of the environment needed for humans and other living things to survive. Wrote about the need for national forest preserves and about the destruction caused by dams. His work raised the environmental consciousness of many generations of readers.

QUICK REVIEW CHECKLIST

☐ **Feeding a Growing Population**
  ☐ human nutritional requirements
  ☐ types of agriculture
    ☐ agroforestry
    ☐ alley cropping
    ☐ crop rotation
    ☐ high-input agriculture
    ☐ industrial agriculture (corporate farming)
    ☐ intercropping
    ☐ interplanting
    ☐ low input agriculture
    ☐ low-till agriculture
    ☐ no-till agriculture
    ☐ conservation-till agriculture
    ☐ monoculture
    ☐ plantation
    ☐ polyculture
    ☐ polyvarietal cultivation
    ☐ subsistence agriculture
    ☐ tillage

☐ **Green Revolution**
  ☐ first green revolution
  ☐ second green revolution

☐ **Genetic Engineering and Crop Production**
  ☐ genetically engineered crops

☐ **Irrigation**

☐ **Sustainable Agriculture**
  ☐ efficient use of inputs
  ☐ selection of site, species, and variety
  ☐ soil management
  ☐ species diversity

☐ **Controlling Pests**
  ☐ types of pesticides
    ☐ biological
    ☐ carbamates
    ☐ chlorinated hydrocarbons
    ☐ fumigants
    ☐ inorganic
    ☐ organic or natural
    ☐ organophosphates
  ☐ pros and cons of using pesticides

QUICK REVIEW CHECKLIST (continued)

☐ **Integrated Pest Management**

☐ **Forestry**
  ☐ ecological services
  ☐ tree plantations (pros and cons)
  ☐ old-growth forests

☐ **Forest Fires**
  ☐ crown fires
  ☐ ground fires
  ☐ surface fires
  ☐ methods to control fires

☐ **Deforestation**
  ☐ even-age management
  ☐ uneven-age management
  ☐ selective cutting
  ☐ high grading
  ☐ shelterwood cutting
  ☐ seed tree cutting
  ☐ clear-cutting
  ☐ strip cutting

☐ **Case Study**
  ☐ Hubbard Brook Experimental Forest

☐ **Forest Management**
  ☐ Responsibilities of the U.S. Forest Service

☐ **Rangelands**
  ☐ overgrazing
  ☐ desertification
  ☐ remediation

☐ **Federal Rangeland Management**
  ☐ factors that affect resource utilization

☐ **Urban Land Development**
  ☐ planned development
  ☐ suburban sprawl and urbanization

☐ **Transportation Infrastructure**
  ☐ Federal Highway System

☐ **Canals and Channels**
  ☐ Case Study (Gatún Lake)

QUICK REVIEW CHECKLIST (continued)

☐ **Roadless Areas and Ecosystem Impacts**
  ☐ Roadless Area Conservation Rule (2001)

☐ **Public and Federal Lands**
  ☐ management
  ☐ Federal Land Policy and Management Act (1976)
  ☐ National Parks
    ☐ Wilderness Act (1964)
    ☐ Wild and Scenic Rivers Act (1968)

☐ **Wildlife Refuges**

☐ **Wetlands**

☐ **Land Conservation Options**

☐ **Mining**
  ☐ extraction
  ☐ site development
  ☐ extraction
  ☐ processing

☐ **Fishing**
  ☐ bottom trawling
  ☐ drift net
  ☐ long line
  ☐ purse seine
  ☐ overfishing

☐ **Aquaculture**
  ☐ Anadromous Fish Conservation Act (1965)
  ☐ Marine Mammal Protection Act (1972)
  ☐ Magnuson Fishery Conservation and Management Act (1976)
  ☐ United Nations Treaty on the Law of the Sea (1982)

☐ **Global Economics**
  ☐ World Bank

☐ **Tragedy of the Commons**

MULTIPLE-CHOICE QUESTIONS

1. Chronically undernourished people are those who receive approximately
   _____ calories or less per day.

   (A) 500
   (B) 1,000
   (C) 1,500
   (D) 2,000
   (E) 3,000

2. The greatest threat to the success of a species is

   (A) environmental pollution
   (B) loss of habitat
   (C) poaching
   (D) hunting
   (E) introduction of new predators into the natural habitat

3. The second law of thermodynamics would tend to support

   (A) people eating more meat than grain
   (B) people eating more grain than meat
   (C) people eating about the same amount of grain as meat
   (D) people being undernourished
   (E) people eating a balanced diet from all food groups

4. The majority of nutrients and calories in the average human diet
   come from

   (A) potatoes, corn, and rice
   (B) wheat, rice, and soybeans
   (C) wheat, corn, and rice
   (D) wheat, corn, and oats
   (E) wheat, corn, and soybeans

5. Planting trees and/or shrubs between rows of crops is known as

   (A) contour farming
   (B) planting windbreaks
   (C) strip cropping
   (D) alley cropping
   (E) tillage farming

6. The one area of the world that is NOT expected to increase food production soon is

    (A) Asia
    (B) Latin America
    (C) sub-Saharan Africa
    (D) China
    (E) India

7. Most foods derived from genetically modified crops contain
    (A) the same number of genes as food produced from conventional crops
    (B) the same number of genes as foods produced from hybrid crops
    (C) one or two additional genes
    (D) hundreds of additional genes
    (E) no genes at all

8. Which region of the developing world is currently as urbanized as the developed world?

    (A) Africa
    (B) Asia
    (C) Latin America
    (D) Europe
    (E) North America

9. Malnutrition primarily refers to

    (A) lack of carbohydrates in the diet
    (B) lack of protein in the diet
    (C) lack of fat in the diet
    (D) a deficiency of micronutrients in the diet
    (E) a deficiency of vitamins in the diet

10. Adding more fertilizer does NOT necessarily increase crop production is an example of

    (A) law of supply and demand
    (B) Leibig's law of minimum
    (C) limiting factors
    (D) second law of thermodynamics
    (E) Gaia hypothesis

11. Soil that is transported by the wind is

    (A) alluvial
    (B) feral
    (C) gangue
    (D) rill
    (E) aeolian

12. Most of Earth's land area is

    (A) urban
    (B) forest
    (C) desert
    (D) rangeland
    (E) agricultural

13. Cities experiencing the greatest urban growth are found in

    (A) Asia
    (B) Africa
    (C) Europe
    (D) North America
    (E) both choices (A) and (B)

14. Which of the following is NOT a concept designed to create a sustainable city?

    (A) Conserve natural habitats
    (B) Focus on energy and resource conservation
    (C) Design affordable and fuel-efficient automobiles
    (D) Provide ample green space
    (E) All are correct

15. Approximately what percentages of the world depends upon wood or charcoal for heating and/or cooking?

    (A) Less than 10%
    (B) Between 20% and 30%
    (C) Between 50% and 60%
    (D) More than 75%
    (E) Has not been determined

16. Of all the jobs in the U.S. Forest Service, the majority are concerned with

    (A) lumber management
    (B) fire management
    (C) mining management
    (D) recreation
    (E) administrative functions

17. In which of the following states or areas would you NOT find significant amounts of old-growth forests?

    (A) New England
    (B) Appalachia
    (C) British Columbia
    (D) Washington State
    (E) California

18. Most grain that is grown in the United States is used

    (A) for export to countries that need grain
    (B) for cereals and baked goods
    (C) to feed cattle
    (D) for trade with other countries
    (E) to create fuel and liquor

19. What is the number one source of soil erosion?

    (A) Physical degradation
    (B) Chemical degradation
    (C) Water erosion
    (D) Wind erosion
    (E) All forms contribute equally

20. A process in which small holes are drilled into Earth and water-based chemical solvents are used to flush out desired minerals is known as

    (A) chemical leaching
    (B) *ex-situ* leaching
    (C) beneficiation
    (D) heap leaching
    (E) *in situ* leaching

FREE-RESPONSE QUESTION

by: Sarah E. Utley
    College Board APES Reader; Princeton, NJ
    Environmental Content Specialist, Center for Digital Innovation
    AP Environmental Science Division, UCLA

TIP

The Free-Response section of the APES exam consists of four required questions: one data-set question, one document-based question, and two synthesis and evaluation questions. Be sure you are comfortable with and know what is required for each type of question.

Deposits of coal were recently discovered in a small valley in the eastern foothills of the Rocky Mountains. These deposits were found both near the surface and underneath the ground.

(a) Describe a method that could be used to mine the coal.

(b) Using coal as an energy source can have serious environmental consequences. Discuss current technological methods employed to reduce the impact of using coal as an energy source.

(c) The residents of this small town have mixed opinions about the impact of the mine on their community. Some residents support the mine because of the increased employment opportunities and increased tax base that supports projects such as schools and roads. Others are concerned about current or possible future environmental impacts on this region of the state. Discuss two of these possible environmental impacts.

MULTIPLE-CHOICE ANSWERS AND EXPLANATIONS

1. **(D)** Chronic malnutrition is defined as receiving fewer than 2,000 calories per day. People who live in developed nations receive an average of 3,340 calories per day.

2. **(B)** There are many reasons why certain species decline and become endangered; however, most environmentalists agree that the most significant factor is habitat loss and degradation.

3. **(B)** The second law of thermodynamics states that there is about a 10% loss in each successive higher trophic level in the energy pyramid.

4. **(C)** Wheat and rice supply approximately 60% of worldwide human calories.

5. **(D)** See the table in the section "Types of Agriculture" for a more detailed explanation about alley cropping.

6. **(C)** Droughts, pestilence, AIDS, and civil strife in this area have had serious impacts on food production in this part of the world.

7. **(C)** Combining genes from different organisms is known as recombinant DNA technology, and the resulting organism is said to be "genetically modified," "genetically engineered," or "transgenic." Genetically modified products include medicines and vaccines, foods and food ingredients, feeds, and fibers.

8. **(C)** Despite the rapid growth in urbanization in Asian and African countries, their current percentage of people living in urban areas is still half of that of Latin America. About 75% of Latin America's population currently lives in urban areas.

9. **(B)** Protein malnutrition, more commonly referred to as protein-energy malnutrition, is the most lethal form of malnutrition. Protein is necessary for key body functions, including provision of essential amino acids and development and maintenance of muscles.

10. **(C)** There is a limit to crop yield versus fertilizer application. Additional fertilizer past a certain point can actually harm plants.

11. **(E)** Aeolian soils are sand-sized particles transported by wind action.

12. **(B)** About one-third of Earth's surface is covered by forests.

13. **(E)** In 2015, the largest cities in the world are projected to be Tokyo (Japan), Bombay (India), Lagos (Nigeria), and Dhaka (Bangladesh).

14. **(C)** Designing affordable and fuel-efficient cars would increase the number of automobiles already in cities since more people could afford them. More private vehicles results in congestion, pollution, and so on.

15. **(C)** As late as the 1850s, wood supplied over 90% of the energy requirements in the United States. Half of the energy currently used in Africa is fuel wood.

16. **(D)** The number of recreational visitors in all National Forests in 1993 was 730 million. By the year 2045, the number is projected to rise by approximately 63 percent.

17. **(A)** Only 20% of the worlds' original forests remain intact. In New England, the first area to be colonized by Europeans, 99% of the frontier forests have been destroyed.

18. **(C)** There are about 1.5 billion cattle on Earth. They graze on 25% of the landmass and consume enough grain to feed hundreds of millions of people. In the United States, cattle consume 70% of all grain produced.

19. **(C)** Water erosion is about half of the problem.

20. **(E)** See the section "Extraction" for a more detailed explanation.

FREE-RESPONSE ANSWER

Ten points necessary for full-credit.
Section (a): worth 2 points. Describe either surface or subsurface mining.
Section (b): worth 4 points. Explain at least two remediation techniques.
Section (c): worth 4 points. Two points awarded for each completely described environmental impact.

(a) 1. Coal that is located close to the surface is frequently extracted by strip-mining. In surface mining, the rock, soil, and vegetation layers that are on top of the coal are scraped away and removed. The removed rock and soil are called overburden. Specifically, in strip mining, a long trench is dug, removing the overburden and the coal. Then a second trench is dug parallel to the first, and the overburden from the second is used to fill the first trench. The hill of loose overburden that fills the previous trench is called a spoil bank. The process continues until the coal is depleted or the cost of future extraction is prohibitive.

2. Coal that is located deep underneath the surface of Earth is removed through underground or subsurface mining techniques. A deep shaft must be dug or blasted into the area. An extensive network of tunnels (which are supported with wooden or metal beams) is then dug, and these extend out from the main shaft. Sometimes part of the coal vein is left intact to support the shafts and tunnels. If the mine is large enough, electricity is wired, an elevator is installed to take workers to the active part of the mine, and tracks are laid on which the loaded bins of coal travel. The coal itself can be removed with special mining machines equipped with large drills. The mined coal is then loaded into the mining bins and taken to the surface.

(b) Coal can contain contaminants such as rocks, ash, mineral dust, and sulfur. These contaminants decrease the efficiency of the coal and can increase the environmental impact of burning the fossil fuel. When coal with high sulfur content is burned, the sulfur can enter the atmosphere as sulfur dioxide—a component of acid rain. Several techniques can decrease the environmental impact of coal use. When coal is initially mined, it is usually washed. Washing involves separating the impurities from the coal based on density. The contaminants are generally denser than the coal and sink in the separation fluid. The process involves first grinding the coal into small pieces. Then the pieces are placed onto a separation table with a fluid. The coal is skimmed off the top. The contaminants, such as pyrite (a sulfur derivative), sink to the bottom. In addition to the initial washing of coal, remediation techniques can decrease the amount of pollutants that result from the burning or use of coal.

The Clean Air Act (revised in 1990) requires that all new coal-burning plants built after 1978 must be equipped with air quality structures that remove the sulfur from the combustion gases before they are released through the smokestack. These structures are called scrubbers. In a scrubber, a mixture of water and crushed limestone is sprayed into the gases. The sulfur from the emissions and the limestone chemically combine to form a solid (calcium sulfate). This solid is then disposed in a solid waste disposal facility or can be used in the production of some building products.

Notable points:

Description of mining type.

Specific description of coal removal.

or

Description of mining type.

Description of coal removal.

Though an answer may contain more than two remediation techniques, at a minimum, a full-credit answer should contain two techniques.

Introduction of techniques.

Purpose of washing.

Description of washing (performed after excavation and prior to combustion).

Purpose of scrubbers.

Description of scrubbers.

Description of fluidized bed combustion.

Purpose of fluidized bed combustion.

In addition to scrubbers, there is additional technology that can decrease the environmental impact of coal burning. In a fluidized bed combustion plant, the coal particles are suspended within the boiler on jets of air. This tumbling of the burning coal allows limestone to be introduced into the plant prior to the smokestack. The limestone combines with the sulfur and chemically bonds with it to capture a greater amount of sulfur pollutants. In addition to removing the sulfur prior to entering the smokestack, a fluidized bed boiler burns at a cooler temperature than a conventional boiler. This cooler temperature prevents the smog-forming NO_x pollutants from becoming volatile and being released into the air.

(c) (Give two implications.)

(1) Water pollution is frequently a concern with the mining of coal. Water is used throughout the mining process, from cooling machinery and smelters to washing coal. If this contaminated water is not contained and purified, it can seep into the local waterways or water table. In addition, precipitation that percolates through the area can leach soil contaminants from the mining process into the local water table or waterways. The water table of the areas located close to the mines must be carefully monitored for quality.

The question specifically asks for two environmental implications.

Name of each concern.

Cause of each impact.

Environmental result of each.

(2) Air pollution is also a concern in the mining industry. The mining machinery, which is frequently powered by diesel engines, contributes to air pollution in the mine vicinity. In addition, the burning of coal often releases air toxins such as sulfur dioxide, nitrogen oxides, and particulate matter into the atmosphere. New combustion units are now mandated by the Clean Air Act to be fitted with scrubbers and other clean air apparatuses. Many states are now requiring that even older equipment must meet new clean air requirements.

(3) There is also the concern of habitat destruction and wildlife–native plant displacement. The process of surface mining, which requires the removal of many tons of soil, disturbs the natural ecosystem and displaces native animals. Subsurface mining is often less disruptive to soil and the soil surface but has its own concerns, such as subsidence and land instability. Though mining corporations are required by law to return the land to the original topography, since the soil structure has been altered, native plant revegetation techniques often fail. When the native plants fail to thrive, reintroducing native wildlife is difficult.

(4) Mining can also be harmful to human health. With subsurface mining, workers must be protected against unstable mine shafts, generally dangerous working conditions, and health concerns such as poor air quality. Safely ventilating the mines below the ground can be difficult, and care must be taken to protect the workers from the respiratory diseases such as emphysema and lung cancer. With the large amount of mining equipment (large cranes, bulldozers, and so on) required in surface mining, human safety is always a concern.

UNIT V: ENERGY RESOURCES AND CONSUMPTION (10–15%)

Areas on Which You Will Be Tested

A. Energy Concepts—energy forms, power units, conversions, and laws of thermodynamics.

B. Energy Consumption
  1. History—Industrial Revolution, exponential growth, and energy crisis.
  2. Present global energy use.
  3. Future energy needs.

C. Fossil Fuel Resources and Use—formation of coal, oil, and natural gas, extraction/purification methods, world reserves and global demand, synfuels, and environmental advantages/disadvantages of sources.

D. Nuclear Energy—nuclear fission process, nuclear fuel, electricity production, nuclear reactor types, environmental advantages/disadvantages, safety issues, radiation and human health, radioactive wastes, and nuclear fusion.

E. Hydroelectric Power—dams, flood control, salmon, silting, and other impacts.

F. Energy Conservation—energy efficiency, CAFE standards, hybrid electric vehicles, and mass transit.

G. Renewable Energy—solar energy, solar electricity, hydrogen fuel cells, biomass, wind energy, small-scale hydroelectric, ocean waves and tidal energy, geothermal, and environmental advantages/disadvantages.

Energy

CHAPTER 8

ENERGY CONCEPTS

Energy is the ability to do work. In the SI system, the unit of energy is the joule. Power is work divided by time. The unit of power is the joule/s, which is also called a watt. The following table describes the different energy forms.

Forms of Energy	
Form	**Description**
Mechanical	There are two types of mechanical energy: potential energy (a book sitting on a table) and kinetic energy (a baseball flying through the air).
Thermal	Heat is the internal energy in substances—the vibration and movement of the atoms and molecules within substances.
Chemical	Chemical energy is stored in bonds between atoms in a molecule.
Electrical	Electrical energy results from the motion of electrons.
Nuclear	Nuclear energy is stored in the nuclei of atoms. It is released by either splitting or joining of atoms.
Electromagnetic	Electromagnetic energy travels by waves.

Power and Units

Power is the amount of work done per time. The most common unit of power is the kilowatt-hour (kWh).

Units of Energy/Power

Unit or Prefix	Description
Btu (British Thermal Unit)	Btu is a unit of energy used in the United States. In most countries it has been replaced with the joule. A Btu is the amount of heat required to raise the temperature of 1 pound of water by 1°F. 1 watt is approximately 3.4 Btu/hr. 1 horsepower is approximately 2,540 Btu/hr. 12,000 Btu/hr. is referred to as a "ton" in many air-conditioning applications.
Horsepower	Primarily used in the automobile industry. 1 horsepower (HP) is equivalent to 746 watts.
Kilo-	1,000 or 10^3. 1 kW = 10^3 watts.
Mega- (M)	1,000,000 or 10^6. 1 MW = 10^6 watts.
Kilowatt hour (kWh)	Unit of energy equal to 1,000 watt hours or 3.6 megajoules. The kilowatt hour is most commonly known as a billing unit for energy delivered to consumers by electric utilities. Examples: A heater rated at 1,000 watts (1 kilowatt), operating for one hour uses one kilowatt hour (equivalent to 3.6 megajoules) of energy. Using a 60 watt lightbulb for one hour consumes 0.06 kilowatt hours of electricity. Using a 60 watt lightbulb for 1,000 hours consumes 60 kilowatt hours of electricity.

Conversions

Let's do a sample conversion problem that you might experience on the APES exam. We will break it down into steps and show you how it might be graded.

TIP

Be sure to practice these types of conversion problems. They appear very frequently in the Free-Response Question on the APES test. Also, be sure you are comfortable with scientific notation and the factor-label method.

EXAMPLE PROBLEM #1

Thorpeville is a rural community with a population of 8,000 homes. It gets its electricity from a small, municipal coal-burning power plant just outside of town. The power plant's capacity is rated at 20 megawatts with the average home consuming 10,000 kilowatt hours (kWh) of electricity per year. Residents of Thorpeville pay the utility $0.12 per kWh. A group of entrepreneurs is suggesting that the residents support a measure to install 10 wind turbines on existing farmland. Each wind turbine is capable of producing 1.5 MW of electricity. The cost per wind turbine is $2.5 million dollars to purchase and operate for 20 years.

(a) The existing power plant runs 8,000 hours per year. How many kWh of electricity is the current plan capable of producing?

2 points. 1 point for correct setup. 1 point for correctly calculating the amount of electricity generated. You must correctly convert MW to kW. No points will be awarded without showing your work. Alternative setups are acceptable.

$$\frac{20\,\text{MW}}{1} \times \frac{(1\times10^6 \text{ watts})}{1 \text{ MW}} \times \frac{1\,\text{kW}}{10^3 \text{ watts}} = 2 \times 10^4 \text{ kW}$$

$$\frac{(2\times10^4 \text{ kW})}{1} \times \frac{8{,}000 \text{ hours}}{1 \text{ yr.}} = 16{,}000 \times 10^4 \text{ kWh/yr}$$

$$= 1.6 \times 10^8 \text{ kWh/yr.}$$

(b) How many kWh of electricity do the residents of Thorpeville consume in one year?

2 points. 1 point for correct setup. 1 point for correctly calculating the amount of electricity generated. You must correctly convert MW to kW. No points will be awarded without showing your work. Alternative setups are acceptable.

$$\frac{8\times10^3 \text{ homes}}{1} \times \frac{1\times10^4 \text{ kWh/home}}{1 \text{ yr.}} = 8 \times 10^7 \text{ kWh/yr.}$$

(c) Compare answers (a) and (b). What conclusions can you make?

2 points, plus 1 possible elaboration point. 1 point for comparing answers (a) with (b) with an explanation of why the numbers in parts (a) and (b) would be the same or different (must be a viable reason).

OR

1 point for a solid or accurate explanation of why (a) and (b) are different even if the calculations were not attempted.

1 possible elaboration point for explanations that go into great detail about why the numbers differ.

Note: If you say that (a) and (b) are the same, you must state that this can only occur if the households have backup systems that will produce energy for them if they exceed the power generated by the plant.

The power plant produces 1.6×10^8 kWh per year. The residents, however, only use 8×10^7 kWh per year. This leaves a surplus of $1.6 \times 10^8 - 8 \times 10^7 = 8 \times 10^7$ kWh in one year that can be sold to other towns. At a rate of $0.12 per kWh, this provides a surplus of 8×10^7 kWh $\times$ $0.12/kWh = $0.96 \times 10^7 = $9,600,000.

Additional revenues could be generated by running the power plant 24 hours a day, 365 days a week.

Differences between Thorpeville's consumption and the power plant's output could be attributed to the following:

- The power plant needs to produce higher amounts of power to compensate for line loss.
- The power plant needs to produce higher amounts of power to supply energy to the town during peak hours, not just the average usage.
- The power plant needs to plan for possible future growth of the town.
- The power plant was built over capacity to provide a source of income to the town.

(d) Assuming that the population of Thorpeville remains the same for the next 20 years, and that electricity consumption remains stable per household, what would be the cost (expressed in $ per kWh) of electricity to the residents over the next 20 years if they decided to go with wind turbines?

2 points. 1 point for correct setup. 1 point for correct answer with calculations. Alternative setups are acceptable. If your answer in part (b) is incorrect but you appropriately use it as the basis for the calculations for answering the question in part (d), you will receive full credit for answering part (d) if the setup and calculations are correct, even if the answer is not correct.

Based on current community consumption of 8×10^7 kWh/yr from part (b).

$$\text{kWh for 20 years} = \frac{8 \times 10^7 \text{ kWh}}{\text{year}} \times 20 \text{ years} = 1.6 \times 10^9 \text{ kWh}$$

$$\text{Direct cost for 20 years} = 10 \text{ turbines} \times \frac{\$2.5 \times 10^6}{\text{turbine}} = \$2.5 \times 10^7$$

$$\text{cost/kWh} = \frac{\$2.5 \times 10^7}{1.6 \times 10^9 \text{ kWh}} = \$1.6 \times 10^{-2}/\text{kWh} = \$0.016/\text{kWh}$$

(e) What are the pros and cons of the existing coal-burning plant compared with the proposed wind farm?

Pros	Cons
WIND: The electricity produced from the wind turbines costs $0.016 per kWh, but each homeowner would also have to pay $25,000,000/8,000 homes = $3,125.00 over 20 years ($156.25/yr.) to pay for the wind turbines. 10,000 kWh at $0.016 per kWh for electricity produced from wind turbines = $160 plus $156.25 per year to pay for the wind turbines = $316.25 per year.	COAL: Electricity costs $0.12 per kWh produced from the coal-burning plant. 10,000 kWh of electricity per year from the coal-burning plant at $0.12 per kWh = $1,200 per year. Clearly, electricity produced from wind turbines is much cheaper.
WIND: Zero emissions. Acid rain would be reduced as well as photochemical smog.	COAL: Produces air pollution, specifically SO_2 and NO_x.
WIND: The wind is free.	COAL: As labor prices increase, the price of coal would be expected to increase over the next 20 years.
WIND: No heavy-metal or radioactive emissions.	COAL: Coal-burning plants produce heavy metals such as mercury, lead, and cadmium pollution along with radioactive contaminants.
WIND: No thermal pollution.	COAL: Can produce thermal pollution to local streams. However, cooling towers can be installed to reduce this form of pollution.
WIND: Multiple use of the land.	COAL: Cannot utilize the concept of multiple use of land.

EXAMPLE PROBLEM #2

(a) An electric water heater requires 0.30 kWh to heat a gallon of water. The thermostat is set to 150°F. The cost of electricity is $0.20 per kWh. A washing machine with a flow rate of 6.0 gallons per minute runs four times each Saturday. Each time it runs it takes in water for a total of 15 minutes. How much total water does the washing machine use in one year?

$$\frac{4 \text{ cycles}}{\text{Saturday}} \times \frac{15 \text{ minutes}}{1 \text{ cycle}} \times \frac{6.0 \text{ gallons}}{1 \text{ minute}} \times \frac{52 \text{ Saturdays}}{1 \text{ year}} = 18{,}720 \text{ gallons / year}$$

2 points. 1 point for correct setup. 1 point for correct answer with calculations. Alternative setups are acceptable.

(b) Calculate the annual cost of the electricity for the washing machine, assuming that 3.0 gallons per minute of the water used by the machine comes from the hot-water heater.

18,720 gallons / 2 = 9,360 gallons of hot water per year

$$\frac{9{,}360 \text{ gallons}}{1 \text{ year}} \times \frac{0.30 \text{ kWh}}{1 \text{ gallon}} \times \frac{\$0.20}{\text{kWh}} = \$561.60 \text{ / year}$$

2 points. 1 point for correct setup. 1 point for correct answer with calculations. Alternative setups are acceptable. If your answer in part (b) is incorrect but you appropriately used information from part (a) as the basis for answering the question, you will receive full credit, even if the numerical answer is wrong.

Laws of Thermodynamics

FIRST LAW

Energy cannot be created or destroyed.

SECOND LAW

When energy is converted from one form to another, a less useful form results (energy quality). Energy cannot be recycled to a higher quality. Only 20% of the energy in gasoline is converted to mechanical energy. The rest is lost as heat and is known as low-quality energy.

ENERGY CONSUMPTION

Wood (a renewable energy source) served as the predominant form of energy up until the Industrial Revolution. During the Industrial Revolution, coal (a nonrenewable energy source) surpassed wood's usage. Coal was overtaken by petroleum during the middle of the 20th century, with petroleum continuing to be the primary source of energy worldwide. Natural gas and coal experienced rapid development in the second half of the 20th century.

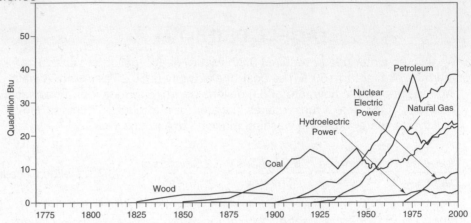

Figure 8.1 Energy consumption by source

Source: Energy Information Administration

The United States was self-sufficient in energy until the late 1950s. At that time, energy consumption began to outpace domestic production, which then led to oil imports.

The industrial sector in the United States has traditionally used the largest share of energy, followed by transportation and then residential and commercial uses. While coal was once the predominant form of energy in the industrial sector, it gave way to natural gas and petroleum in the late 1950s, with rapid increases occurring through the 1970s.

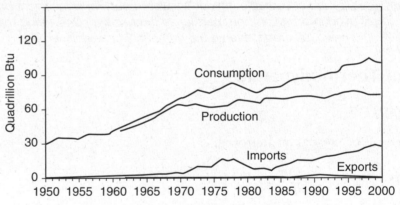

Figure 8.2 U.S. energy overview

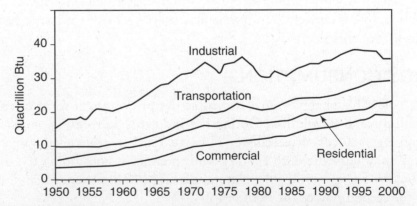

Figure 8.3 Energy consumption in the United States by end use

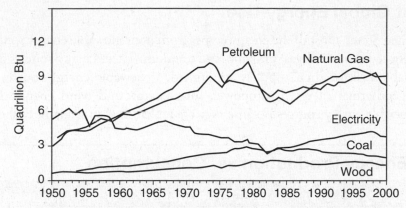

Figure 8.4 Industrial energy consumption

Beginning in 1998, net imports of oil surpassed the domestic oil supply in the United States. The United States accounts for 25% of the world consumption of petroleum.

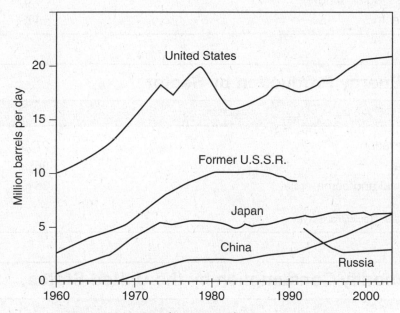

Figure 8.5 Leading petroleum consumers

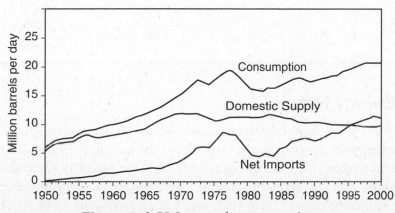

Figure 8.6 U.S. petroleum overview

Present Global Energy Use

In the United States, most of the energy comes from nonrenewable energy sources such as coal, petroleum, natural gas, propane, and uranium. These energy sources are called nonrenewable because their supplies are limited. Renewable energy sources include biomass, geothermal energy, hydropower, solar energy, and wind energy. They are called renewable energy sources because they are replenished in a relatively short time.

U.S. Energy Production vs. Consumption

Commodity	U.S. Production	U.S. Consumption
Oil	18%	39%
Natural gas	27%	23%
Coal	33%	23%
Nuclear	10%	7%
Renewable (geothermal, biomass, solar, wind)	9%	3.6%
Hydroelectric	5%	4%

U.S. Energy Production by Sector

Sector	%
Transportation	27%
Industrial	38%
Residential and commercial	36%

*Source: U.S. Geological Survey

Commodity Consumption by the United States

Commodity	%
(% of Total World Usage)	
Oil	40%
Natural gas	23%
Coal	23%

Future Energy Needs

Although the United States's energy history is one of large-scale change as new forms of energy were developed, the outlook for the next few decades is for continued growth and reliance on the three major fossil fuels: petroleum, natural gas, and coal. The most realistic, economical, and viable resources of future energy needs for the immediate future are clean coal, methane hydrates, oil shale, and tar sands.

CLEAN COAL

The world's supply of coal is substantial and can be expected to meet the world's energy needs for many years to come. Clean-coal technology refers to processes that reduce the negative environmental effects of burning coal. The processes include washing the coal to remove minerals and impurities and capturing the sulfur dioxide and carbon dioxide from the flue gases. Other clean-coal technology is focusing on natural gas or microbial fuel cells charged from biomass or sewage.

In the following diagram, oxygen is introduced to burn coal completely at Step 1. At Step 2, coal is pulverized in order to burn more completely. Coal is also washed at this point to remove contaminants. At Step 3, ash is removed using electrostatic precipitators. At Step 4, steam is condensed and returned to the boiler. And at Step 5, CO_2 is recovered using lime and then sequestered.

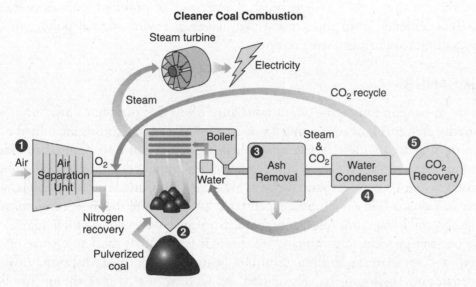

Figure 8.7 Steps used in producing clean coal

METHANE HYDRATES

Methane hydrates (methane locked in ice) are a recently discovered source of methane that form at low temperature and high pressure. They are found in two types of geologic settings: on land in permafrost regions where cold temperatures persist in shallow sediments and beneath the ocean floor at water depths greater than 1,640 feet (500 m) where high pressures dominate and where the hydrate deposits themselves may be several hundred meters thick. Some believe there is enough methane in the form of hydrates to supply energy for hundreds or thousands of years.

Natural gas is expected to take on a greater role in power generation. This is largely because of increasing pressure for clean fuels and the relatively low capital costs of building new natural gas–fired power equipment. The United States will consume increasing volumes of natural gas well into the 21st century. U.S. natural gas consumption is expected to increase 40%. Also, natural gas demand is expected to grow because of its expanded use as a transportation fuel and potentially, in the long-term, as a source of alternative liquid fuels and a source of hydrogen for fuel cells. The primary waste product of burning natural gas is carbon dioxide—a greenhouse gas that contributes to global warming.

OIL SHALE

Oil shales contain an organic material called kerogen. If the oil shale is heated in the absence of air, the kerogen converts to oil. There are approximately 3 trillion barrels of recoverable oil from oil shale in the world, with 750 billion of it being located in the United States. Oil shale can be extracted through either surface mining or *in situ* methods that consist of heating the oil shale underneath the ground and extracting the oil and gases through pumping. Most of the oil shale in the United States is found in Wyoming, Utah, and Colorado. The largest world reserves are found in Estonia, Australia, Germany, Israel, and Jordan.

Surface mining of oil shale negatively impacts the environment. The net energy yield of producing oil through oil shale is moderate since energy is required for blasting, drilling, crushing, heating the material, disposing of waste material, and environmental restoration. *In situ* methods have the potential of affecting aquifers. Even though the world has large oil shale reserves, the problem remains that once the oil is obtained from shale, traditional issues of environmental pollution, acid rain, and global warming will continue.

TAR SANDS

Tar sands contain bitumen—a semisolid form of oil that does not flow. Specialized refineries are capable of converting bitumen to oil. Tar sand deposits are mined using strip-mining techniques. *In situ* methods using steam can also be used to extract bitumen from tar sands. The sulfur content of oil obtained from tar sands is high, about 5%. Most of the tar sand deposits are located in Canada and Venezuela, with those in Canada being the most concentrated and therefore the most economical to mine. The oil in tar sands represents about two-thirds of the world's total oil reserves. The net-energy yield of producing oil through tar sands is moderate since energy is required for blasting, drilling, crushing, heating the material, disposing of waste material, and environmental restoration. As with oil shale, once the oil has been extracted from tar sands, the problems of environmental pollution, acid rain, and global warming continue.

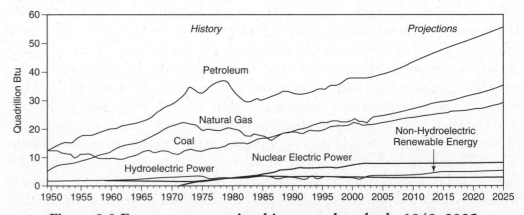

Figure 8.8 Energy consumption history and outlook, 1949–2025

Energy Crisis

In a free-market economy, the price of energy is driven by the principle of supply and demand. Sudden changes in the price of energy can occur if either supply or demand changes. In some cases, an energy crisis is brought on by a failure of world markets to adjust prices in response to shortages. Oil supply is largely controlled by nations with significant reserves of easily extractable oil, such as Saudi Arabia and Venezuela, who belong to an association of oil-producing countries known as OPEC (Organization of Petroleum Exporting Countries). When OPEC reduces the output quotas of its member countries, the price of oil increases as the supply diminishes. Similarly, OPEC can boost oil production in order to increase supplies, which drives down the price. When OPEC raises the price of oil too high, demand decreases and the production of oil from alternative sources becomes profitable. Historically, there have been several energy crises.

Most of the world's energy is supplied by burning oil. At current rates of consumption, world oil reserves are predicted to last 50 years, with oil reserves in the United States predicted to last 25 years. The industrialization of China will significantly decrease this predicted number of years. As supply decreases, prices will increase. Higher prices for oil may make other sources (shale oil and tar sands) more economical.

FOSSIL FUEL RESOURCES AND USE

Coal is produced by decomposition of ancient (286 million-year-old) organic matter under high temperature and pressure. Sulfur from the decomposition of hydrogen sulfide (H_2S) by anaerobic bacteria became trapped in coal. There are three types of coal: lignite, bituminous, and anthracite. Lignite or brown is the softest and has the lowest heat content. Bituminous is soft, has a high sulfur content, and constitutes 50% of the U.S. reserve. Anthracite is hard, has a high heat content and low sulfur content, and makes up 2% of the U.S. reserve. Peat is precoal and is used in some countries for heat but it has low heat content. Coal supplies 25% of the world's energy, with China and the United States consuming the most. In the United States, 87% of the coal is used for power plants to produce electricity. The Clean Air Act requires up to a 90% reduction in the release of sulfur-containing gases.

Oil is a fossil fuel produced by the decomposition of deeply buried organic material (plants) under high temperatures and pressures for millions of years. Compounds derived from oil are known as petrochemicals. They are used in the manufacture of paints, drugs, plastics, etc.

Natural gas (known as methane or CH_4) is produced by the decomposition of ancient organic matter under high temperatures and pressure. Conventional sources of methane are found associated with oil deposits. Unconventional sources include coal beds, shale, gas hydrates, and tight sands. Methane can be liquefied (LNG), which allows for worldwide distribution.

Extraction-Purification Methods

COAL

There are two primary methods of mining coal: surface mining and underground mining. Coal that is going to be burned in solid form may go through a variety of preparation processes. These include removing foreign material, screening for size, crushing, and washing to remove contaminants. It is also possible to turn solid coal into a gas or liquid fuel through clean-coal technologies.

OIL

Oil occurs in certain geologic formations at varying depths in Earth's crust. In many cases, elaborate, expensive equipment is required to extract it. Oil is usually found trapped in a layer of porous sandstone, which lies just beneath a dome-shaped or folded layer of some nonporous rock such as limestone. In other formations, the oil is trapped at a fault, or break in the layers of the crust. Natural gas is usually present just below the nonporous layer and immediately above the oil. Below the oil layer, the sandstone is usually saturated with salt water. The oil is released from this formation by drilling a well and puncturing the limestone layer. The oil is usually under such great pressure that it flows naturally, and sometimes with great force, from the well. However, in some cases, this pressure later diminishes so that the oil must be pumped. Once the oil has been collected, it is sent to a refinery, where it is cracked. Cracking involves separating the components of oil by their boiling points. Refining crude oil produces gasoline, heating oil, diesel oil, asphalt, etc.

CASE STUDY

Arctic National Wildlife Refuge (ANWR): The largest national wildlife refuge in the United States, the ANWR is located in northeastern Alaska and consists of 19 million acres (78,000 km^2). The question of whether to drill for oil in the ANWR has been an ongoing political controversy in the United States since 1977. Much of the debate over whether to drill in ANWR rests on the amount of economically recoverable oil, as it relates to world oil markets, weighed against the potential harm oil exploration might have on the wildlife.

The Keystone Pipeline System is a pipeline concept to transport synthetic crude oil and diluted bitumen from oil sands in Canada to multiple destinations in the United States, which include refineries in Illinois, a distribution hub in Oklahoma, and proposed connections to refineries along the Gulf Coast of Texas. As of October 2012, the plan has not been approved by Congress.

NATURAL GAS

Natural gas typically flows from wells under its own pressure. It is collected by small pipelines that feed into the large gas transmission pipelines. In the United States, about 20 trillion cubic feet (560 billion m^3) of gas is produced each year.

Hydrofracking

Natural gas from dense shale rock formations that have previously been uneconomical to extract has become the fastest-growing source of natural gas in the United States and could become a significant new global energy source. In hydraulic fracturing, or hydrofracking, chemicals are mixed with large quantities of water and sand and injected into wells at extremely high pressure to create fractures in rock that allow oil and natural gas to escape and flow out of the well. Studies estimate that up to 80% of natural gas wells drilled in the next decade will require hydraulic fracturing.

Hydrofracking

Pros	Cons
Process of bringing a well to completion is short, after which the well can be in production for 20 to 40 years.	Dangerous chemicals are used in the process and can enter the water table.
Makes it possible to produce oil and natural gas in places where conventional technologies were ineffective or cost-prohibitive.	Toxic, radioactive, and caustic liquid waste by-products pose storage, treatment, and disposal problems. There are no adequate safeguards or regulations currently in place for this process.
Allows greater independence from foreign sources of energy and helps create new jobs, which in turn stimulates the economy.	Hydrofracking results in contaminated water supplies, air pollution, destroyed streams, and negative environmental impacts on local flora and fauna.

World Reserves and Global Demand

Coal, oil, and natural gas are nonrenewable energy resources. Following is a brief description of the known world reserves of these energy sources and their expected demands in the future.

COAL

Coal is currently the world's single largest source of fuel used to produce electricity. The United States has the largest proven recoverable coal reserves in the world, whereas China is the world's largest producer of coal. Global coal reserves are estimated to last about 300 years at current levels of extraction with continuing issues of CO_2 emissions and global warming.

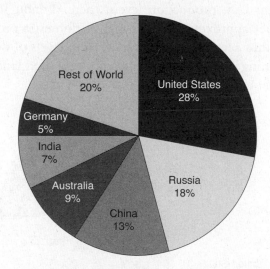

Figure 8.9 World recoverable coal reserves by region

OIL

A large portion of the Earth's global crude oil reserve—45% to 70%—has already been depleted. It is estimated that there is a 50-year supply left on Earth, with countries in the Middle East owning about half of what's left. The United States owns

3% of the world's oil reserves but uses 30% of the oil extracted in the world each year. Increased competition for foreign oil by China and India increases the world's cost of oil. Two-thirds of the oil used in the United States is used for transportation (gasoline, diesel, jet fuel, etc.). About one-fourth of it is used in industries such as plastics, medicines, etc. Oil imports in the United States have decreased slightly in the last few years due to improvements in energy efficiency and higher fuel economy standards. Of all known oil reserves, 65% is found in 1% of all fields—primarily in the Middle East.

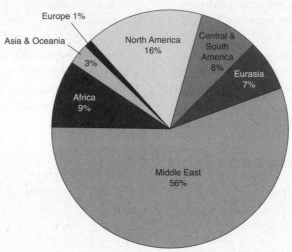

Figure 8.10 World oil reserves by region

NATURAL GAS

Most of the world's natural gas reserves are located in the Middle East (34% of the world total). Russia and Kazakhstan combined have approximately 40% of the world's natural gas reserves, the Middle East has about 25%, and the United States has about 3%. Given U.S. production levels, there is enough natural gas in the United States to meet approximately 75 years of domestic production. This estimate does not take into account expected increasing levels of domestic production or the potential opening up of access to currently restricted land such as the Arctic National Wildlife Refuge.

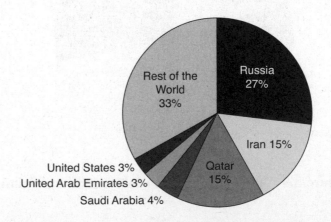

Data from CIA *World Factbook 2008*

Figure 8.11 World natural gas reserves by region

Synfuels

A synthetic fuel (synfuel) is any fuel produced from coal, natural gas, or biomass through chemical conversion. The conversion process creates substances that are chemically the same as crude oil or processed fuels, but were synthesized through artificial means. The raw materials used to make synfuels have to be subjected to intense chemical and physical changes to be usable as crude oil or processed fuel. An example is synthetic natural gas (SNG), which is chemically produced from coal liquefaction.

SOLID COAL TO SYNTHETIC NATURAL GAS (SNG), METHANOL, OR SYNTHETIC GASOLINE

Pros
- Easily transported through pipelines.
- Produces less air pollution.
- Large supply of raw materials available worldwide to meet current demands for hundreds of years.
- Can produce gasoline, diesel, or kerosene directly without reforming or cracking.

Cons
- Low net energy yield and requires energy to produce SNG.
- Plants are expensive to build.
- Would increase depletion of coal due to inherent inefficiencies.
- Product is more expensive than petroleum products.

Environmental Advantages/Disadvantages of Sources

COAL

Pros
- Abundant, known world reserves will last approximately 300 years at current rate of consumption.
- Unidentified world reserves are estimated to last 1,000 years at current rate of consumption.
- United States reserves are estimated to last about 300 years at current rate of consumption.
- Relatively high net-energy yield.
- U.S. government subsidies keep prices low.
- Stable; nonexplosive; not harmful if spilled.

Cons
- Most extraction in the United States is done through either strip mining or underground mining. These methods cause disruption to the land through erosion, runoff, and decrease in biodiversity.
- Up to 20% of coal ends up as fly ash, boiler slag, or sludge. Burning coal releases mercury, sulfur, and radioactive particles into the air. Thirty-five percent of all CO_2 releases are due to the burning of coal, with 30% of all pollution due to NO_x.

- Underground mining is dangerous and unhealthy.
- Expensive to process and transport. Cannot be used effectively for transportation needs.
- Pollution causes global warming. Scrubbers and other antipollution control devices are expensive.

OIL

Pros

- Inexpensive; however, prices are increasing, making alternatives more attractive.
- Easily transported through established pipelines and distribution networks.
- High net-energy yield.
- Ample supply for immediate future.
- Large U.S. government subsidies in place.
- Versatile—used to manufacture many products (paints, medicines, plastics, etc.).

Cons

- World oil reserves are limited and declining.
- Produces pollution (SO_2, NO_x, and CO_2). Production releases contaminated wastewater and brine.
- Causes land disturbances in the drilling process, which accelerates erosion.
- Oil spills both on land and in the ocean from platforms and tankers.
- Disruption to wildlife habitats (e.g., Arctic Wildlife Refuge).
- Supplies are politically volatile.

NATURAL GAS

Pros

- Pipelines and distribution networks are in place. Easily processed and transported as LNG over rail or ship.
- Relatively inexpensive, but prices are increasing. Viewed by many as a transitionary fossil fuel as the world switches to alternative sources.
- World reserves are estimated to be 125 years at current rate of consumption.
- High net energy yield.
- Produces less pollution than any other fossil fuel.
- Extraction is not as damaging to the environment as either coal or oil.

Cons

- H_2S and SO_2 are released during processing.
- LNG processing is expensive and dangerous, and results in lower net energy.
- Leakage of CH_4 has a greater impact on global warming than does CO_2.
- Disruption to areas where it is collected.
- Extraction releases contaminated wastewater and brine.
- Land subsidence.

Fuel Type	Btu Value per Unit	Units Req'd to Produce 1,000,000 Btus	Fuel Price per Unit	Cost to Produce 1,000,000 Btus	Appliance Efficiency[3]	Effective Cost per 1,000,000 Btus
Electricity	3,413 per kWh	293 kWh	$0.14	$41.84	100%	$41.84[1]
Natural Gas	1,000 per ft³ 1 therm = 100 ft³	10 therms	$1.35 per therm	$13.50	80%	$16.88[1]
Fuel Oil	139,000 per gallon	7.1 gallons	$2.86 per gallon	$20.30	80%	$25.38
LP Gas	91,690 per gallon	11 gallons	$2.95 per gallon	$32.45	80%	$40.56
Wood	22,000,000 per cord	0.0607 cords	$225 per cord	$13.66	60%	$22.77[2]
Wood Pellets	8,000 per pound	125 pounds	$0.12 per pound	$14.69	80%	$18.36
Kerosene	135,000 per gallon	7.41 gallons	$3.30 per gallon	$24.45	95%	$25.68

[1] Does not include a fixed monthly charge.

[2] Includes delivery.

[3] Electricity is set at 100% for comparative purposes.

Figure 8.12 Comparisons of U.S. Fuel Prices (as of March 2010)

NUCLEAR ENERGY

During nuclear fission, an atom splits into two or more smaller nuclei along with by-product particles (neutrons, photons, gamma rays, and beta and alpha particles). The reaction gives off heat (exothermic). If controlled, the heat that is produced is used to produce steam that turns generators that then produces electricity. If the reaction is not controlled, a nuclear explosion can result.

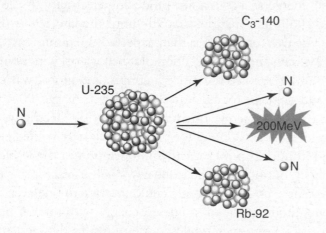

Figure 8.13 Nuclear fission

The amount of potential energy contained in nuclear fuel is 10 million times more than that of more traditional fuel sources such as coal and petroleum. The downside is that nuclear wastes remain highly radioactive for thousands of years and are difficult to dispose. The most common nuclear fuels are U-235, U-238, and Pu-239.

Nuclear Fuel

U-235

U-235 differs from U-238 in its ability to produce a fission chain reaction. The minimum amount of U-235 required for a chain reaction is called the critical mass. Low concentrations of U-235 can be used if the speed of the neutrons is slowed down through the use of a moderator. Less than 1% of all natural uranium on Earth is U-235. Uranium that has been processed to separate out U-235 is known as enriched uranium and is a subject of current controversy with Iran and North Korea. Nuclear weapons contain 85% or more U-235; nuclear power plants contain about 3% U-235. The half-life of U-235 is 700 million years.

U-238

U-238 is the most common (99.3%) isotope of uranium and has a half-life of 4.5 billion years. When hit by a neutron, it eventually decays into Pu-239, which is used as a fuel in fission reactors. Most depleted uranium is U-238.

Pu-239

Pu-239 has a half-life of 24,000 years. It is produced in breeder reactors from U-238. Plutonium fission provides about one-third of the total energy produced in a typical commercial nuclear power plant. Control rods in nuclear power plants need to be changed frequently due to the buildup of Pu-239 that can be used for nuclear weapons and due to the buildup of Pu-240, a contaminant. International inspections of nuclear power plants regulate the amount of Pu-239 produced by power plants.

Electricity Production

The use of nuclear energy as a source for producing electricity in the United States started during the 1960s and increased rapidly until the late 1980s. Reasons for its decline included cost overruns, higher-than-expected operating costs, safety issues, disposal of nuclear wastes, and the perception of it being a risky investment. However, due to electricity shortages, fossil fuel price increases, and global warming, there is a renewed interest and demand for nuclear power plants. As of 2005, nuclear power provided 6% of the world's energy and 15% of the world's electricity, with the U.S., France, and Japan together accounting for 57% of nuclear generated electricity. As of 2007, there were 439 nuclear power reactors in operation in the world, operating in 31 countries. Globally, during the 1980s, one new nuclear reactor started up every 17 days on average; by the year 2015 this rate could increase to one every 5 days.

The United States produces the most nuclear energy, with nuclear power providing 19% of the electricity it consumes, while France produces the highest percentage of its electrical energy from nuclear reactors—78% as of 2006. In the European Union as a whole, nuclear energy provides 30% of the electricity.

Proponents of nuclear energy assert that nuclear power is a sustainable energy source that reduces carbon emissions and increases energy security by decreasing dependence on foreign oil. Proponents also claim that the risks of storing waste are small and can be further reduced by the technology in the new reactors. Furthermore, the operational safety record is already good when compared to the other major kinds of power plants. Critics believe that nuclear power is a potentially dangerous and declining energy source, with decreasing proportions of nuclear energy in power production, and dispute whether the risks can be reduced through new technology. Critics also point out the problem of storing radioactive waste, the potential for possibly severe radioactive contamination by accident or sabotage, the possibility of nuclear proliferation, and the disadvantages of centralized electrical production.

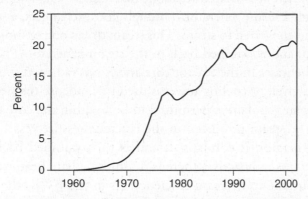

Figure 8.14 Nuclear share of electricity in the United States

Nuclear Reactor Types

Several different nuclear reactor types are in use: light-water reactors, heavy-water reactors, graphite-moderated reactors, and exotic reactors. However, they all have several features in common (see Figure below).

A. The *core* contains up to 50,000 fuel rods. Each fuel rod is stacked with many fuel pellets, each pellet having the energy equivalent of 1 ton (0.9 m.t.) of coal, 17,000 cubic feet (481 m³) of natural gas, or 149 gallons (564 L.) of oil.

B. Enriched or concentrated U-235 is usually the *fuel*. The fission of an atom of uranium produces 10 million times the energy produced by the combustion of an atom of carbon from coal.

C. *Control rods* (usually made of boron) move in and out of the core to absorb neutrons and slow down the reaction.

D. A neutron *moderator* is a medium that reduces the velocity of fast neutrons, thereby turning them into thermal neutrons capable of sustaining a nuclear chain reaction. Moderators can be water, graphite (can produce plutonium for weapons), or deuterium oxide (heavy water).

E. *Coolant* removes heat and produces steam to generate electricity.

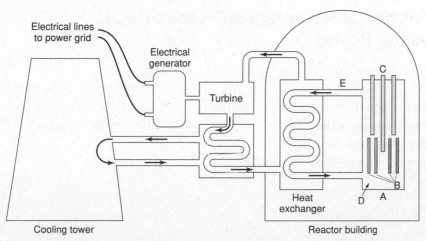

Figure 8.15 Diagram of nuclear power plant

LIGHT-WATER REACTORS

Both the moderator and coolant are light or normal water (H_2O). To this category belong the pressurized-water reactors (PWR) and boiling-water reactors (BWR).

In PWR, the water coolant operates at a high pressure and is then pumped through the reactor core, where it is heated to about 620°F (325°C). The superheated water is pumped through a steam generator. Through heat exchangers, a secondary loop of water is heated and converted to steam. This steam drives one or more turbine generators, is condensed, and is pumped back to the steam generator. The secondary loop is isolated from the water in the reactor core and is not radioactive. A third stream of water from a lake, river, or cooling tower is used to condense the steam.

In BWR, the water coolant is permitted to boil within the core by operating at a lower pressure. The steam produced in the reactor pressure vessel is piped directly to the turbine generator, is condensed, and is then pumped back to the reactor. Although the steam is radioactive, there is no intermediate heat exchanger between the reactor and turbine to decrease efficiency. As in the PWR, the condenser cooling water has a separate source such as a lake or river.

HEAVY-WATER REACTORS

Heavy water, also known as deuterium oxide (D_2O), is a form of water in which each hydrogen in the water molecule contains one proton, one electron, and one neutron. Heavy water is used in certain types of nuclear reactors where it acts as a neutron moderator to slow down neutrons so that they can react with the uranium in the reactor. The use of heavy water essentially increases the efficiency of the nuclear reaction.

GRAPHITE-MODERATED REACTORS

This category uses light water for cooling, graphite for moderation, and uranium for fuel. These reactors require no separated isotopes such as enriched uranium or heavy water. This type of reactor, built by the Russians, was very unstable and is no longer being produced (see the Chernobyl case study).

EXOTIC REACTORS

Fast-breeder reactors and other experimental installations are in this group. Breeder reactors produce more fissionable material than they consume.

Environmental Advantages/Disadvantages of Nuclear Power

PROS

- No air pollutants if operating correctly.
- Releases about 1/6 the CO_2 as fossil fuel plants, thus reducing global warming.
- Water pollution is low.
- Disruption of land is low to moderate.

CONS

- Nuclear wastes take millions of years to degrade. Problem of where to store them and keeping them out of hands of terrorists.
- Nuclear power plants are licensed in the United States by the Nuclear Regulatory Commission (NRC) for 40 years. After that, they can ask to renew their license, or they can shut down the plant and decommission it. Decommissioning means shutting down the plant and taking steps to reduce the level of radiation so that the land can be used for other things. Since it may cost $300 million or more to shut down and decommission a plant, the NRC requires plant owners to set aside money when the plant is still operating to pay for future shutdown costs.
- Low net-energy yield—energy required for mining uranium, processing ore, building and operating plant, dismantling plant, and storing wastes.
- Safety and malfunction issues.

Safety Issues (Radiation and Human Health)

The U.S. Department of Energy (DOE) estimates that up to 50,000 radioactive contaminated sites within the United States require cleanup with a projected cost of $1 trillion dollars. The situation is many times worse in the former Soviet Union.

Estimated Health Risks per Year in the United States

	Nuclear	Coal
Premature death	6,000	65,000
Genetic defects/damage	4,000	200,000

CASE STUDY

Chernobyl, Ukraine (1986): Explosion in a nuclear power plant sent highly radioactive debris throughout northern Europe. Estimates run as high as 32,000 deaths, and 62,000 square miles (161,000 sq km) remain contaminated. About 500,000 people were exposed to dangerous levels of radiation. According to the World Health Organization, the Chernobyl nuclear disaster will cause 50,000 new cases of thyroid cancer among young people living in the areas most affected by the nuclear disaster, and the incidence of thyroid cancer in children rose tenfold in the Ukraine region. In certain parts of Belarus, 36% percent of children who were under four when the accident occurred can expect to develop thyroid cancer. Cost estimates run as high as $400 billion. The cause was determined to be both design and human error.

Nuclear Fusion

Nuclear fusion can occur when extremely high temperatures are used to force nuclei of isotopes of lightweight atoms to fuse together, which causes large amounts of energy to be released. A coal-fed electrical generating plant producing 1,000 megawatts of electricity in one day produces 30,000 tons of CO_2 gases, 600 tons of SO_2 gas, and 80 tons of NO_2 gas. In contrast, a fusion plant producing the same amount of electricity would produce 4 pounds of harmless helium as a waste product.

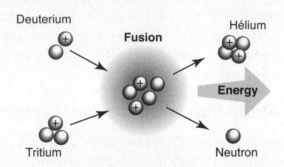

Figure 8.16 Nuclear fusion

HYDROELECTRIC POWER

Dams are built to trap water, which in turn is then released and channeled through turbines that generate electricity. Hydroelectric power supplies about 10% of the electricity in the United States and approximately 3% worldwide.

Pros

- Dams control flooding.
- Low operating and maintenance costs.
- No polluting waste products.
- Long life spans.
- Moderate to high net-useful energy.
- Areas of water recreation.

Cons

- Dams create large flooded areas behind the dam from which people are displaced. Water is slow moving and can breed pathogens.
- Dams destroy wildlife habits and keep fish from migrating.
- Sedimentation requires dredging. Prevents sedimentation from reaching downstream and enriching farmland.
- Expensive to build.
- Destroys wild rivers.
- Large-scale projects are subject to earthquakes.
- Water loss due to increased water surface areas.

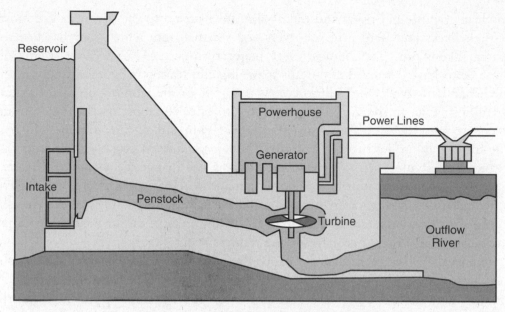

Figure 8.17 Typical hydroelectric dam

Flood Control

Methods to control floods include:

CHANNELIZATION

Straighten and deepen streams. Cons: removes bank vegetation and increases stream velocity, which causes erosion; may increase downstream flooding and sedimentation, which negatively impacts aquatic habitats.

DAMS

Dams store water in reservoirs. During periods of excessive rainfall, dams can be overwhelmed and excess water needs to be released.

IDENTIFY AND MANAGE FLOOD-PRONE AREAS

By identifying flood-prone areas, precautionary building practices such as floodways, building elevation, and pumping stations can be adopted.

LEVEES OR FLOODWALLS

Levees are raised embankments to prevent a river from overflowing. Levees contain river and stream flows but increase water velocity. Levees can break as they did in New Orleans during Hurricane Katrina in 2005.

PRESERVE WETLANDS

This technique preserves natural flood plains and maintains biodiversity.

Salmon

There are an estimated 74,993 dams in America, blocking 600,000 miles (965,000 km) of what had once been free-flowing rivers. Salmon are migratory fish that hatch in streams and rivers and then swim downstream to the ocean to live most of their lives. They return to the rivers and streams from which they hatched to spawn. Dams now block almost every major river system in the West. Many of those dams have destroyed important spawning and rearing habitats for salmon. In the Sacramento Valley in California, less than 5% of the salmon's original habitat is still available. In the Columbia River Basin, once the most productive salmon river system in the world, less than 70 miles (110 km) still remains free-flowing. As a result of habitat destruction, at least 106 major U.S. west coast salmon runs are extinct, and 25 more are now endangered. Dams also change the character of rivers, creating slow-moving, warm-water pools that are ideal for predators of salmon. Low water velocities in large reservoirs can also delay salmon migration and expose fish to higher water temperatures and disease. Cutting trees in forests near the streams and rivers clouds the water with silt and reduces water quality.

Things that have been done to reduce the impacts of dams on fish include fish passage facilities and fish ladders that help juvenile and adult fish migrate over or around many dams. Spilling water at dams over the spillway can help pass juvenile fish downstream because it avoids sending the fish through turbines. Water releases from upstream storage reservoirs have been used to increase water velocities and to reduce water temperatures in order to improve migration conditions through reservoirs. Juvenile fish also are collected and transported downstream in barges and trucks.

Silting and Other Impacts

DISEASE

Dam reservoirs in tropical areas, due to their slow movement, are literally breeding grounds for mosquitoes, snails, and flies—the vectors that carry malaria, schistosomiasis, and river blindness.

DISPLACEMENT

Flooded areas behind dams destroy rich croplands and displace people.

EFFECTS ON WATERSHED

Downstream areas are deprived of the nutrient-rich silt that would revitalize depleted soil profiles.

IMPACT ON WILDLIFE

Migration and spawning cycles are disrupted.

SILTING

Silting occurs when silt (very fine particles intermediate in size between sand and clay) that is dissolved in river water settles out behind dams. Over time, the silt builds up and must be removed (dredged).

WATER LOSS

Large losses of freshwater occur through evaporation and seepage through porous rock beds.

ENERGY CONSERVATION

Energy Star is a joint program of the U.S. Environmental Protection Agency and the U.S. Department of Energy. It is designed to protect the environment through energy-efficient products and practices. Programs coordinated through Energy Star saved enough energy in 2005 to avoid greenhouse gas emissions equivalent to 23 million cars and $12 billion in utility bills. The symbol shown below appears on products that meet Energy Star standards.

CAFE Standards

Transportation needs consume two-thirds of the petroleum consumption in the United States. This sector of energy consumption is increasing faster than any other sector (1.8% growth per year). Imports of crude oil and other petroleum products are expected to increase 66% by 2020. CAFE (Corporate Average Fuel Economy) standards are the average fuel economies of a manufacturer's fleet of passenger cars or light trucks. The testing follows the guidelines established by the Environmental Protection Agency. It is estimated that CAFE standards result in savings of over 55 billion gallons (210 billion L) of fuel annually with a substantial reduction in carbon dioxide emissions of approximately 10%. CAFE standards are achieved through better engine design, efficiency, and weight reduction. The average CAFE standard of 27.5 miles per gallon (11.7 km/L) for automobiles has not increased in the United States since 1996 with SUV standards less than those for light trucks. Significant improvements in fuel mileage could be achieved by expanding CAFE standards to include:

1. Streamlining

2. Reduced tire-rolling resistance

3. Engine improvements, especially transitioning to a hybrid technology

4. Optimized transmission improvements

5. Transition to higher voltage automotive electrical systems

6. Performance-based tax credits

Hybrid Electric Vehicles

Cars should be able to drive at least 300 miles (482 km) between refuelings, be refueled quickly and easily, and keep up with the other traffic on the road. A gasoline-powered car meets these requirements but produces a relatively large amount of pollution and generally gets poor gas mileage. For example, 1 gallon (4 L) of gasoline weighs 6 pounds (2.7 kg). When burned, the carbon in it combines with oxygen from the air to produce nearly 20 pounds (9 kg) of carbon dioxide. An electric car, however, produces almost no pollution but has limited range between charges. A hybrid vehicle attempts to increase the mileage and reduce the emissions of a gas-powered car significantly while overcoming the shortcomings of an electric car.

Gasoline-electric hybrid cars contain five important parts. First, it has an engine. The gasoline engine on a hybrid is smaller than on a gas-only car and uses advanced technologies to reduce emissions and increase efficiency. Second, the fuel tank in a hybrid is the energy storage device for the gasoline engine. Gasoline has a much higher energy density than batteries do. For example, about 1,000 pounds (455 kg) of batteries are needed to store as much energy as 1 gallon (3.7 L) or 6 pounds (2.7 kg) of gasoline. Third, advanced electronics allow the electric motor to act as a generator. For example, when it needs to, the motor can draw energy from the batteries to accelerate the car. When acting as a generator, it can slow down the car and return energy to the batteries. Fourth, the generator is similar to an electric motor, but it acts only to produce electrical power. It is used mostly on series hybrids. Fifth, the batteries in a hybrid car are the energy storage device for the electric motor. Unlike the gasoline in the fuel tank, which can power only the gasoline engine, the electric motor on a hybrid car can put energy into the batteries as well as draw energy from them.

A parallel hybrid has a fuel tank that supplies gasoline to the engine and a set of batteries that supplies power to the electric motor. Both the engine and the electric motor power the car at the same time. In a series hybrid, the gasoline engine turns a generator, which charges the batteries and/or powers an electric motor. The gasoline engine never directly powers the vehicle.

Plug-in hybrid electric vehicles are hybrid cars with an added battery. Plug-in hybrids can be plugged in to a 120-volt outlet and charged. Plug-ins run on the stored energy for much of a typical day's driving—up to 60 miles per charge. When the charge is used up, plug-in hybrids automatically keep running on the fuel in the fuel tank. A plug-in does not entail any sacrifice of vehicle performance or driver amenities. A midsize plug-in can accelerate from 0 to 60 miles per hour (0–96 kph) in less than 9 seconds and sustain a top speed of 97 miles per hour (156 kph). Higher initial costs are partly offset by lower operating costs.

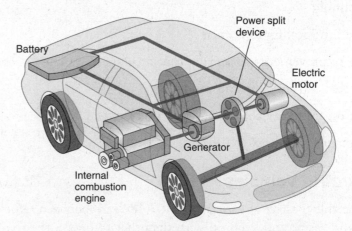

Figure 8.18 Hybrid electric vehicle

Alternative Fuels (LNG and CNG)

A natural gas vehicle or NGV is an alternative fuel vehicle that uses compressed natural gas (CNG). In 2010, there were 12.7 million natural gas vehicles throughout the world, led by Pakistan with 2.7 million vehicles. The Asia-Pacific region leads the world with 6.8 million NGVs, followed by Latin America with 4.2 million vehicles. In the Latin American region, almost 90% of NGVs have bi-fuel engines, allowing these vehicles to run on either gasoline or CNG.

Vehicles running on CNG require high compression and thus thick walled tanks, adding to the material cost and weight. CNG is also lighter than air, dispersing upwards in case of a leak. The cost of CNG is much cheaper than oil based products; some areas of the United States in April 2012 were selling a gallon's equivalent of CNG for $1.09. NGVs and especially CNG tends to corrode and wear the parts of an engine less rapidly than gasoline. Driving 500,000 miles (800,000 km) on an engine designed to run on CNG is not uncommon.

Finally, emissions are cleaner with NGVs, there is generally less wasted fuel, and lower emissions of carbon and lower particulate emissions per equivalent distance traveled.

Electric Cars

An electric car is an automobile that is propelled only by an electric motor and using electrical energy stored in batteries. Electric cars were popular in the late-19th century and early 20th century until advances in internal combustion engine technology and mass production of cheaper gasoline vehicles led to a decline in the use of the electric drive vehicle. A renewed interest in the production of electric cars occurred during the mid 2000s due mainly to concerns about rapidly increasing oil prices and the need to curb greenhouse gas emissions.

Electric cars have several potential benefits compared to conventional internal combustion automobiles, including: a significant reduction of urban air pollution as they do not emit harmful tailpipe pollutants from the onboard source of power at the point of operation, reduced greenhouse gas emissions from the onboard source of power, and less dependence on foreign oil.

Electric cars, however, are significantly more expensive than conventional internal combustion engine vehicles and hybrid electric vehicles due to the additional cost of their lithium-ion battery pack. Other factors discouraging the adoption of electric cars are the lack of public and private recharging infrastructure and the driver's fear of the batteries running out of energy before reaching their destination due to the limited range of existing electric cars.

Mass Transit

Mass transit includes rail, bus services, subways, airlines, ferries, and so on. Mass transit often determines where people live, where they work, and how much air pollution they are subjected to. In the United States, private cars are the primary mode of transportation. In the rest of the world, mass transport is the primary form. For example, in the United States, only 3% of the population utilizes mass transit on a regular basis. In Japan, that figure expands to 47%. Land availability and whether cities expand vertically (land is not available) or expand horizontally (land is available) often determines the preferred mode of transportation. Use of mass transit rises sharply with population density. Mass transit can be faster than private cars when either a separate infrastructure is reserved for mass transit or special lanes on shared highways are designated for mass transit vehicles, which results in higher speeds and less delay. In some areas, public transport systems are poorly developed and take significantly longer than an equivalent trip in a private vehicle. Perhaps the most efficient method to promote mass transit is to adopt a user-pay approach, where all external costs of operating a private vehicle are factored into license fees and/or vehicle taxes. However, this approach would be met with fierce private and political opposition.

LIGHT RAIL

Consists of trains that share space with road traffic and trains that have their own right-of-way and are separated from road traffic.

BUS RAPID TRANSIT

Includes bus-dedicated and grade-separated right-of-ways, bus lanes, bus signal preference and preemption, bus turnouts, bus-boarding islands, curb realignment, off-bus fare collection, and level boarding.

CAR SHARING

Car sharing is a model of car rental where people rent cars for short periods of time, often by the hour. It is often promoted as an alternative to owning a car where public transit, walking, and bicycling can be used most of the time and a car is only necessary for out-of-town trips, moving large items, or special occasions. It can also be an alternative to owning multiple cars for households with more than one driver. Car sharing differs from traditional car rentals in the following ways: Car sharing is not limited by office hours; reservation, pickup, and return is all self-service; vehicles can be rented by the hour, as well as by the day; vehicle locations are distributed throughout the service area, and often located for access by public transport; and insurance and fuel costs are included in the rates.

Recent studies have found that 30% of households that participated in car sharing sold a car; others delayed purchasing one while transit use, bicycling, and walking also increased among members. Car sharing is generally not cost-effective for commuting to a full-time job on a regular basis. Car sharing can also help reduce congestion and pollution. Replacing private automobiles with shared ones directly also reduces demand for parking spaces.

Successful car sharing development has tended to be associated mainly with densely populated areas. Low-density areas are considered more difficult to serve with car sharing because of the lack of alternative modes of transportation and the potentially larger distance that users must travel to reach the cars.

RENEWABLE ENERGY

Several different forms of renewable energy can be used.

Solar Energy

Solar energy consists of collecting and harnessing radiant energy from the sun to provide heat and/or electricity. Electrical power and/or heat can be generated at home and industrial sites through photovoltaic cells, solar collectors or at a central solar-thermal plant.

Active solar collectors use the sun's energy to heat water or air inside a home or business. It requires an electrical input for pumps and fans. Passive solar requires no moving parts. The structure is built to maximize solar capture, such as large, south-facing windows. Photovoltaic cells are used to generate electricity.

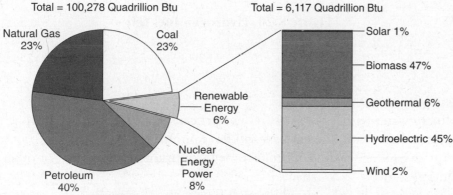

The Role of Renewable Energy Consumption in the Nation's Energy Supply, 2004

Total = 100,278 Quadrillion Btu   Total = 6,117 Quadrillion Btu

Natural Gas 23%
Coal 23%
Renewable Energy 6%
Nuclear Energy Power 8%
Petroleum 40%

Solar 1%
Biomass 47%
Geothermal 6%
Hydroelectric 45%
Wind 2%

Figure 8.19 Renewable energy sources

Pros

• Supply of solar energy is limitless.
• Reduces reliance on foreign imports.
• Only pollution is in manufacture of collectors. Little environmental impact.
• Can store energy during the day and release it at night—good for remote locations.

Cons

• Inefficient where sunlight is limited or seasonal.
• Maintenance costs are high.
• Systems deteriorate and must be periodically replaced.
• Current efficiency is between 10%–25% and not expected to increase soon.

Hydrogen Fuel Cells

Nine million tons of hydrogen is produced in the United States each day—enough to power 20 to 30 million cars or 5 to 8 million homes. Most of this hydrogen is used by industry in refining, treating metals, and processing foods.

The hydrogen fuel cell operates similar to a battery. It has two electrodes, an anode and a cathode, that are separated by a membrane. Oxygen passes over one electrode and hydrogen over the other. The hydrogen reacts with a catalyst on the anode that converts the hydrogen gas into negatively charged electrons and positively charged hydrogen ions. The electrons flow out of the cell to be used as electrical energy. The hydrogen ions move through the electrolyte membrane to the cathode, where they combine with oxygen and the electrons to produce water. Unlike batteries, fuel cells never run out.

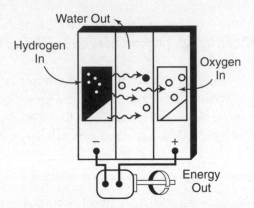

Figure 8.20 Hydrogen fuel cell

Pros

• Waste product is pure water.
• Ordinary water (either ocean or freshwater) can be used to obtain hydrogen.
• Does not destroy wildlife habitats and has minimal environmental impact.
• Energy to produce hydrogen could come from fusion reactor, solar, or other less-polluting source.
• Hydrogen is easily transported through pipelines.
• Hydrogen can be stored in compounds to make it safe to handle. Hydrogen is explosive, but so are methane, propane, butane, and gasoline.

Cons

• Takes energy to produce the hydrogen from either water or methane.
• Changing from a current fossil fuel system to a hydrogen-based system would be very expensive.
• Hydrogen gas is explosive.
• At the current time, it is difficult to store hydrogen gas for personal cars.

Biomass

Biomass is any carbon-based, biologically derived fuel source such as wood, manure, charcoal, or bagasse grown for use as a biofuel. Examples include biodiesel, methanol, and ethanol. Plants that are suitable for biofuel include switch grass, hemp, corn, and sugarcane. Biomass can also be used for building materials and biodegradable plastics and paper. Approximately 15% of the world's energy supply is derived from biomass and is most commonly used in developing nations.

Pros

• Renewable energy source as long as used sustainably.
• Can be sustainable if issue of deforestation and soil erosion are controlled.

- Could supply half of the world's demand for electricity.
- Biomass plantations (cottonwoods, poplars, sycamores, switch grass, and corn) can be located in less desirable locations and can reduce soil erosion and restore degraded land. In the U.S. alone, 200 million acres (80 million ha) are suitable for biomass plantations.

Cons

- Requires adequate water and fertilizer, of which sources are declining.
- Use of inorganic fertilizers, herbicides, and pesticides would harm environment.
- Corn being diverted to produce ethanol raises food prices.
- Would cause massive deforestation and loss of habitat, resulting in a decrease in biodiversity.
- Inefficient methods of burning biomass would lead to large levels of air pollution, especially particulate matter.
- Expensive to transport because it is heavy.
- Not efficient. About 70% of the energy derived from burning biomass is lost as heat.

Biomass can also be burned in large incinerators as an energy source.

Pros

- Crop residues are available (e.g., sugarcane in Hawaii).
- Ash can be collected and recycled.
- Reduces impact on landfills.

Cons

- Net-energy yield is low to moderate. Energy required for drying and transporting material to a centralized facility is prohibitive.
- Severe air pollution if not burned in a centralized facility.
- CO_2 production would have a major impact on global warming.

CASE STUDY

Bagasse is the fibrous matter that remains after sugarcane or sorghum stalks are crushed to extract their juice. It is currently used as a biofuel and in the manufacture of paper and building materials. For each 10 tons of sugarcane crushed, a sugar factory produces nearly 3 tons of bagasse. The high moisture content of bagasse, typically 40 to 50%, is a drawback to its use as a fuel. Approximately 90% of the cars in Brazil run on either alcohol or gasohol (a mixture of gasoline and ethanol) produced from sugarcane grown in Brazil.

Using the by-products of agricultural crops such as bagasse for paper production, rather than wood, does offset commercial forestry practices and reduces the rate of conversion of the rain forest to commercial tree stock; however, this is offset by the clearing of tropical rain forests to areas suitable for growing sugarcane.

Bagasse has been used as a sorbent material to clean up oil spills and to make disposable food containers, replacing materials such as Styrofoam.

Wind Energy

Wind turns giant turbine blades that then power generators. Turbines can be grouped in clusters called wind farms.

Pros

- All electrical needs of the United States could be met by wind in North Dakota, South Dakota, and Texas.
- Wind farms can be quickly built and can also be built out on sea platforms.
- Maintenance is low and the farms are automated.
- Moderate-to-high net-energy yield.
- No pollution. Wind farms are in remote areas so noise pollution is minimal to humans.
- Land underneath wind turbines can be used for agriculture (multiple-use).

Cons

- Steady wind is required to make investment in wind farms economical. Few places are suitable.
- Backup systems need to be in place when the wind is not blowing.
- Visual pollution.
- May interfere with flight patterns of birds.
- May interfere with communication, such as microwaves, TV, and cell phones.
- Noise pollution.

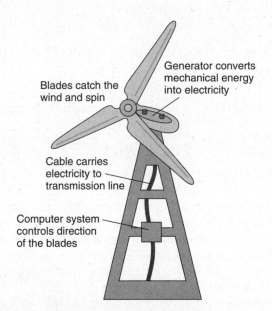

Figure 8.21 Wind turbine

Small-Scale Hydroelectric

Small-scale hydropower utilizes small turbines connected to generators submerged in streams to generate power. Generally, the capacity of small-scale hydropower is 100 kW or less. This technology does not impede stream navigation or fish movement. This technology is especially attractive in remote areas where power lines are not available. Several factors should be considered when installing small-scale hydropower:

- The amount of water flow available on a consistent basis
- The amount of drop (head) the water has between the intake and output of the system
- Regulatory issues such as water rights and easements.

In many cases, there are economic incentives for installing small-scale hydropower systems through grants, loans, and tax incentives.

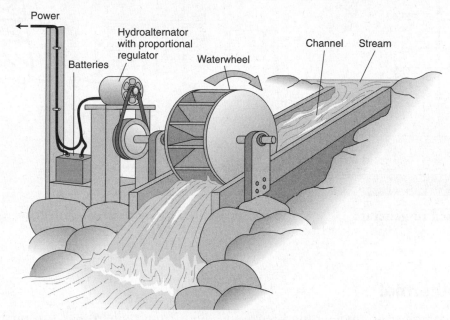

Figure 8.22 Small-scale hydroelectric generator

Ocean Waves and Tidal Energy

The natural movement of tides and waves spin turbines that generate electricity. Only a few plants are currently operating worldwide. They are on the north coast of France and in the Bay of Fundy between the United States and Canada.

Pros

- No pollution.
- Minimal environmental impact.
- Net-energy yield is moderate.

Cons

- Construction is expensive.
- Few suitable sites.
- Equipment can be damaged by storms and corrosion.

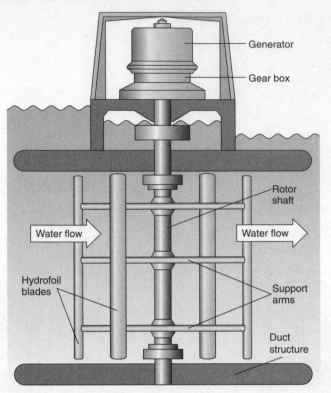

Figure 8.23 Tide turbine used to generate electricity

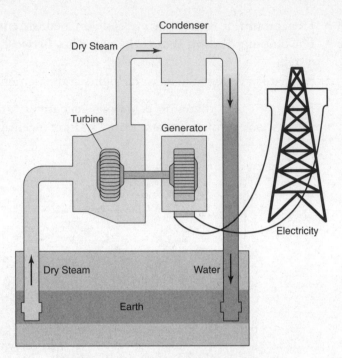

Figure 8.24 Geothermal generator

Geothermal

Heat contained in underground rocks and fluids from molten rock (magma), hot dry-rock zones, and warm-rock reservoirs produce pockets of underground dry steam, wet steam, and hot water. This steam can be used to drive turbines, which can then generate electricity. Geothermal energy supplies less than 1% of the energy needs in the United States. Geothermal energy is currently being used in Hawaii, Iceland, Japan, Mexico, New Zealand, Russia, and California. Areas of known geothermal resources tend to follow tectonic plate boundaries.

Pros

• Moderate net-energy yield.
• Limitless and reliable source if managed properly.
• Little air pollution.
• Competitive cost.

Cons

• Reservoir sites are scarce.
• Source can be depleted if not managed properly.
• Noise, odor, and land subsidence.
• Can degrade ecosystem due to corrosive, thermal, or saline wastes.

RELEVANT LAW

Renewable Energy Law, China (2007): Rapid economic development throughout China has resulted in a significant increase in energy consumption, leading to a rise in harmful emissions and power shortages. China's 2007 Renewable Energy Law requires power grid operators to purchase resources from registered renewable energy producers. The law also offers financial incentives, such as a national fund to foster renewable energy development, and discounted lending and tax preferences for renewable energy projects. The Renewable Energy Law is designed to help protect the environment, prevent energy shortages, and reduce dependence on imported energy. The law includes the use of solar photovoltaics and active solar water heating. Finally, the law includes specific penalties for noncompliance. In 2003, China's renewable energy consumption accounted for only 3% of the country's total energy consumption. The government plans to increase this figure to 10% by 2020. The United States now stands at about 2% usage of renewable energy sources.

CASE STUDY

Bloom Boxes: A Bloom Box is a collection of solid oxide fuel cells that use liquid or gaseous hydrocarbons (such as gasoline, diesel, or propane produced from fossil or bio sources) to generate electricity on the site where it will be used. 20 percent of Bloom Box cost savings result from avoiding transfer losses associated with transmitting energy over an electrical grid. 15 percent of the power at eBay is currently created with Bloom Boxes. Other large companies that are using Bloom Boxes include Google, Staples, Walmart, FedEx, Coca-Cola, and Bank of America.

QUICK REVIEW CHECKLIST

- ☐ **Energy Concepts**
  - ☐ forms of energy
  - ☐ power and units
  - ☐ conversion problems
  - ☐ laws of thermodynamics
- ☐ **Energy Consumption**
  - ☐ present global energy use
  - ☐ energy production vs. consumption
  - ☐ energy production by sector
  - ☐ commodity consumption
- ☐ **Future Energy Needs**
  - ☐ clean coal
  - ☐ methane hydrates
  - ☐ oil shale
  - ☐ tar sands
  - ☐ energy crisis

QUICK REVIEW CHECKLIST (continued)

☐ **Fossil Fuel Resources and Use**
  ☐ extraction-purification methods
    ☐ coal
    ☐ oil
    ☐ natural gas
  ☐ world reserves and global demand
    ☐ coal
    ☐ oil
    ☐ natural gas
    ☐ synfuels

☐ **Environmental Advantages/Disadvantages of Sources**
  ☐ coal
  ☐ oil
  ☐ natural gas

☐ **Nuclear Energy**
  ☐ nuclear fuel
  ☐ electricity production
  ☐ nuclear reactor types
    ☐ light-water reactors
    ☐ heavy-water reactors
    ☐ exotic reactors
  ☐ environmental advantages/disadvantages of nuclear power
  ☐ safety issues
  ☐ Chernobyl
  ☐ nuclear fusion

☐ **Hydroelectric Power**
  ☐ diagram
  ☐ pros/cons
  ☐ flood control
  ☐ channelization
  ☐ levees-floodwalls
  ☐ salmon
  ☐ silting and other impacts

☐ **Energy Conservation**
  ☐ CAFE standards
  ☐ hybrid electric vehicles
    ☐ regenerative braking
    ☐ electric motor/drive assist
    ☐ automatic start/shutoff
    ☐ plug-ins

QUICK REVIEW CHECKLIST (continued)

- [ ] **Mass Transit**
  - [ ] light rail
  - [ ] group or personal rapid transit
  - [ ] automated highway systems
  - [ ] bus rapid transit
  - [ ] maglev
  - [ ] tubular rail
- [ ] **Renewable Energy**
  - [ ] solar energy (pros/cons)
  - [ ] hydrogen fuel cells
    - [ ] diagram
    - [ ] pros/cons
  - [ ] biomass
    - [ ] relation to carbon sink
    - [ ] pros/cons
    - [ ] Brazil and biomass
- [ ] **Wind Energy**
  - [ ] diagram
  - [ ] pros/cons
- [ ] **Small-Scale Hydroelectric**
  - [ ] diagram
  - [ ] pros/cons
- [ ] **Ocean Waves and Tidal Energy**
  - [ ] diagram
  - [ ] pros/cons
- [ ] **Geothermal**
  - [ ] dry steam
  - [ ] hot water
  - [ ] hot dry rock
  - [ ] geopressurized systems
  - [ ] diagram
  - [ ] pros/cons
- [ ] **Renewable Energy Law (China, 2007)**

MULTIPLE-CHOICE QUESTIONS

1. Which of the following forms of energy is a renewable resource?

   (A) Synthetic oil
   (B) Breeder fission
   (C) Biomass
   (D) Oil shale
   (E) Synthetic natural gas

2. Which of the following forms of energy has low, short-term availability?

   (A) Solar energy
   (B) Synthetic oil derived from coal
   (C) Nuclear energy
   (D) Coal
   (E) Petroleum

3. Which of the following is a source of high net energy?

   (A) Tar sands
   (B) Wind
   (C) Fission
   (D) Synthetic natural gas
   (E) Geothermal

4. Which of the following alternatives would NOT lead to a sustainable energy future?

   (A) Phase out nuclear power subsidies.
   (B) Create policies to encourage governments to purchase renewable energy devices.
   (C) Assess penalties or taxes on continuous use of coal and oil.
   (D) Decrease fuel-efficiency standards for cars, appliances, and HVAC (Heating, Ventilation, and Air Conditioning) systems.
   (E) Create tax incentives for independent power producers.

5. At today's rate of consumption, known U.S. oil reserves will be depleted in about

   (A) 100 years
   (B) 50 years
   (C) 25 years
   (D) 10 years
   (E) 3 years

6. Which country currently ranks number one in both coal reserves and use of coal as an energy source?

   (A) Russian Federation
   (B) United States
   (C) China
   (D) India
   (E) Brazil

7. The lowest average generating cost (cents per kWh) comes from what energy source?

   (A) Large hydroelectric facilities
   (B) Geothermal
   (C) Nuclear
   (D) Solar photovoltaic
   (E) Coal

8. The fastest-growing renewable energy resource today is

   (A) nuclear
   (B) coal
   (C) wind
   (D) large-scale hydroelectric
   (E) geothermal

9. The least-efficient energy conversion device listed is

   (A) steam turbine
   (B) fuel cell
   (C) fluorescent light
   (D) incandescent light
   (E) internal combustion engine

10. Which is NOT an advantage of using nuclear fusion?

    (A) Abundant fuel supply
    (B) No generation of weapons-grade material
    (C) No air pollution
    (D) No high-level nuclear waste
    (E) All are advantages

11. Only about 10% of the potential energy of gasoline is used in powering an automobile. The remaining energy is lost as low-quality heat. This is an example of the

    (A) first law of thermodynamics
    (B) second law of thermodynamics
    (C) law of conservation of energy
    (D) first law of efficiency
    (E) law of supply and demand

12. The law of conservation of mass and energy states that matter can neither be created nor destroyed and that the total energy of an isolated system is constant despite internal changes. Which society offers the best long-term solution to the constraints of this law?

    (A) Low-throughput society
    (B) High-throughput society
    (C) Matter-recycling society
    (D) Free-market economy
    (E) Global market economy

13. Which of the following methods CANNOT be used to produce hydrogen gas?

    (A) Reforming
    (B) Thermolysis
    (C) Producing it from plants
    (D) Coal gasification
    (E) All are methods of producing hydrogen gas

14. Energy derived from fossil fuels supplies approximately what percentage of the world's energy needs?

    (A) 10%
    (B) 33%
    (C) 50%
    (D) 85%
    (E) 99%

15. Which of the following terms is NOT a unit of power?

    (A) Watt
    (B) Kilowatt
    (C) Horsepower
    (D) Joule
    (E) All are units of power

16. Automobile manufacturers make money by selling cars. Cars pollute and use fossil fuels. Traditionally, American auto manufacturers have encouraged customers to buy bigger and more powerful cars. This lack of incentive to improve energy efficiency is known as a

    (A) harmful, positive-feedback loop
    (B) beneficial, positive-feedback loop
    (C) harmful, negative-feedback loop
    (D) beneficial, negative-feedback loop
    (E) mutualistic, positive-feedback loop

17. If you were designing a house in the United States, you lived in a cold climate, and you wanted it to be as energy efficient as possible, you would place large windows to capture solar energy on which side of the house?

    (A) North
    (B) South
    (C) East
    (D) West
    (E) It makes no difference which side

18. Which of the following is NOT an advantage of building a hydroelectric power plant?

    (A) Low pollution
    (B) High construction cost
    (C) Relatively low operating cost
    (D) Control flooding
    (E) Moderate-to-high net-useful energy

19. The country that has made the largest commitment to increasing its share of renewable energy resources is

    (A) the United States
    (B) the Russian Federation
    (C) China
    (D) Saudi Arabia
    (E) Iran

20. Which of the following nonrenewable energy sources has the least environmental impact?

    (A) Gasoline
    (B) Coal
    (C) Oil shale
    (D) Tar sands
    (E) Natural gas

FREE-RESPONSE QUESTION

A large, natural gas–fired electrical power facility produces 15 million kilowatt-hours of electricity each day it operates. The power plant requires an input of 13,000 Btus of heat to produce 1 kilowatt-hour of electricity. One cubic foot of natural gas supplies 1,000 Btus of heat energy.

(a) Showing all steps in your calculations, determine the:
   (i) Btus of heat needed to generate the electricity produced by the power plant in 2 hours.
   (ii) Cubic feet of natural gas consumed by the power plant each year.
   (iii) Cubic feet of carbon dioxide gas released by the power plant each day. Assume that methane combusts with oxygen to produce only carbon dioxide and water vapor and that the pressures and temperatures are kept constant.
   (iv) Gross profit per year for the power company. The power company is able to sell electricity at $50 per 500 kWh. They pay a wholesale price of $5.00 per 1,000 cubic feet of natural gas.
(b) What environmental effect might the production of electricity through the burning of natural gas pose?
(c) Describe two other methods of producing electricity, and provide technological, economic, and environmental pros and cons for each method discussed.

MULTIPLE-CHOICE ANSWERS AND EXPLANATIONS

1. **(C)** Renewable resources are those resources that theoretically will last indefinitely either because they are replaced naturally at a higher rate than they are consumed or because their source is essentially inexhaustible.

2. **(E)** At the current rate of consumption, global oil reserves are expected to last another 45–50 years. With projections of increased consumption in the near future, this figure will be even lower.

3. **(B)** Net-useful energy is defined as the total amount of useful energy available from an energy resource over its lifetime minus the amount of energy used in extracting it and delivering it to the end user.

4. **(D)** The key word in this question is "NOT." To foster a sustainable energy future, fuel efficiency standards would have to increase.

5. **(C)** World oil demand is increasing at a rate of about 2% per year. Known U.S. oil reserves are projected to last another 25 years.

6. **(C)** China gets approximately 75% of its energy from coal.

7. **(A)** Nonrenewable resources of energy, such as natural gas or oil, have had recent and dramatic price increases that have resulted in major increases in the cost of electricity. Renewable resources of energy, especially hydroelectric and wind, provide the least-expensive method of producing electricity.

8. **(C)** During the 1990s, wind power experienced an annual growth of 22%. Wind power supplies less than 2% of the energy used in the United States. The country with the largest sector of its energy needs met through wind power is Denmark at 8%.

9. **(D)** An incandescent lightbulb is only 5% efficient as compared with a fluorescent light at 22%. The typical internal combustion engine is 10% efficient, and a hydrogen fuel cell is 60% efficient. The United States wastes as much energy each day as two-thirds of the world consumes.

10. **(E)** The major fuel used in fusion reactors, deuterium, could be readily extracted from ordinary water. The tritium required would be produced from lithium, which can be extracted from seawater and land deposits.

11. **(B)** The second law of thermodynamics states that when energy changes from one form to another, some of the useful energy is always degraded into a lower-quality, more dispersed (higher entropy), and less-useful form.

12. **(A)** A low-throughput society, also known as a low-waste society, focuses on matter and energy efficiency through reusing and recycling, using renewable resources at a rate no faster than can be replenished, reducing unnecessary consumption, emphasizing pollution prevention rather than waste reduction, and controlling population growth.

13. **(E)** Reforming is a chemical process of splitting water molecules. Thermolysis is breaking water molecules apart at high temperatures. Hydrogen gas can be produced from algae by depriving the algae of oxygen and sulfur. Coal gasification is the conversion of coal into synthetic natural gas (SNG), which can then be converted into hydrogen.

14. **(D)** Oil supplies about 36%, coal about 26%, and natural gas about 23%.

15. **(D)** A joule is a unit of energy.

16. **(A)** Pollution is harmful, which is the first part of the answer. A negative-feedback loop tends to slow down a process, while a positive-feedback loop tends to speed it up. Speeding up or increasing sales would be a positive feedback, which is the second half of the answer.

17. **(B)** In cold climates, large south-facing windows allow significant solar energy into the house and also provide daylighting. Properly sized overhangs can prevent overheating in the summer. In hot climates, north-facing windows can provide daylighting without heating the house. East- and west-facing windows generally cause excessive heat gains in the summer and heat losses in the winter, so they are usually small. Although overhangs are impractical for east- and west-facing windows, vertical shading can be used or trees and shrubs can be strategically located to shade the windows.

18. **(B)** The construction of hydroelectric plants is initially expensive but the operating costs are low since there are no fuel costs.

19. **(C)** China has passed legislation to increase its percentage of renewable energy to 10%. The United States' percentage is about 2%.

20. **(E)** The combustion of natural gas releases very small amounts of sulfur dioxide and nitrogen oxides, virtually no ash or particulate matter, and lower levels of carbon dioxide, carbon monoxide, and other reactive hydrocarbons when compared with coal or oil. The following chart compares emission levels of various pollutants for three forms of energy: natural gas, oil, and coal.

Fossil Fuel Emission Levels

Pollutant	Natural Gas	Oil	Coal
(pounds of pollutant per billion Btu of energy input)			
Carbon dioxide	117,000	164,000	208,000
Carbon monoxide	40	33	208
Nitrogen oxides	92	448	457
Sulfur dioxide	1	1,122	2,591
Particulates	7	84	2,744
Mercury	0.000	0.007	0.016

FREE-RESPONSE ANSWER

(a)

(i)

$$\frac{24 \text{ hours}}{1 \text{ days}} \times \frac{1.5 \times 10^7 \text{ kWh}}{24 \text{ hours}} \times \frac{13,000 \text{ Btu}}{1 \text{ kWh}} = 2.0 \times 10^{11} \text{ Btu per day}$$

(ii)

$$\frac{2 \times 10^{11} \text{ Btu}}{\text{day}} \times \frac{1 \text{ ft}^3}{1,000 \text{ Btu}} \times \frac{1 \text{ day}}{24 \text{ hours}} = 8.3 \times 10^6 \text{ ft}^3 \text{ natural gas/hr}$$

(iii) $CH_{4(g)} + 2O_{2(g)} \rightarrow CO_{2(g)} + 2H_2O_{(g)}$. If everything is a gas, and if temperature and pressure are kept constant, then the coefficients can represent volume. Therefore, we could say that 1 ft³ of methane gas combines with 2 ft³ of oxygen gas to produce 1 ft³ of carbon dioxide gas and 2 ft³ of water vapor.

$$\frac{9.3 \times 10^6 \ \cancel{ft^3 \ CH_4}}{\cancel{hr}} \times \frac{1 \ ft^3 \ CO_2}{1 \ \cancel{ft^3} \ \cancel{CH_4}} \times \frac{24 \ \cancel{hours}}{1 \ day} = 2.2 \times 10^8 \ ft^3 \ CO_2 \ per \ day$$

Income:

$$\frac{1.57 \ \cancel{kWh}}{\cancel{day}} \times \frac{365 \ \cancel{days}}{year} \times \frac{\$50}{500 \ \cancel{kWh}} = \$5.48 \times 10^8 \ (\$548 \ million) \ per \ year$$

Costs that the power company spends on natural gas:

$$\frac{8.36 \times 10^6 \ \cancel{ft^3 \ natural \ gas}}{\cancel{hour}} \times \frac{24 \ \cancel{hours}}{1 \ \cancel{day}} \times \frac{365 \ \cancel{days}}{year} \times \frac{\$5.00}{1,000 \ \cancel{ft^3} \ \cancel{natural \ gas}} =$$

$$\$3.66 \times 10^8 \ (\$366 \ million) \ per \ year$$

Gross profit: $(\$5.48 \times 10^8) - (\$3.66 \times 10^8) = \$1.82 \times 10^8 \ (\$182 \ million)$

(b) An environmental effect that results from the burning of natural gas to produce electricity is the production of carbon dioxide gas—a greenhouse gas. As a greenhouse gas, the carbon dioxide gas molecules absorb radiant energy reflected from Earth's surface and re-radiate this energy as long-wave infrared radiation back to Earth. This trapping of Earth's heat has serious environmental consequences in terms of its effect on global weather patterns, local climate conditions, and the glaciers and polar ice caps.

Other environmental effects produced by burning natural gas include the production of hydrogen sulfide gas (H_2S) and sulfur dioxide gas (SO_2) as production by-products. Sulfur dioxide gas when combined with atmospheric water vapor produces what is known as acid rain. Acid rain has serious environmental effects primarily on aquatic organisms in lower trophic levels and on the reproduction of developing aquatic organisms. Hydrogen sulfide has serious environmental effects as a toxic pollutant and respiratory irritant. Leakage of CH_4 during processing adds methane to the atmosphere, which has a more deleterious effect as a greenhouse gas than does carbon dioxide.

Drilling platforms to extract natural gas, the production of contaminated wastewater and brine, the possibility of land subsidence, and the disruption of wildlife and habitat refuges are other deleterious side effects of using natural gas. However, these negative side effects have less environmental impacts than does using other forms of fossil fuels.

(c) Other methods of producing electricity include burning coal and using dams. The advantages of using coal are that it is plentiful and relatively inexpensive. Currently, the world used 4 billion metric tons of coal per year. At this rate, the world has enough coal to last for the next 200 years. Drawbacks to using coal are that current extraction methods are environmentally damaging to the land and natural habitat and are potentially dangerous to miners. Burning coal also releases radioactive particles and mercury into the air. Carbon dioxide produced from burning coal adds to the problem of global warming. The nitrogen oxides released add to the problems of acid precipitation.

Dams create large flooded areas that destroy natural habitats and disrupt the lives of the people who inhabit the area. The collection of sediment by the dam, the loss of this valuable sediment downstream and thereby making the land less productive, the obstructions that dams cause to migrating fish populations, and the aesthetic loss of a wild river are detractors of hydroelectric power. On the positive side, hydroelectric power does not significantly add to global pollution.

TIP

When answering essay and short answer questions, use the analyze, organize, and respond model.

Analyze—Write down key phrases. Even if a key phrase pops into your mind for another question, write it down immediately or you risk forgetting it.

Organize—Sequence your ideas. This ensures that your answer is logically developed and will be clear to the grader.

Respond—Try to provide support for claims that you make in your answer using Case Studies when possible. Star or underline important points. Write legibly. If your handwriting is poor, then print. Use proper grammar and be neat.

UNIT VI: POLLUTION
(25–30%)

Areas on Which You Will Be Tested

A. Pollution Types
1. **Air pollution**—primary and secondary sources, major air pollutants, measurement units, smog, acid deposition—causes and effects, heat islands and temperature inversions, indoor air pollution, remediation and reduction strategies, Clean Air Act, and other relevant laws.
2. **Noise pollution**—sources, effects, and control measures.
3. **Water pollution**—types, sources, causes and effects, cultural eutrophication, groundwater pollution, maintaining water quality, water purification, sewage treatment/septic systems, Clean Water Act, and other relevant laws.
4. **Solid waste**—types, disposal, and reduction.

B. Impacts on the Environment and Human Health
1. **Hazards to human health**—environmental risk analysis, acute and chronic effects, dose-response relationships, air pollutants, smoking, and other risks.
2. **Hazardous chemicals in the environment**—types of hazardous waste, treatment/disposal of hazardous waste, cleanup of contaminated sites, biomagnification, and relevant laws.

C. Economic Impacts—cost-benefit analysis, externalities, marginal costs, and sustainability.

Pollution

CHAPTER 9

AIR POLLUTION

Primary pollutants are emitted directly into the air from natural sources such as volcanoes, mobile sources such as cars, or stationary sources such as industrial smokestacks. Examples include: particulate matter or soot (PM_{10}), nitric oxide (NO), nitrogen dioxide (NO_2), sulfur dioxide (SO_2), carbon dioxide (CO_2), and carbon monoxide (CO).

Secondary pollutants result from the reaction of primary pollutants in the atmosphere to form a new pollutant. Examples include sulfur trioxide (SO_3), sulfuric acid (H_2SO_4), ozone (O_3), and chemicals found in photochemical smog such as PANS and peroxyacyl nitrates.

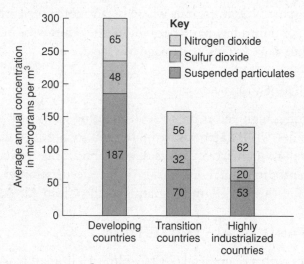

Figure 9.1 Particulate matter pollution in developing countries
Source: World Resources Institute.

Major Air Pollutants

NITROGEN DIOXIDE (NO_2)

Forms when fuels are burned at high temperatures. Also results from forest fires, volcanoes, lightning, and bacterial action in soil. Forms nitric acid (HNO_3) in the air and contributes to acid deposition and cultural eutrophication. Results in lung irritation and damage, suppresses plant growth, and may be a carcinogen.

CRITERIA AIR POLLUTANTS

Criteria air pollutants are a set of air pollutants that cause smog, acid rain, and other health hazards and are typically emitted from many sources in industry, mining, transportation, power generation, and agriculture. They were the first set of pollutants recognized by the U.S. Environmental Protection Agency as needing standards on a national level. In most cases they are the products of the combustion of fossil fuels or industrial processes. They include: ozone; particulate matter; carbon monoxide; sulfur dioxide; nitrogen oxides; and lead.

OZONE (O_3)

Major component of photochemical smog. Formed by sunlight reacting with NO_x and VOCs in the air. Causes lung irritation and damage, bronchial constriction, coughing, wheezing, and eye irritation. Damages plants, rubber, and plastics.

PEROXYACYL NITRATES (PANs)

PANs are secondary air pollutants formed from the reaction of various hydrocarbons combining with oxygen and nitrogen dioxide and being catalyzed by ultraviolet radiation from the sun: hydrocarbons + O_2 + NO_2 + light → $CH_3COOONO_2$ (PAN). Since they dissociate quite slowly in the atmosphere into radicals and NO_2, PANs are able to transport these unstable compounds far away from the urban and industrial origin. PANs transport NO_x to regions where it can more efficiently produce ozone. At concentrations of only a few parts per billion, they cause eye irritation. At higher concentrations, they cause extensive damage to vegetation.

SULFUR DIOXIDE (SO_2)

Produced by burning high-sulfur oil or coal, smelting of metals, and paper manufacturing. Combines with water vapor in the air to produce acid precipitation, which reduces the productivity of plants. Causes breathing difficulties. Significant decreases in concentrations and emissions in SO_2 concentration in the United States reflect the success of the Acid Rain Program and the Clean Air Act.

SUSPENDED PARTICULATE MATTER (PM_{10})

PM_{10}s are particles with a diameter 1/7 the width of a human hair or less (< 10 μm) and include smoke, dust, diesel soot, lead, and asbestos. PM_{10}s cause lung irritation and damage. Many are known mutagens, teratogens, and carcinogens. Reduction in PM_{10}s would produce health benefits 10 times greater than similar reductions in all other air pollutants combined.

VOLATILE ORGANIC COMPOUNDS (VOCs)

Include organic compounds that have a high vapor pressure. Over 600 compounds have been identified. Examples of VOCs include toluene, xylene, formaldehyde, benzene, and acetone. These are found in paints, aerosol sprays, dry-cleaning fluids, and industrial solvents. Causes respiratory irritation and damage. Most are carcinogenic and cause liver, kidney, and central nervous system damage. Concentration of VOCs may be 1,000 times higher indoors than outdoors.

Measurement Units

The most common form of expressing air pollutants is parts per million (ppm); 1 ppm represents one particle of a pollutant for every 999,999 particles of air. The symbol µ is often used to represent a millionth. This concentration is equivalent to one drop of ink in 40 gallons (150 L) of water or one second in 280 hours. To change ppm to a percentage, move the decimal place four places to the left and add a % sign. For example, a concentration of 400 ppm (0.04%) of carbon monoxide may be fatal. Two other common measurements are parts per billion (ppb or nano) and parts per trillion (ppt or pico).

Smog

There are two forms of smog: industrial smog and photochemical smog. Industrial smog tends to be sulfur-based and is also called grey-air smog. Photochemical smog is catalyzed by UV radiation and tends to be nitrogen-based. Photochemical smog is also called brown-air smog.

Formation of Industrial Smog

Step	Chemical Reaction
1. Carbon in coal or oil is burned in oxygen gas to produce carbon dioxide and carbon monoxide gas.	$C + O_2 \rightarrow CO_2$ $C + O_2 \rightarrow CO$
2. Unburned carbon ends up as soot or particulate matter (PM).	C
3. Sulfur in oil and coal reacts with oxygen gas to produce sulfur dioxide.	$S + O_2 \rightarrow SO_2$
4. Sulfur dioxide reacts with oxygen gas to produce sulfur trioxide.	$SO_2 + O_2 \rightarrow SO_3$
5. Sulfur trioxide reacts with water vapor in the air to form sulfuric acid.	$SO_3 + H_2O \rightarrow H_2SO_4$
6. Sulfuric acid reacts with atmospheric ammonia to form brown, solid ammonium sulfate.	$H_2SO_4 + NH_3 \rightarrow (NH_4)_2SO_4$

FORMATION OF PHOTOCHEMICAL SMOG

Net result: $NO + VOCs + O_2 + uv \rightarrow O_3 + PANs$

- **6 A.M.–9 A.M.**
  As people drive to work, concentrations of nitrogen oxides and VOCs increase:

$$N_2 + O_2 \rightarrow 2NO$$

$$NO + VOCs \rightarrow NO_2$$

$$NO_2 \overset{uv}{\rightarrow} NO + O$$

- **9 A.M.–11 A.M.**

As traffic begins to decrease, nitrogen oxides and VOCs begin to react, forming nitrogen dioxide:

$$2NO + O_2 \rightarrow 2NO_2$$

- **11 A.M.–4 P.M.**

As the sunlight becomes more intense, nitrogen dioxide is broken down and the concentration of ozone increases:

$$NO_2 \overset{uv}{\rightarrow} NO + O$$

$$O_2 + O \rightarrow O_3$$

Nitrogen dioxide also reacts with water vapor to produce nitric acid and nitric oxide:

$$3NO_2 + H_2O \rightarrow 2HNO_3 + NO$$

Nitrogen dioxide can also react with VOCs released by vehicles, refineries, gas stations, and so on to produce toxic PANs (peroxacyl nitrates):

$$NO_2 + VOCs \rightarrow PANs$$

- **4 P.M.–Sunset**

As the sun goes down, the production of ozone is halted.

CASE STUDY

Great Smog of '52: A period of cold weather, combined with windless conditions, collected airborne pollutants mostly from the use of coal to form a thick layer of (gray) smog over London in 1952. It lasted from December 5–9,1952, and then dispersed quickly after a change of weather. Estimates run as high as 100,000 people becoming ill and over 12,000 deaths.

Catalytic Converters

A catalytic converter is an exhaust emission control device that converts toxic chemicals in the exhaust of an internal combustion engine into less noxious substances. Inside a catalytic converter, a catalyst stimulates a chemical reaction in which noxious by-products of combustion are converted to less toxic substances by way of catalysed chemical reactions. Most present-day vehicles that run on gasoline are fitted with a "three way" converter, since it converts the three main pollutants:

(a) Reduction of nitrogen oxides to nitrogen and oxygen: $NO_x \rightarrow O_2 + N_2$
(b) Oxidation of carbon monoxide to carbon dioxide: $CO + O_2 \rightarrow CO_2$
(c) Oxidation of unburned hydrocarbons to carbon dioxide and water:
$C_xH_{2x} + O_2 \rightarrow CO_2 + H_2O$

Although catalytic converters are effective at removing hydrocarbons and other harmful emissions, they do not reduce the emission of carbon dioxide produced when fossil fuels are used for fuel. Additionally, the Environmental Protection Agency (EPA) has stated catalytic converters are a significant and growing cause of global warming, because of their release of nitrous oxide (N_2O), a greenhouse gas over 300 times more potent than carbon dioxide. The EPA states that motor vehicles contribute approximately 50% of nitrous oxide emissions and nitrous oxide makes up 7% of greenhouse gases.

Acid Deposition—Causes and Effects

Acid rain is a broad term used to describe several ways that acids fall out of the atmosphere. A more precise term is acid deposition, which has two parts: wet and dry. Wet deposition refers to acidic rain, fog, and snow. As this acidic water flows over and through the ground, it affects a variety of plants and animals. The strength of the effects depends on many factors, including how acidic the water is, the chemistry and buffering capacity of the soils involved, and the types of organisms, such as fish, macroinvertebrates, trees, and other living things that rely on water and soil.

Dry deposition refers to acidic gases and particles. About half of the acidity in the atmosphere falls back to Earth through dry deposition. Wind blows these acidic particles and gases onto buildings, cars, homes, and trees. Dry-deposited gases and particles can also be washed from trees and other surfaces by rainstorms. When that happens, the runoff water adds those acids to the acid rain, making the combination more acidic than the falling rain alone.

Acid deposition due to sulfur dioxide (SO_2) begins with sulfur dioxide being introduced into the atmosphere by burning coal and oil, smelting metals, organic decay, and ocean spray. It then combines with water vapor to form sulfurous acid ($SO_2 + H_2O \rightarrow H_2SO_3$). Finally, the sulfurous acid reacts with oxygen to form sulfuric acid ($H_2SO_3 + \frac{1}{2}O_2 \rightarrow H_2SO_4$).

Acid deposition due to nitrogen oxides (NO_x) begins with nitrogen oxides formed by burning oil, coal, or natural gas. They also are found in volcanic vent gases and formed by forest fires, bacterial action in soil, and lightning-induced atmospheric reactions. Nitrogen monoxide, also known as nitric oxide (NO), reacts with oxygen gas to produce nitrogen dioxide gas ($NO + \frac{1}{2}O_2 \rightarrow NO_2$). Finally, nitrogen dioxide reacts with water vapor in the atmosphere to produce nitrous and nitric acids ($2NO_2 + H_2O \rightarrow HNO_2 + HNO_3$).

Acid rain causes acidification of lakes and streams. It contributes to the damage of trees at high elevations and many sensitive forest soils through nitrogen saturation and creating acidic conditions that are unhealthy for decomposers and mycorrhizal fungi. Acid shock, which is caused by the rapid melting of snow pack that contains dry acidic particles, results in acid concentrations in lakes and streams 5 to 10 times higher than acidic rainfall. In addition, acid rain accelerates the decay of building materials and paints, including irreplaceable statues and sculptures. Acid rain also leaches essential plant nutrients from the soil such as Ca^{2+}, K^+ and Mg^{2+}. Heavy metal ions such as Pb^{2+}, Cd^{2+}, and Hg^{2+} that are contained within rock structures may be leached out of the rocks and into the soil structure. Prior to falling to Earth, SO_2 and NO_x gases and their particulate matter derivatives (sulfates and nitrates) contribute to breathing difficulties and other health matters.

Heat Islands and Temperature Inversions

Urban heat islands occur in metropolitan areas that are significantly warmer than their surroundings. Urban air can be 10°F (6°C) warmer than the surrounding area. Since warmer air can hold more water vapor, rainfall can be as much as 30% greater downwind of cities when compared with areas upwind. One of the main reasons for higher-than-normal nighttime air temperatures in urban areas are buildings that reduce the radiation of urban heat to the night sky. Thermal properties of surface materials (bricks, concrete, and asphalt) store heat longer. The lack of vegetation and standing water in many urban areas also increase urban temperatures. The canyon effect results from buildings reflecting and absorbing heat and blocking winds that reduce heat through convection. Human activities that increase the heat island effect include the operation of automobiles, air conditioners, and industry. Since demand for air-conditioning rises during summer months, problems associated with energy availability and pricing become compounded.

High levels of pollution in urban areas can also create a localized greenhouse effect. Urban heat islands can directly influence the health and welfare of urban residents who cannot afford air-conditioning. As many as 1,000 people per year die in the United States due to excessive temperatures. Urban heat islands can produce secondary effects on local meteorology, including the altering of local wind patterns, the development of clouds and fog, the number of lightning strikes, and the rate of precipitation. The heat island effect can be slightly reduced by using white or reflective building materials and increasing the amount of landscaping and parks.

Temperature inversions occur when air temperature increases with height above the ground, as opposed to the normal decrease in temperature with height. This effect can lead to pollution such as smog being trapped close to the ground, with possible adverse effects on human health (asthma, emphysema, and increases in lung cancer). Temperature inversions commonly occur at night when solar heating ceases and the surface cools, which then cools the atmosphere immediately above it. A warm air mass then moving over a colder one keeps the cooler air mass trapped below, and the air becomes still. This results in dust and pollutants being trapped and their concentrations increasing. A nearly permanent temperature inversion occurs over Antarctica.

CASE STUDY

In 1948, 20 people were asphyxiated and over 7,000 were hospitalized or became ill as the result of severe air pollution over Donora, PA, a town of 14,000. Smog from the local zinc and steel smelting plants settled in the valley where the town was located. Four days later, winds finally cleared the toxins from the town. The investigation of this incident by state and federal health officials resulted in the first meaningful federal and state laws to control air pollution and marked the beginning of modern efforts to assess and deal with the health threats from air pollution.

Indoor Air Pollution (Sick Building Syndrome)

Many people spend the majority of their lives indoors sleeping, working, eating, and relaxing, where air circulation may be restricted and where indoor air pollutant levels may be 25% to 60% greater than outdoor levels. Sick building syndrome (SBS) is a term used to describe a combination of ailments (a syndrome) associated with an individual's place of work or residence. Up to a third of new and remodeled buildings worldwide may be linked to symptoms of SBS. The most common pollutants found indoors include molds, bacteria, carbon monoxide, radon, allergens, asbestos, tobacco smoke, and formaldehyde and other volatile organic compounds (VOCs) released from carpeting, adhesives, and particleboard. SBS is frequently pinned down to flaws in heating, ventilation, and air conditioning (HVAC) systems. Symptoms of indoor air pollution can range from headaches, breathing difficulties, and allergies to asthma, cancer, emphysema, and various nerve disorders.

Remediation and Reduction Strategies

Strategies designed to improve air quality in general include:

1. Emphasizing tax incentives for pollution control rather than fines and penalties

2. Setting legislative standards for energy efficiency

3. Increasing funding for research into renewable energy sources

4. Incorporating incentives for reducing air pollution into trade policies

5. Distributing solar cookstoves to developing countries to replace coal and firewood

6. Phasing out two-cycle gasoline engines

7. For issues involving sick building syndrome: (a) modify building codes to control materials used in construction; (b) replace and repair areas that have received water damage in order to control mold (carpeting, ceiling tiles, inside walls, etc.); (c) use paints, adhesives, solvents, cleaning products, and pesticides in well-ventilated areas and during periods of non-occupancy; (d) increase the number of complete air exchanges in the building; and (e) ensure proper maintenance of HVAC systems.

8. Providing incentives to use mass transit

Several strategies have been designed to reduce the effects of acid rain. They include designing more efficient engines to reduce NO_x emissions and increasing the efficiency of coal-burning plants to reduce SO_2, NO_x, and particulates through washing coal and using scrubbers, advanced filtration on smokestacks, electrostatic precipitators, and staged catalytic burners with afterburners. Other strategies include increasing penalties on stationary sources that do not reduce emissions, providing tax incentives to companies that do reduce pollutants, providing incentives to consumers to purchase Energy Star appliances and products that conserve energy, and increasing CAFE standards.

The EPA's Acid Rain Program is designed to achieve significant environmental and public health benefits through reductions in emissions of sulfur dioxide (SO_2) and nitrogen oxides (NO_x), the primary causes of acid rain. To achieve this goal at the lowest cost to society, the program employs both traditional and innovative market-based approaches for controlling air pollution. In addition, the program encourages energy efficiency and pollution prevention. Specific strategies employed include an allowance trading system, an opt-in program that allows nonaffected industrial and small utility units to participate in allowance trading, setting new NO_x emissions standards for existing coal-fired utility boilers, and allowing emissions averaging to reduce costs. Another method is a permit process that affords sources maximum flexibility in selecting the most cost-effective approach to reducing emissions. Continuous emission monitoring (CEM) requirements provide credible accounting of emissions to ensure the integrity of the market-based allowance system and to verify the achievement of the reduction goals. The excess emissions provision provides incentives to ensure self-enforcement. Another strategy is an appeals procedure that allows the regulated community to appeal decisions with which it may disagree.

The Clean Air Act was originally signed into law in 1963. Its goal was to protect public health from air pollution and limit the effects of air pollution on the environment. Early versions allowed individual states to set their own standards. Later versions of the act switched responsibility of setting uniform standards to the federal government. Primary standards protect human health. Secondary standards protect materials, crops, climate, visibility, and personal comfort. The 1990 version addressed acid rain, urban smog, air pollutants, ozone protection, marketing pollution rights, and VOCs. Estimates are that the Clean Air Act is responsible each year for saving 15,000 lives, reducing bronchitis cases by 60,000, and reducing 9,000 hospital admissions due to respiratory illnesses. The following chart shows some of the progress that the Clean Air Act has been responsible for since it was passed into law. Notice however, the increase in PM_{10} and NO_x levels.

Pollutant	% Change
Pb	−98%
VOC	−42%
SO_2	−37%
CO	−31%
NO_x	+17%
PM_{10}	+266 %

The 1997 Kyoto Protocol would have required the United States to reduce greenhouse emissions by 7% when compared with 1990 levels over a five-year period. Under this agreement, the United States would have faced penalties if it did not meet its emission cuts. The United States saw this as an unattainable target since carbon dioxide and greenhouse gases continue to increase and are projected to increase for the next 20 years. The United States felt that the protocol held developed nations responsible for meeting the cuts but did not apply the same standards to developing nations. Reasons given for not agreeing with the rest of the signing members of the Kyoto Protocol were the cost of meeting the emission targets would be too high, the time frame was too short for implementation, and there was no evidence of a correlation between greenhouse gases and global warming.

In 2012, the year the 1997 Kyoto Protocol expired, Canada, Japan, and Russia joined the United States stating that they will not agree to an extension of the Kyoto Protocol unless the unbalanced requirements of developing and developed countries are changed.

RELEVANT LAWS AND PROTOCOLS

Air Pollution Control Act (1955): The nation's first piece of federal legislation regarding air pollution. Identified air pollution as a national problem and announced that research and additional steps to improve the situation needed to be taken. It was an act to make the nation more aware of this environmental hazard.

Clean Air Act (1963): The Clean Air Act is designed to control air pollution on a national level. It requires the Environmental Protection Agency (EPA) to develop and enforce regulations to protect the general public from exposure to airborne contaminants that are known to be hazardous to human health. The act (amended in 1967, 1977, and 1990):

- required comprehensive federal and state regulations for both stationary (industrial) and mobile sources of air pollution.
- expanded federal enforcement authority.
- addressed acid rain, ozone depletion, and toxic air pollution.
- established new auto gasoline reformulation requirements.
- was the first major environmental law in the United States to include a provision for citizen suits.

National Environmental Policy Act (1970): Created the Environmental Protection Agency (EPA). Also mandated creation of Environmental Impact Statements.

Montreal Protocol (1989): An agreement among nations requiring the phaseout of chemicals that damage the ozone layer.

Pollution Prevention Act (1990): Requires industries to reduce pollution at its source. Reduction can be in terms of volume and/or toxicity.

Kyoto Protocol (1997 and 2001): An agreement among 150 nations requiring greenhouse gas reductions.

NOISE POLLUTION

Noise pollution is unwanted human-created sound that disrupts the environment. The dominant form of noise pollution is from transportation sources, primarily motor vehicles, aircraft noise, and rail transport noise. Besides transportation noise, other prominent sources are office equipment, factory machinery, appliances, power tools, and audio entertainment systems. Noise regulation by governmental agencies effectively began in the United States with the 1972 Federal Noise Control Act.

Effects

Normal hearing depends on the health of the inner, middle, and outer ear. Three kinds of hearing loss occur: conductive, sensory, and neural. Sensory hearing loss is caused by damage to the inner ear and is the most common form associated with noise.

In addition to contributing to hearing loss, too much noise can affect health in other ways too. Immediate effects may be temporary or may become permanent. They may include cardiovascular problems with an accelerated heartbeat and high blood pressure, gastric-intestinal problems, a decrease in alertness and ability to memorize, nervousness, pupil dilation, and a decrease in the visual field. Effects that may be longer lasting include insomnia, nervousness, bulimia, chronically high blood pressure, anxiety, depression, and sexual dysfunction.

Control Measures

Roadway noise can be reduced through the use of noise barriers, limitations on vehicle speed, newer roadway surface technologies, limiting times for heavy-duty vehicles, computer-controlled traffic flow devices that reduce braking and acceleration, and changes in tire design. Aircraft noise can be reduced through developing quieter jet engines and rescheduling takeoff and landing times. Industrial noise can be reduced through new technologies in industrial equipment and installation of noise barriers in the workplace. Residential noise such as power tools, garden equipment, and loud radios can be controlled through local laws and enforcement.

RELEVANT LAW

Noise Control Act (1972): Establishes a national policy to promote an environment for all Americans free from noise that jeopardizes their health and welfare. To accomplish this, the act establishes a means for the coordination of federal research and activities in noise control, authorizes the establishment of federal noise emissions standards for products distributed in commerce, and provides information to the public respecting the noise emission and noise reduction characteristics of such products.

WATER POLLUTION

Water pollution can originate from either a point or a nonpoint source. A point source occurs when harmful substances are emitted directly into a body of water. An example of a point source of water pollution is a pipe from an industrial facility discharging effluent directly into a river. Point source pollution is usually monitored and regulated in developed countries.

Nonpoint sources deliver pollutants indirectly through transport or environmental change. An example of a nonpoint source of water pollution is when fertilizer from a farm field is carried into a stream by rain (run off). Nonpoint sources are much more difficult to monitor and control, and they account for the majority of contaminants in streams and lakes.

Sources of Water Pollution

The following sections describe the varied sources of water pollution. They include air pollution, chemicals, microbiological sources, mining, noise, nutrients, oxygen-depleting substances, suspended matter, and thermal sources.

AIR POLLUTION

Pollutants like mercury, sulfur dioxide, nitric oxides, and ammonia fall out of the air and into the water. They can then cause mercury contamination in fish and acidification and eutrophication of lakes. The oceans have absorbed enough carbon dioxide to have already caused a slight increase in ocean acidification and which may be causing the carbonate structures of corals, algae, and marine plankton to dissolve. These organisms form the base of the food pyramid in the ocean.

METHYL MERCURY

Mercury has been well known as an environmental pollutant for several decades. There are many sources of mercury in the environment, both natural and human made. Natural sources include volcanoes, natural mercury deposits, and volatilization from the ocean. The primary human-related sources include: coal combustion, waste incineration, and metal processing. Best estimates to date suggest that human activities have about doubled or tripled the amount of mercury in the atmosphere, and the atmospheric burden is increasing by about 1.5% per year. Like many environmental contaminants, mercury undergoes bioaccumulation. The bioaccumulation effect is generally compounded the longer an organism lives, so that larger predatory game fish will likely have the highest mercury levels. Adding to this problem is the fact that mercury concentrates in the muscle tissue of fish. Humans generally uptake mercury in two ways: (1) as methylmercury (CH_3Hg^+) from fish consumption, or (2) by breathing vaporous mercury in the ambient air. The ultimate source of mercury to most aquatic ecosystems is deposition from the atmosphere, primarily associated with rainfall. Once in surface water, mercury enters a complex cycle in which one form can be converted to another. Studies have shown that bacteria that process sulfate (SO_4^{2-}) in the environment take up mercury in its inorganic form, and through metabolic processes convert it to methylmercury. The conversion of inorganic mercury to methylmercury is important for two reasons: (1) methylmercury is much more toxic than inorganic mercury, and (2) organisms require considerably longer to eliminate methylmercury than elemental mercury. At this point, the methylmercury-containing bacteria may be consumed by the next higher level in the food chain, or the bacteria may release the methylmercury to the water where it can quickly adsorb to plankton, which are also consumed by the next level in the food chain.

OTHER CHEMICALS

A variety of chemicals from industrial and agricultural sources can cause water pollution. Examples include metals, solvents and oils, detergents, and pesticides. These can accumulate in fish and shellfish, poisoning the people, animals, and birds that consume them. On a square-foot basis, homeowners apply more chemicals to their lawns than farmers do to their fields. Each year, road runoff and other nonspill sources impart an amount of oil to the oceans that is more than 5 times greater than the *Exxon Valdez* spill—about 21 million barrels. Discharge of oily wastes and oil-contaminated ballast water and wash water are all significant sources of marine pollution. Drilling and extraction operations for oil and gas can also contaminate coastal waters and groundwater. The EPA estimates that about 100,000 gasoline storage tanks are leaking chemicals into groundwater. In Santa Monica, California, wells supplying half the city's water have been closed because of dangerously high levels of the gasoline additive MTBE. New evidence strongly suggests that components of crude oil, called polycyclic aromatic hydrocarbons (PAHs), persist in the marine environment for years and are toxic to marine life at concentrations in the low parts per billion (ppb) range. Chronic exposure to PAHs can affect the development of marine organisms, increase susceptibility to disease, and jeopardize normal reproductive cycles in many marine species.

Studies have shown that up to 90% of drug prescriptions pass through the human body unaltered. Animal farming operations that use growth hormones and antibiotics also send large quantities of these chemicals into the water. Most wastewater treatment facilities are not equipped to filter out personal care products, household products, or pharmaceuticals. As a result, a large portion of these chemicals pass directly into local waterways. Studies on the effects of these chemicals have discovered fragrance molecules inside fish tissues, ingredients from birth control pills causing gender-bending hormonal effects in frogs and fish, and the chemical nonylphenol, a remnant of detergent, disrupting fish reproduction and growth.

MICROBIOLOGICAL SOURCES

Disease-causing (pathogenic) microorganisms such as bacteria, viruses, and protozoa can result in swimmers getting sick and fish and shellfish becoming contaminated. Examples of waterborne diseases include cholera, typhoid, shigella, polio, meningitis, and hepatitis. In developing countries, an estimated 90% of the wastewater is discharged directly into rivers and streams without treatment. In the United States, 850 billion gallons (3 trillion L) of raw sewage are dumped into rivers, lakes, and bays each year by leaking sewer systems and inadequate combined sewer/storm systems that overflow during heavy rains. Leaking septic tanks and other sources of sewage can also cause groundwater and stream contamination. Beaches suffer the effects of water pollution from sewage. About 25% of all beaches in the United States annually have water pollution advisories or are closed each year due to bacterial buildup caused by sewage.

MINING

Mining causes water pollution in a number of ways. The mining process exposes heavy metals and sulfur compounds that were previously locked away in Earth. Rainwater leaches these compounds out of the exposed Earth, resulting in acid mine drainage and heavy-metal pollution that can continue long after the mining operations have ceased. Second, the action of rainwater on piles of mining waste (tailings) transfers pollution to freshwater supplies. In the case of gold mining, cyanide is intentionally poured on piles of mined rock (a leach heap) to extract the gold from the ore chemically. Some of the cyanide ultimately finds its way into nearby water. Additionally, huge pools of mining waste slurry are often stored behind containment dams that often leak or infiltrate ground water supplies. Fourth, mining companies in developing countries often dump mining waste directly into rivers or other bodies of water as a method of disposal.

The U.S. government in 2003 reclassified mining waste from mountaintop removal (a type of coal mining) so it could be dumped directly into valleys and burying streams altogether. The Iron Mountain mine in California has been closed since 1963 but continues to drain sulfuric acid and heavy metals into the Sacramento River. Experts say the pollution from this particular mine may continue for another 3,000 years.

NOISE

Many marine organisms, including marine mammals, sea turtles, and fish, use sound to communicate, navigate, and hunt. Because of oceanic water noise pollution caused by commercial shipping, military sonar, and recreational boating, some species may have a harder time hunting or detecting predators. They may also not being able to navigate properly.

NUTRIENTS

Phosphorus and nitrogen are necessary for plant growth and are plentiful in untreated wastewater. When added to lakes and streams, they can cause the growth of aquatic weeds that block waterways as well as algal blooms. If the source is from humans, it is called cultural eutrophication. Deposition of atmospheric nitrogen (from nitrogen oxides) also causes nutrient-type water pollution. Nutrient pollution is also a problem in estuaries and deltas, where the runoff that was aggregated by watersheds is finally dumped at the mouths of major rivers.

OIL SPILLS

Oil is one of the world's main sources of energy, but because it is unevenly distributed worldwide, it must be transported by ship across oceans and by pipelines across land. This can result in accidents when transferring oil to vessels, when transporting oil, and when pipelines break, as well as when drilling for oil. Oil accidentally released into a marine environment drastically affects wildlife. The oil penetrates the feathers of seabirds, reducing the feathers' insulating ability and making the birds more vulnerable to temperature fluctuations and much less buoyant in the water. It also impairs seabirds' flight, and thus their abilities to forage and escape from predators. As they attempt to preen, birds typically ingest oil that covers their feathers, causing kidney and liver damage. This, along with the limited foraging ability, quickly results in dehydration. Marine mammals exposed to oil spills are affected in similar ways as seabirds. Because oil floats on top of water, less sunlight penetrates into the water, limiting the photosynthesis of marine plants and phytoplankton and affecting the food web in the ecosystem.

Recovering the oil is difficult and depends on many factors, including the type of oil spilled, the temperature of the water, and the types of shorelines and beaches involved. Methods for cleaning up include the use of microorganisms to break down oil; chemical agents, dispersants, sorbents, and detergents that act to disperse the oil, absorb it, or cause it to clump into gel-like agglomerations that sink; controlled burning; and booming, skimming, and/or vacuuming the oil from the surface or shoreline.

OXYGEN-DEPLETING SUBSTANCES

Biodegradable wastes are used as nutrients by bacteria and other microorganisms. Excessive biodegradable wastes can cause oxygen depletion in receiving waters. This can result in increases in anaerobic bacteria that produce ammonia, amines, sulfides, and methane (swamp gas) and decreases of aerobic organisms such as fish.

PLASTIC

The Great Pacific Garbage Patch, or Pacific Trash Vortex, is a large system of rotating ocean currents (gyre) of marine litter in the central North Pacific Ocean and is characterized by high concentrations of floating plastics, chemical sludge, and other debris that have been trapped by the currents of the North Pacific Gyre. It was formed gradually as a result of marine pollution gathered by oceanic currents. The gyre's rotational pattern draws in waste material from across the North Pacific Ocean, including coastal waters off North America and Japan. As material is captured in the currents, wind-driven surface currents gradually move floating debris toward the center, trapping it in the region. Estimates of its size range from 0.5% to 8.1% of the size of the Pacific Ocean. As the plastic photodegrades into smaller and smaller pieces, it remains as plastic polymers with some leaching toxic chemicals such as bisphenol A and PCBs, which then concentrate in the upper water column. As it disintegrates, the plastic ultimately becomes small enough to be ingested by aquatic organisms that reside near the ocean's surface and enters the marine food chain. Some of these long-lasting plastics end up in the stomachs of marine birds and animals, and their young. The floating debris can also absorb organic pollutants from sea water, which ultimately results in bioaccumulation of these toxins throughout the food chain. Marine plastics also facilitate the spread of invasive species that attach to floating plastic in one region and drift long distances to colonize other ecosystems. On the macroscopic level, the physical size of the plastic kills birds and turtles as the animals' digestion cannot break down the plastic inside their stomachs. Another effect of the macroscopic plastic is that it makes it much more difficult for animals to see and detect their normal sources of food through the water column. Research has shown that this plastic marine debris affects over 250 species worldwide.

SUSPENDED MATTER

Suspended wastes eventually settle out of water and form silt or mud at the bottom. Toxic materials can also accumulate in the sediment and affect organisms throughout the food web. When forests are clear-cut, the root systems that previously held soil in place die and the sediment is free to run off into nearby streams, rivers, and lakes.

Plastics and other plastic-like substances (such as nylon from fishing nets and lines) can entangle fish, sea turtles, and marine mammals, causing injury and death. Certain types of plastic can break down into microparticles and become ingested by tiny marine organisms and move up the marine food chain. Plastic remains in the ecosystem and will continue to harm marine organisms far into the future.

THERMAL SOURCES

Produced by industry and power plants. Heat reduces the ability of water to hold oxygen and causes death to organisms that cannot tolerate heat and/or low oxygen levels. Global warming is also imparting additional heat to the oceans, rivers, and streams with unknown consequences.

> **CASE STUDIES**
>
> **Minamata disease:** Twenty-seven tons of mercury-containing compounds from industrial processes were dumped into Minamata Bay in Japan between 1932 and 1968. The mercury collected in fish and shellfish caught from the bay. Symptoms included blurred vision, hearing loss, loss of muscular coordination, and reproductive disorders.
>
> ***Exxon Valdez* (1989):** In 1989, the oil tanker *Exxon Valdez* spilled 11 to 30 million gallons (42 to 110 million L) of crude oil into Prince William Sound, Alaska. As a result, 250,000 sea birds, 3,000 otters, 300 seals, 300 bald eagles, and 22 whales died along with billions of salmon and herring eggs. The oil also destroyed the majority of the plankton in the sound.
>
> **Gulf of Mexico Oil Spill (2010):** In April 2010, a massive oil spill followed an explosion on the *Deepwater Horizon* offshore drilling rig operated by British Petroleum, becoming the most significant environmental disaster to occur in the United States. As the oil from the well site reached the Gulf coast, billions of dollars in damage was done to the Gulf of Mexico fishing industry, the tourism industry, and the habitat of hundreds of bird, fish, and other wildlife species.

Cultural Eutrophication

Cultural eutrophication is defined as the process whereby human activity increases the amount of nutrients entering surface waters. The two most important nutrients that cause cultural eutrophication are nitrates (NO_3^-) and phosphates (PO_4^{3-}) that come from fertilizer, sewage discharge, and animal wastes.

Nitrates are water soluble. Nitrates found in fertilizers can remain on fields and accumulate, leach into groundwater, end up in surface runoff, and/or volatize and enter the atmosphere where they contribute to acid precipitation. Nitrates cause nitrate poisoning in water supplies, reduce the effectiveness of hemoglobin, and may be responsible for worldwide declines in amphibians.

Phosphates are also a component of inorganic fertilizers. However, they are not water soluble and adhere to soil particles. Soil erosion contributes to the buildup of phosphates in water supplies. Phosphate levels in water supplies are 75% higher than they were during preindustrial times. Phosphate buildup is more damaging in freshwater systems. In contrast, nitrate pollution is more damaging in wetlands where nitrogen is the limiting factor.

Nitrates and phosphates are algal nutrients. Increased concentrations of these nutrients increase the carrying capacity of lakes and streams. Explosions in the amount of algae as a result of cultural eutrophication are called algal blooms. The steps involved in algal bloom include the following:

* Increased algae due to increased nitrate and/or phosphate concentrations result in decreased light penetration and killing off of deeper plants and their supply of oxygen to water.
* Oxygen concentration decreases in the water due to the consequences of increased material for decomposers.
* Lower oxygen concentrations cause fish and other aquatic organisms to die and contaminate the water at a high rate.
* Decaying fish and algae produce toxins in the water.

Several methods can control cultural eutrophication. Planting vegetation (buffer zones) along streambeds slows erosion and absorbs some of the nutrients. Controlling the application and timing of applying fertilizer, controlling runoff from feedlots, and using biological controls such as denitrifying bacteria that convert nitrates into atmospheric nitrogen are other methods.

Groundwater Pollution

About 50% of the people in the United States depend on groundwater for their water supplies. In some countries, it may reach as high as 95%. Almost half of the water used for agriculture in the United States comes from groundwater. The Environmental Protection Agency (EPA) estimates that each day, 4.5 trillion liters of contaminated water seep into groundwater supplies in the United States. In the United States, 34 billion liters per year (60%) of the most hazardous liquid waste solvents, heavy metals, and radioactive materials are injected directly into deep groundwater via thousands of injection wells. Although the EPA requires that these effluents be injected below the deepest source of drinking water, some pollutants have already entered underground water supplies in Florida, Texas, Ohio, and Oklahoma.

Water entering an aquifer remains there for approximately 1,400 years compared with 16 days for water entering a river system. Once an aquifer is contaminated, it is practically impossible to remove the pollutants. For example, in Denver, Colorado, just 80 liters of organic solvents contaminated 4.5 trillion liters of groundwater. Initial cleanup of contaminated groundwater locations in the United States could cost up to $1 trillion over the next 30 years.

Maintaining Water Quality and Water Purification

DRINKING WATER TREATMENT METHODS

- **Adsorption**—Contaminants stick to the surface of granular or powdered activated charcoal.
- **Disinfection**—Chlorine, chloramines, chlorine dioxide, ozone, and UV radiation.
- **Filtration**—Removes clays, silts, natural organic matter, and precipitants from the treatment process. Filtration clarifies water and enhances the effectiveness of disinfection.
- **Flocculation-Sedimentation**—Process that combines small particles into larger particles that then settle out of the water as sediment. Alum, iron salts, or synthetic organic polymers are generally used to promote coagulation.
- **Ion Exchange**—Removes inorganic constituents. It can be used to remove arsenic, chromium, excess fluoride, nitrates, radium, and uranium.

WATER TREATMENT REMEDIATION TECHNOLOGIES

- **Adsorption/absorption**—Solutes concentrate at the surface of a sorbent (an absorbing surface), thereby reducing their concentration.
- **Aeration**—Bubbling air through water increases rates of oxidation.
- **Air stripping**—VOCs are separated from groundwater by exposing water to air (the VOCs evaporate due to their high vapor pressure).
- **Bioreactors**—Groundwater is acted upon by microorganisms.
- **Constructed wetlands**—Uses natural geochemical and biological processes that parallel natural wetlands. Also known as living machines.

- **Deep-well injection**—Uses injection wells to place treated or untreated liquid waste into geologic formations that do not pose a potential risk to groundwater.
- **Enhanced bioremediation**—The natural rate of bioremediation is enhanced by adding oxygen and nutrients into groundwater.
- **Fluid-vapor extraction**—A vacuum system is applied to low-permeable soil to remove liquids and gases.
- **Granulated activated carbon (GAC)**—Groundwater is pumped through a series of columns containing activated carbon.
- **Hot water or steam flushing**—Steam or hot water is forced into an aquifer to vaporize volatile contaminants and is then treated through fluid-vapor extraction.
- **In-well air stripping**—Air is injected into wells—the air picks up various contaminants, particularly VOCs. Vapors are drawn off by vapor extraction.
- **Ion exchange**—Involves exchange of one ion for another.
- **Phytoremediation**—Uses plants to remove contamination.
- **UV oxidation**—Uses ultraviolet light, ozone, or hydrogen peroxide to destroy microbiological contaminants.

Sewage Treatment/Septic Systems

Sewage treatment incorporates physical, chemical, and biological processes to remove contaminants from wastewater. There are three stages of wastewater treatment.

A septic system consists of a tank and a drain field. Wastewater enters the tank, where solids settle. Anaerobic digestion using bacteria treats the settled solids and reduces their volume. Excess liquid leaves the tank and moves through a pipe with holes in it to a leach field where the water then percolates into the soil.

Some pollutants, especially nitrogen, do not decompose in a septic system and may contaminate the groundwater. Approximately 25% of Americans rely on septic systems.

PRIMARY TREATMENT—SEPARATION OF SOLIDS

Primary treatment is to reduce oils, grease, fats, sand, grit, and coarse solids. Specific steps include sand catchers, screens, and sedimentation. This is a physical method of cleaning.

SECONDARY TREATMENT—BREAKDOWN BY BACTERIA

Secondary treatment is designed to degrade substantially the biological content of the sewage derived from human waste, food waste, soaps, and detergent. Specific steps include filters, activated sludge, filter (oxidizing) beds, trickling filter beds using plastic media, and secondary sedimentation. This is a biological method of cleaning.

TERTIARY TREATMENT—DISINFECTION

Tertiary treatment provides a final stage to raise the effluent quality to the standard required before it is discharged to the receiving environment (sea, river, lake, or ground). Specific steps may include sand filtration, lagooning, constructed wetlands, nutrient removal through biological or chemical precipitation, denitrification using bacteria, phosphorous removal using bacteria, microfiltration, and disinfection using UV light, chlorine, or ozone.

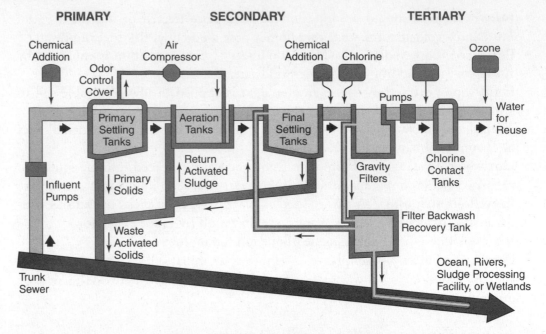

Figure 9.2 Sewage treatment plant

RELEVANT LAWS

Clean Water Act (1972): Established the basic structure for regulating discharges of pollutants into the waters of the United States. It gave the EPA the authority to implement pollution control programs such as setting wastewater standards for industry. The Clean Water Act also continued requirements to set water quality standards for all contaminants in surface waters. The act made it unlawful for any person to discharge any pollutant from a point source into navigable waters unless a permit was obtained under its provisions. It also funded the construction of sewage treatment plants under the construction grants program and recognized the need for planning to address the critical problems posed by nonpoint source pollution.

Safe Drinking Water Act (1974): Established standards for safe drinking water in the United States.

Ocean Dumping Ban Act (1988): Made it unlawful for any person to dump or transport for the purpose of dumping sewage, sludge, or industrial wastes into the ocean.

Oil Spill Prevention and Liability Act (1990): Strengthened the EPA's ability to prevent and respond to catastrophic oil spills.

SOLID WASTE

Types of solid waste include:

- **Organic**—Kitchen wastes, vegetables, flowers, leaves, or fruits. Usually decomposes within 2 weeks. Wood can take 10 to 15 years to decompose.
- **Radioactive**—Spent fuel rods and smoke detectors. Radioactive wastes can take hundreds of thousands of years to decompose.
- **Recyclable**—Paper, glass, metals, and some plastics. Paper decomposes in 10 to 30 days. Glass does not decompose. Metals decompose in 100 to 500 years. Some plastics can take up to 1 million years to decompose.

- **Soiled**—Hospital wastes. Cotton and cloth can take 2 to 5 months to decompose.
- **Toxic**—Paints, chemicals, pesticides, and so on. Toxic wastes can take hundreds of years to decompose.

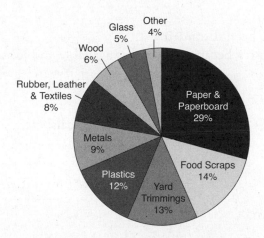

Figure 9.3 Amounts and types of municipal solid wastes (MSW) in the U.S.

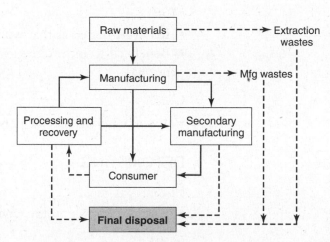

Figure 9.4 Solid waste flow diagram

Disposal and Reduction

BURNING, INCINERATION, OR ENERGY RECOVERY

Pros

- Heat can be used to supplement energy requirements.
- Reduces impact on landfills.
- Mass burning is inexpensive.
- What is left is 10% to 20% of original volume.
- U.S. incinerates 15% of its wastes.
- France, Japan, Sweden, and Switzerland incinerate > 40% of their wastes and use the heat to generate electricity.

Cons

- Air pollution including lead, mercury, NO_x, cadmium, SO_2, HCl, and dioxins.
- Sorting out batteries, plastics, etc., is expensive.
- No way of knowing toxic consequences.
- Ash is more concentrated with toxic materials.
- Initial costs of incinerators are high.
- Adds to acid precipitation and global warming.

COMPOSTING

Pros

- Creates nutrient-rich soil additive.
- Aids in water retention.
- Slows down soil erosion.
- No major toxic issues.

Cons

- Public reaction to odor, vermin, and insects.
- Not in my backyard (NIMBY).

REMANUFACTURING

Pros

- Recovers materials that would have been discarded.
- Beneficial to inner cities as an industry because material is available and jobs are needed.

Cons

- Toxic materials may be present (CFCs, heavy metals, toxic chemicals, etc.).

DETOXIFYING

Pros

- Reduces impact on the environment.

Cons

- Expensive.

EXPORTING

Pros

- Gets rid of problem immediately.
- Source of income for poor countries.

Cons

- Garbage imperialism or environmental racism.
- Long-term effects not known.
- Expensive to transport.

LAND DISPOSAL—SANITARY LANDFILLS

Pros

- Waste is covered each day with dirt to help prevent insects and rodents.
- Plastic liners, drainage systems, and other methods help control leaching material into groundwater.
- Geologic studies and environmental impact studies are performed prior to building.
- Collection of methane and use of fuel cells to supplement energy demand.
- Use of anaerobic methane generators reduces dependence on other energy sources.

Cons

- Rising land prices. Current costs are $1 million per hectare.
- Transportation costs to the landfill.
- High cost of running and monitoring landfill.
- Legal liability.

- Suitable areas are limited.
- NIMBY.
- Degradable plastics do not decompose completely.

LAND DISPOSAL—OPEN DUMPING

Pros

- Inexpensive.
- Provides a source of income to the poor by providing recyclable products to sell.

Cons

- Trash blows away in the wind.
- Vermin and disease.
- Leaching of toxic materials into the soil.
- Aesthetics.

OCEAN DUMPING

Pros

- Inexpensive.

Cons

- Debris floats to unintended areas.
- Marine organisms and food webs are impacted.
- Illegal in the United States.

RECYCLING

Pros

- Turns waste into an inexpensive resource.
- Reduces impact on landfills.
- Reduces need for raw materials and the costs associated with it.
- Reduces energy requirements to produce product. For example: recycling aluminum cuts energy use 95% and producing steel from scrap reduces energy requirements 75%.
- Reduces dependence on foreign oil.
- Reduces air and water pollution.
- Bottle bills provide economic incentive to recycle.

Cons

- Poor regulation.
- Fluctuations in market price.
- Throwaway packaging is more popular.
- Current policies and regulations favor extraction of raw materials. Energy, water, and raw materials are sold below real costs to stimulate new jobs and the economy.

REUSE

Pros

- Most efficient method of reclaiming materials.
- Industry models already in place—auto salvage yards, building materials, and so on.
- Refillable glass bottles can be reused approximately 15 times.
- Cloth diapers do not impact landfills.

Cons

- Cost of collecting materials on a large scale is expensive.
- Cost of washing and decontaminating containers is expensive.
- Only when items are expensive and labor is cheap is reuse economical.

RELEVANT LAWS

Resource Conservation and Recovery Act (RCRA) (1976): Encouraged states to develop comprehensive plans to manage nonhazardous industrial solid and municipal wastes. Set criteria for municipal landfills and disposal facilities, and prohibited open dumping of solid wastes.

Toxic Substances Control Act (TOSCA) (1976): Gave the EPA the authority to track industrial chemicals produced within or imported into the United States. Allows the EPA to ban the manufacture or importation of chemicals that pose risks.

Comprehensive Environmental Response, Compensation, and Liability Act (CERCLA—Superfund) (1980): Provided authority for the federal government to respond to releases or possible releases of hazardous substances that could threaten public health and/or the environment. Established rules for closed and abandoned hazardous waste sites. Established liability for corporations responsible for hazardous waste sites. Created a trust fund for cleanup if responsible parties for contaminated sites could not be found.

Nuclear Waste Policy Act (1982): Established federal authority to provide locations for permanent disposal of high-level radioactive wastes and required the operators of nuclear power plants to pay the costs of permanent disposal.

QUICK REVIEW CHECKLIST

- ☐ **Air Pollution**
  - ☐ major air pollutants
    - ☐ nitrogen dioxide
    - ☐ ozone
    - ☐ sulfur dioxide
    - ☐ suspended particulate matter (PM$_{10}$)
    - ☐ volatile organic compounds (VOCs)
  - ☐ measurement units
  - ☐ smog
    - ☐ formation of industrial smog
    - ☐ formation of photochemical smog
  - ☐ acid deposition
    - ☐ causes and effects
  - ☐ heat islands
  - ☐ temperature inversions
  - ☐ indoor air pollution
  - ☐ remediation and reduction strategies
  - ☐ relevant laws and protocols
    - ☐ Air Pollution Control Act (1955)
    - ☐ Clean Air Act (1963)
    - ☐ National Environmental Policy Act (1969)
    - ☐ Montreal Protocol (1989)
    - ☐ Pollution Prevention Act (1990)
    - ☐ Kyoto Protocol (1997)

- ☐ **Noise Pollution**
  - ☐ causes
  - ☐ effects
  - ☐ control measures
  - ☐ Noise Control Act (1972)

- ☐ **Water Pollution**
  - ☐ sources
    - ☐ air pollution
    - ☐ chemicals
    - ☐ microbiological
    - ☐ mining
    - ☐ noise
    - ☐ nutrients
    - ☐ oxygen-depleting substances
    - ☐ suspended matter
    - ☐ thermal sources
  - ☐ Minamata disease
  - ☐ *Exxon Valdez* (1989)

- ☐ **Cultural Eutrophication**

QUICK REVIEW CHECKLIST (continued)

☐ **Groundwater Pollution**

☐ **Maintaining Water Quality and Water Purification**
- ☐ drinking water treatment methods
- ☐ water treatment remediation technologies

☐ **Sewage Treatment/Septic Systems**
- ☐ primary treatment
- ☐ secondary treatment
- ☐ tertiary treatment

☐ **Relevant Laws**
- ☐ Clean Water Act (1972)
- ☐ Safe Drinking Water Act (1974)
- ☐ Ocean Dumping Ban Act (1988)
- ☐ Oil Spill Prevention and Liability Act (1990)

☐ **Solid Wastes**
- ☐ different types
  - ☐ organic
  - ☐ radioactive
  - ☐ recyclable
  - ☐ soiled
  - ☐ toxic
- ☐ amounts and type of MSW
- ☐ disposal and reduction methods
  - ☐ incineration
  - ☐ composting
  - ☐ remanufacturing
  - ☐ detoxifying
  - ☐ exporting
  - ☐ land disposal—sanitary landfills
  - ☐ land disposal—open dumping
  - ☐ ocean dumping
  - ☐ recycling
  - ☐ reuse
  - ☐ solid waste flow diagram

☐ **Relevant Laws**
- ☐ Resource Conservation and Recovery Act (1976)
- ☐ Toxic Substances Control Act (TOSCA) (1976)
- ☐ Comprehensive Environmental Response, Compensation, and Liability Act (CERCLA-Superfund) (1980)
- ☐ Nuclear Waste Policy Act (1982)

MULTIPLE-CHOICE QUESTIONS

1. _____ contributes to the formation of _____ and thereby compounds the problem of _____.

    (A) ozone, carbon dioxide, acid rain
    (B) carbon dioxide, carbon monoxide, ozone depletion
    (C) sulfur dioxide, acid deposition, global warming
    (D) nitrous oxide, ozone, industrial smog
    (E) nitric oxide, ozone, photochemical smog

2. Photochemical smog does NOT require the presence of

    (A) nitrogen oxides
    (B) ultraviolet radiation
    (C) peroxyacyl nitrates
    (D) volatile organic compounds
    (E) ozone

3. Which of the following is generally NOT considered to be a teratogen?

    (A) Ethanol or drinking alcohol
    (B) Benzene
    (C) Radiation
    (D) PCBs or polychlorinated biphenyls
    (E) All are considered teratogens

4. Which of the following steps is NOT involved in the production of industrial smog?

    (A) $C + O_2 \rightarrow CO_2$
    (B) $C + O_2 \rightarrow CO$
    (C) $S + O_2 \rightarrow SO_2$
    (D) $NO_2 \rightarrow NO + O$
    (E) $SO_2 + O_2 \rightarrow SO_3$

5. Household water is most likely to be contaminated with radon in homes that

    (A) are served by public water systems that use a groundwater source
    (B) are served by public water systems that use a surface water source
    (C) have private wells
    (D) use bottled water
    (E) are served by water agencies that use ozone to disinfect the water

6. Which reaction is NOT involved in the formation of acid deposition?

(A) $O_3 + C_xH_y \rightarrow PANs$
(B) $SO_2 + H_2O \rightarrow H_2SO_3$
(C) $H_2SO_3 + \frac{1}{2}O_2 \rightarrow H_2SO_4$
(D) $NO + \rightarrow \frac{1}{2}O_2 \rightarrow NO_2$
(E) $2NO_2 + H_2O \rightarrow HNO_2 + HNO_3$

7. Normal rainfall has a pH of about

(A) 2.3
(B) 5.6
(C) 7.0
(D) 7.6
(E) 8.3

8. According to the Environmental Protection Agency, about _____ of all commercial buildings in the United States are classified as sick.

(A) 5%
(B) 15%
(C) 50%
(D) 75%
(E) 100%

9. In developing countries, the most likely cause of respiratory disease would be

(A) photochemical smog
(B) industrial smog
(C) smoking
(D) PM_{10}
(E) asbestos

10. Humans LEAST susceptible to the effects of air pollution are

(A) newborns
(B) children between the age of 2 and 10
(C) teenagers
(D) adult males
(E) the elderly

11. Acid precipitation, leaching out the metal _____ , causes fish and other aquatic organisms to die from acid shock.

(A) Al
(B) Pb
(C) Hg
(D) Cd
(E) Fe

12. Which pollutant best illustrates the effectiveness of legislation?

    (A) NO_2
    (B) SO_2
    (C) CO_2
    (D) O_3
    (E) Pb

13. The diagram below shows the range of organisms found within certain sections of a river in an industrial area. Which section of the river most likely has the LOWEST level of dissolved oxygen?

Effects of Sewage Discharge in a River
Organisms commonly found in discharge zone

Sewage pipe				
Clean zone	**Decomposition zone**	**Septic zone**	**Recovery zone**	**Clean zone**
Trout	Carp	Worms	Carp	Trout
Perch	Catfish	Fungi	Catfish	Perch
Carp	Few perch	Bacteria	Blue-green algae	Carp
Catfish	Blue-green algae		Green algae	Catfish
Green algae	Green algae			Green algae

Direction of water flow ⟶

    (A) Clean zone
    (B) Decomposition zone
    (C) Septic zone
    (D) Recovery zone
    (E) None of the above

14. Which country listed below did NOT agree to the Kyoto Protocol to reduce its greenhouse gas emissions in 1997 and the extension in 2012?

    (A) Canada
    (B) United States
    (C) Russia
    (D) Australia
    (E) Brazil

15. The major source of solid waste in the United States comes from what source?

    (A) Homes
    (B) Factories
    (C) Agriculture
    (D) Petroleum refining
    (E) Mining wastes

16. What is the largest type of domestic solid waste in the United States?

   (A) Yard wastes
   (B) Paper
   (C) Plastic
   (D) Glass
   (E) Metal

17. Which of the following is most readily recyclable?

   (A) Plastic
   (B) Paper
   (C) Metal
   (D) Glass
   (E) All are equally and readily recyclable

18. Which of the following statements is TRUE?

   (A) Recycling is more expensive than trash collection and disposal.
   (B) Landfills and incinerators are more cost effective and environmentally sound than recycling options.
   (C) The marketplace works best in solving solid waste management problems; no public sector intervention is needed.
   (D) Landfills are significant job generators for rural communities.
   (E) None of the above are true.

19. In the 1970s, houses were built over a toxic chemical waste disposal site. This case study is known as

   (A) Love Canal
   (B) Bet Trang
   (C) Bhopal
   (D) Brownfield
   (E) Chernobyl

20. Which of the following methods of handling solid wastes is against the law in the United States?

   (A) Incineration
   (B) Dumping it in open landfills
   (C) Burying it underground
   (D) Exporting the material to foreign countries
   (E) Dumping the material in the open ocean

FREE-RESPONSE QUESTION

By Dr. Ian Kelleher
Brooks School
North Andover, MA

(a) Study the following graph, which shows projected trends in annual carbon dioxide emissions, and then answer the following questions.

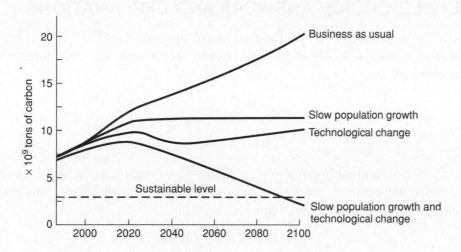

    (i) In the "business as usual" model, what factors do you think might contribute to the increase in carbon dioxide emissions?

    (ii) Given the shape of the graph, what do you think is meant in this case by "technological change"?

    (iii) What is meant by a "sustainable level" of carbon dioxide emissions? According to these predictions, what needs to happen for this level to be brought about?

(b) Use the example of acid deposition to illustrate the difference between remediation and alleviation of an environmental problem.

(c) Look at the graph of CFC production and account for the trends you observe.

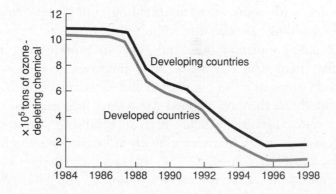

(d) In 1992, the World Bank designated indoor air pollution in developing countries as one of the four most critical global environmental problems. Use examples of major sources of indoor pollutants to illustrate how the issue of indoor air quality in developing countries differs from that in developed countries.

MULTIPLE-CHOICE ANSWERS AND EXPLANATIONS

1. **(E)** As the sunlight becomes more intense, nitrogen dioxide (NO_2) is broken down into nitric oxide (NO) and oxygen atoms, and the concentration of ozone increases:

$$NO_2 \xrightarrow{uv} NO + O$$

$$O_2 + O \rightarrow O_3$$

2. **(C)** Nitrogen dioxide can also react with volatile organic compounds (VOCs) released by vehicles, refineries, gas stations, and the like to produce toxic chemicals such as peroxyacyl nitrates (PANs):

$$NO_2 + VOCs \rightarrow PANs$$

3. **(E)** Teratogens are substances or environmental agents that cause abnormalities in a fetus during pregnancy. Choices (A)–(D) are all known teratogens.

4. **(D)** This step is present in the formation of photochemical smog.

5. **(C)** Radon is a naturally occurring radioactive gas that causes cancer and may be found in drinking water and indoor air. The Safe Drinking Water Act requires the EPA to develop regulations to reduce radon in drinking water.

6. **(A)** Review the reactions required to form acid deposition.

7. **(B)** Rain is naturally acidic due to the presence of carbon dioxide in the air, which forms carbonic acid with rainwater.

8. **(B)** The World Health Organization (WHO) estimates that up to 30% of commercial buildings worldwide may be unhealthy with up to one-third of the people who work in these buildings experiencing health effects. In the United States, with more stringent building codes, approximately 15% of commercial buildings are classified as unhealthy.

9. **(D)** Developing countries around the world are experiencing increased levels of particulate matter as a result of rapid increases in energy consumption, motor vehicle use, and rapid population and economic growth.

10. **(D)** The elderly are the most affected group since they may have compromised respiratory systems based on many years of breathing pollutants. Newborns are a very close second. Children are highly affected due to high activity levels and increased cases of asthma.

11. **(A)** Aluminum reduces the ion exchange through the gills and causes fish to excrete excess mucus, which reduces oxygen uptake.

12. **(E)** From 1976 to 1980, the average amount of lead in children was 15 µg/dL. From 1991 to 1994, the average level had been reduced to 2.7 µg/dL.

13. **(C)** The level of dissolved oxygen in the septic zone is too low to support organisms that live by aerobic respiration. In this region, anaerobic bacteria flourish and produce waste products such as hydrogen sulfide (H_2S) and methane (CH_4).

14. **(B)** The United States is the only country listed to have failed to agree to the Kyoto Protocol in 1997 to reduce its greenhouse gases and its extension in 2012. In 2011, Canada withdrew from the Protocol.

15. **(E)** Mining wastes, along with oil and gas production, constitute about 75% of all solid wastes generated in the United States.

16. **(B)** Paper accounts for approximately 35% of domestic solid waste in the United States.

17. **(B)** Paper can be broken down into fibers and reused generally without requiring a sorting process. For some refined paper products, recycled paper must be sorted based on ink types.

18. **(E)** In choice (A), when designed properly, recycling programs are cost-competitive with trash collection and disposal. In choice (B), properly designed recycling programs are cost competitive with landfills and incinerators. In choice (C), solid waste systems generally operate under public sector guidelines. In choice (D), recycling creates many more jobs in rural and urban communities than landfill employment.

19. **(A)** Love Canal is an abandoned canal in New York where a huge amount of toxic waste was buried. The waste was composed of at least 300 different chemicals, totaling an estimated 20,000 metric tons.

20. **(E)** The Ocean Dumping Act prohibits dumping of wastes into territorial ocean waters.

FREE-RESPONSE ANSWER

(a) (i) Increased carbon dioxide emissions primarily come from increased burning of fossil fuels worldwide. The "business as usual" model would include an increasing global population, which would mean an increase in the demand for energy. In the "business as usual" model, this need would be met primarily by increased use of fossil fuels, by far the most common source of energy in the world today. Increasing rates of development, particularly of developing countries in this model, would also lead to an increase in fossil fuel use.

Notable points:

Cover all of the relevant aspects of "business as usual," such as increasing size and development of populations.

(a) (ii) "Technological change" reduces carbon dioxide emissions on the graph. Increased use of alternative sources of energy such as nuclear power, hydro-electric, solar, and tidal energy would all decrease fossil fuel emissions. Technologies that increase energy efficiency and save power, such as more efficient car engines, would also decrease the use of fossil fuels and thus carbon dioxide emissions.

"Technological change" is a vague phrase that could mean many things. In this case, the curve thus labeled shows a dramatic reduction in carbon dioxide emissions, so the question is asking about technologies that reduce fossil fuel use.

(a) (iii) A "sustainable level" of carbon dioxide emissions is one in which the amount absorbed by Earth's natural systems, such as oceans and plants, equals the amount released into the atmosphere. A combination of both technological change and slowed population growth is needed to bring carbon dioxide emissions to sustainable levels. As the graph shows, neither factor can produce sustainable levels on its own.

Explaining the difference between these two similar terms would be a good way to start. These are basic terms that you should know. Starting like this should lessen the chance of mixing them up.

Thus, cause and effect must both be covered in the answer, and you must take care to distinguish the two. Acid deposition is a secondary pollutant. Explaining the sequence of events that leads to its formation might be helpful in distinguishing cause from effect.

Anything that reduces the use of fossil fuels or limits the emissions of sulfur and nitrogen oxides would be an example of alleviation. Anything that neutralizes the effects of acid in the environment would be an example of remediation. Be careful in giving just an appropriate level of detail.

When asked to explain the trends shown in a graph, a good starting point is to describe what they are. Writing this down first should also help you compose your explanation.

This sample essay may provide more detail than necessary. The important point is that the amounts for both developed and developing countries fell dramatically at the same time and basically at the same rate. This suggests that the drops come as a result of the implementation of new laws. Since we are dealing with a global situation, it is likely to be in the form of an international agreement.

(b) Alleviation of an environmental problem means stopping or lessening its cause. Remediation means cleaning up the effects of the problem.

Acid deposition forms in the atmosphere mainly from sulfur oxides (SO_x) and nitrogen oxides (NO_x). Both are present primarily as a consequence of the combustion of fossil fuels.

Methods of alleviation may concentrate on decreasing the amount of sulfur and nitrogen oxides released into the atmosphere. Any method of reducing the rate of consumption of fossil fuels, such as increased use of nuclear power and alternative energy sources, or laws and education to help conserve energy would decrease the amount of sulfur and nitrogen oxides released. Clean-fuel technologies, such as fluidized-bed combustion of coal, would result in less of these gases being produced on combustion. Scrubbers in smokestacks can remove much of what is produced. Perhaps the most important step in reducing emissions in the United States, however, was the passing of the Clean Air Act in 1963 and its subsequent amendments.

Methods of remediation may concentrate on neutralizing the acid deposited in an environment. Acids can be neutralized by adding a base. Since an abundant, low-cost, and nontoxic material is often needed, limestone, $CaCO_3$, is commonly used. For example, limestone might be added to a lake to increase its pH. Powdered limestone could also be spread over agricultural lands to increase the pH of the soil. Many nutrients are more soluble in acidic soils and therefore might be washed away by rain. As a result, the addition of fertilizer is also required. Deforestation caused by acid deposition may be addressed by treating the soil and then replanting.

(c) Between 1984 and 1988, both developed and developing countries used ozone-depleting chemicals at a fairly consistent rate of approximately 1 million tons per year. After 1988, the amount used by both developed and developing countries decreased sharply and at a fairly constant rate for the next seven years. By 1996, developed countries used 50,000 tons per year and developing countries used 150,000 tons. Use remained at approximately these levels for the next two years.

The reduction is likely to be due to countries implementing technologies to comply with the Montreal Protocol (1989), which set limits for the emission of chemicals that cause depletion of the ozone layer. The biggest reduction has come from using alternatives to chlorofluorocarbons (CFCs)—the principal ozone-depleting agents. Alternatives include HFCs and HCFCs.

(d) Homes are perhaps the most important factor when considering indoor air quality since people tend to spend more time there than anywhere else. The majority of people in developing countries live in homes with much simpler technologies. Many of the pollutants found in houses in developed countries are not present. The major source of indoor air pollutants in developing countries, therefore, is the combustion of poor-quality fuels for heating, lighting, and cooking. Such dirty fuels include animal wastes, kerosene, and low-grade coal that may release large amounts of particulates, carbon monoxide, sulfur oxides, and other toxins on combustion. As fires are often burned indoors in places with inadequate ventilation, the levels of these pollutants and carbon monoxide can be greatly concentrated.

Energy sources are much more technologically advanced in developed countries, so this is not such an important source of indoor air pollution. For example, poorly maintained furnaces may produce some carbon monoxide. However, this problem is not nearly as widespread or generally as serious as the energy issue in developing countries. Instead, major sources of air pollution include lead (from lead paints), asbestos, and fumes from volatile organic compounds (VOCs) in paints, glues, plastics, and furniture. These things will be less common in houses in developing countries. Houses in developed countries are often tightly sealed, more so than in developing countries. This leads to increased levels of concentration. For example, radon gas, which occurs naturally from radioactive decay in certain rocks, might accumulate to dangerous levels in a modern air-conditioned house but not in a simpler hut with no glass in the windows.

The wording of the World Bank designation suggests that the question is referring to the general masses of the population in developing countries rather than the small, technologically developed percentage. The answer is thus strongly focused on this difference in lifestyle.

Impacts on the Environment and Human Health

HAZARDS TO HUMAN HEALTH

Environmental risk analysis is the comparing of the risk of a situation to its related benefits. It is the overall process that allows one to evaluate and deal with the consequences of events, based on their probability. There are four classes of risk:

1. High risk—such as smoking or driving while intoxicated.

2. Low risk—infrequent events that may have a large consequence, such as an earthquake on the East Coast of the United States.

3. Very low risk—events that have never occurred in recorded history, such as a major meteor striking the North American continent.

4. Mixed risk—outcomes that increase in frequency against a background of occurrences, such as additional cases of cancer beyond that normally expected.

Understanding how people accept risk requires an understanding of how preferences are accepted and measured. The three types of preferences are revealed, expressed, and natural standards. Revealed preferences are observations on the risks people actually take. Expressed preferences are often measured through public opinion polls. Natural standards are levels of risk humans have lived with in the past.

Risk estimation is a scientific question, while acceptability of a given level of risk is a political question. We can see this distinction by comparing the risk of smoking with working in a coal mine. If the United States spent as much money on premature deaths and illnesses caused by smoking tobacco as we do on coal mine safety, there would be little money left in the United States for any other purpose. This is the political reality of risk acceptance that goes beyond risk estimation.

External influences are factored into decisions regarding environmental risk. These influences include public concern, economic interest, and legislative actions that affect the possible choices available.

Risk analysis is divided into risk assessment and risk management. Risk assessment is an objective estimation of risk. It includes the identification of hazards, dose-response assessment, exposure assessment, and risk characterization.

Risk management is the process of determining what to do about risk. This includes risk identification and use of mitigating measures to reduce risk.

Acute and Chronic Effects

Acute health effects are characterized by sudden and severe exposure and rapid absorption of the substance. Normally, a single large exposure is involved. Acute health effects are often reversible, such as carbon monoxide poisoning.

Chronic health effects are characterized by prolonged or repeated exposures over many days, months, or years. Symptoms may not be immediately apparent. Chronic health effects are often irreversible. Examples include lead or mercury poisoning, asbestosis, or cancer.

Dose-Response Relationships

Dose-response relationships describe the change in effect on an organism or a population caused by differing levels of exposures to a substance. These relationships are used to determine whether various environmental risks are safe or hazardous.

A dose-response curve is a graph that relates the amount of drug or toxin given (plotted on the x-axis and usually the logarithm of the dose) compared with the response (plotted on the y-axis and usually provided in percentages). The point on the graph where the response is first observed is known as the threshold dose. For most drugs, the desired effect is found slightly above the threshold dose. When past this point, negative side effects begin to appear.

LD_{50} (lethal dose, 50%) is the median lethal dose of a pollutant or drug that kills half the members of a tested population within 14 days and is the most common indicator of toxicity. EC_{50} is the concentration of a compound where 50% of its effect is observed.

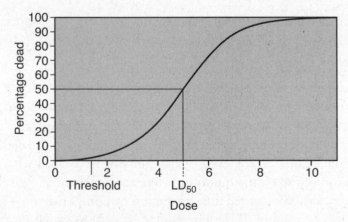

Figure 10.1 A dose-response curve

Air Pollutants and Their Effect on Human Health

AIR TOXICS

Air toxics are a group of air pollutants that are known or suspected to cause serious health problems. Examples of air toxics include asbestos, benzene, chloroform, formaldehyde, lead, mercury and nickel compounds, and perchlorocthylene. People exposed to air toxics at sufficient concentrations and durations may have an increased chance of developing cancer or other serious health problems. These include damage to the immune system as well as neurological, reproductive (reduced fertility), developmental, and respiratory problems.

ASBESTOS

Studies of people who were exposed to asbestos in factories and shipyards have shown that breathing high levels of asbestos fibers can lead to an increased risk of lung cancer, mesothelioma (a cancer of the lining of the chest and the abdominal cavity), and asbestosis (the lungs become scarred with fibrous tissue). Researchers have not yet determined a safe level of exposure. However, they know that the greater and longer the exposure, the greater the risk of contracting an asbestos-related disease. Risks of lung cancer and mesothelioma increase with the number of fibers inhaled. The risk of lung cancer from inhaling asbestos fibers is also greater for smokers. People who develop asbestosis have usually been exposed to high levels of asbestos for a long time. The symptoms of these diseases do not usually appear until about 20 to 30 years after the first exposure.

CARBON MONOXIDE (CO)

Carbon monoxide enters the bloodstream through the lungs and binds chemically to hemoglobin, the substance in blood that carries oxygen to cells. In this way, CO interferes with the ability of the blood to transport oxygen to organs and tissue throughout the body. This can cause slower reflexes, confusion, and drowsiness. It can also reduce visual perception and coordination and can decrease the ability to learn. People with cardiovascular disease, such as angina, are most at risk from exposure to CO. These individuals may experience chest pain and other cardiovascular symptoms if they are exposed to CO, particularly while exercising.

INDOOR AIR POLLUTANTS

Health effects from indoor air pollutants may be acute and experienced soon after exposure or may be chronic. Immediate effects may show up after a single exposure or repeated exposures. These include headaches, dizziness, fatigue, and irritation of the eyes, nose, and throat. Such immediate effects are usually short term and treatable. The likelihood of immediate reactions to indoor air pollutants depends on several factors: age, preexisting medical conditions, and individual sensitivity. Certain immediate effects are similar to those from colds or other viral diseases. Other health effects from exposure to indoor air pollutants may show up either years after exposure has occurred or only after long or repeated periods of exposure. These effects, which include some respiratory diseases, heart disease, and cancer, can be severely debilitating or fatal.

> **Remember**
>
> Environmental problems have a cultural and social context. Understanding the role of cultural, social, and economic factors is vital to the development of solutions.

LEAD (Pb)

Exposure to lead can occur through inhalation of air and ingestion of lead in food, water, soil, or dust. Excessive lead exposure can cause seizures, brain and kidney damage, mental retardation, and/or behavioral disorders. Children six and under are most at risk because their bodies are growing quickly.

NITROGEN DIOXIDE (NO_2)

Health effects of exposure to nitrogen dioxide include coughing, wheezing, and shortness of breath in children and adults with respiratory disease such as asthma. Even short exposures to nitrogen dioxide can affect lung function and may cause permanent structural changes in the lungs.

OZONE (O_3)

The reactivity of ozone causes health problems because it damages lung tissue, reduces lung function, and sensitizes the lungs to other irritants. Exposure to ozone for several hours at relatively low concentrations has been found to reduce lung function significantly and to induce respiratory inflammation in normal, healthy people during exercise. This decrease in lung function is generally accompanied by symptoms including chest pain and pulmonary congestion.

PM_{10}

Coarse particles can aggravate respiratory conditions such as asthma. Exposure to fine particles is associated with several serious health effects, including premature death. When exposed to PM_{10}, people with existing heart or lung diseases such as asthma, chronic obstructive pulmonary disease, and congestive or ischemic heart disease are at increased risk of premature death. Older persons are especially sensitive to PM_{10} exposure. When exposed to PM_{10}, children and people with existing lung disease may not be able to breathe as deeply or as vigorously as they normally would. They may also experience symptoms such as coughing and shortness of breath. PM_{10} can increase susceptibility to respiratory infections and can aggravate existing respiratory diseases, such as asthma and chronic bronchitis.

RADON

Radon is an invisible, radioactive gas that results from the radioactive decay of radium, which may be found in rock formations beneath buildings or in certain building materials. Radon is probably the most pervasive serious hazard for indoor air in the United States and Europe, probably responsible for tens of thousands of deaths from lung cancer each year. Radon is a heavy gas and thus will tend to accumulate at the floor level, and radon accumulation is greatest for well-insulated homes. Radon mitigation methods include sealing concrete slab floors, basement foundations, water drainage systems, or by increasing ventilation.

SULFUR DIOXIDE (SO_2)

High concentrations of sulfur dioxide affect breathing and may aggravate existing respiratory and cardiovascular disease. Sensitive populations include asthmatics, individuals with bronchitis or emphysema, children, and the elderly.

Smoking and Other Preventable Risks

Cigarette smoke contains over 4,700 chemical compounds including 60 known carcinogens. No threshold level of exposure to cigarette smoke has been defined. However, conclusive evidence indicates that long-term smoking greatly increases the likelihood of developing numerous fatal conditions.

Cigarette smoking is responsible for more than 85% of lung cancers and is also associated with cancers of the mouth, pharynx, larynx, esophagus, stomach, pancreas, kidney, bladder, and colon. Cigarette smoking has also been linked to leukemia. Apart from the carcinogenic aspects of cigarette smoking, links to increased risks of cardiovascular diseases (including stroke), sudden death, cardiac arrest, peripheral vascular disease, and aortic aneurysm have also been established. Many components of cigarette smoke have also been characterized as ciliotoxic materials. These irritate the lining of the respiratory system, resulting in increased bronchial mucus secretion and chronic decreases in pulmonary and mucociliary function.

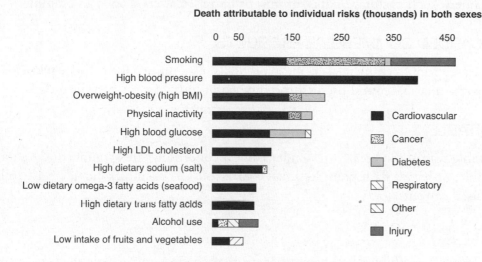

Source: U.S. National Library of Medicine; National Institutes of Health

Figure 10.2 Preventable risks and death rates in the United States

---

RELEVANT LAWS

Federal Hazardous Substances Act (1960): Required that certain hazardous household products bear cautionary labels to alert consumers to the potential hazards of these products.

Federal Environmental Pesticides Control Act (1972): Required registration of all pesticides in the United States.

Hazardous Materials Transportation Act (HAZMAT) (1975): Governs the transportation of hazardous materials and wastes. Covers containers, labeling, and marking standards.

Toxic Substances Control Act (1976): Gives the Environmental Protection Agency (EPA) the ability to track the 75,000 industrial chemicals currently produced in or imported into the United States.

Comprehensive Environmental Response, Compensation and Liability (CERCLA or SUPERFUND) (1980): Established federal authority for emergency response and cleanup of hazardous substances that have been spilled, improperly disposed of, or released into the environment.

Nuclear Waste Policy Act (1982): Established a study to find a suitable site for disposal of spent fuel from nuclear reactors. Yucca Mountain, Nevada, seems most feasible at this time.

HAZARDOUS CHEMICALS IN THE ENVIRONMENT

A hazardous waste is a waste with properties that make it dangerous or potentially harmful to human health or the environment. Hazardous wastes can be liquids, solids, contained gases, or sludges. The Environmental Protection Agency has separated hazardous wastes into the following categories:

CORROSIVE

Corrosive wastes are strong acids or strong bases that are capable of corroding metal containers, such as storage tanks, drums, or barrels. Battery acid is an example.

DISCARDED COMMERCIAL PRODUCTS

These are specific commercial chemical products in an unused form. Some pesticides and some pharmaceutical products become hazardous wastes when discarded.

IGNITABLE

Ignitable wastes can create fires under certain conditions, are spontaneously combustible, or have a flash point less than 140°F (60°C). Examples include waste oils and used solvents.

MUTAGENS

A mutagen is a physical or chemical agent that changes the genetic material, usually DNA, of an organism and thus increases the frequency of mutations above the natural background level. Many mutations cause cancer, as mutagens are typically also carcinogens. Examples of common mutagens include nitrous acid, bromine and bromine-containing compounds, sodium azide (as found in car safety air bag systems), and benzene.

NONSPECIFIC SOURCE

These include wastes from common manufacturing and industrial processes, such as solvents that have been used in cleaning or degreasing operations.

REACTIVE

Reactive wastes are unstable under normal conditions. They can cause explosions, toxic fumes, gases, or vapors when heated, compressed, or mixed with water. Examples include lithium batteries and explosives.

SOURCE SPECIFIC

These are wastes from specific industries, such as petroleum refining or pesticide manufacturing. Examples include certain sludges and wastewaters from treatment and production processes.

TERATOGENS

Teratogens are substances found in the environment that can cause a birth defect. Examples of teratogens include drinking alcohol (ethanol), radioactive compounds, dioxin, certain pharmaceuticals (e.g., Dilantin, which is used to treat seizures and certain heart arrhythmias), lithium, mercury, tetracyclines, ethers, tobacco, and excessive caffeine.

TOXIC

Toxic wastes are harmful or fatal when ingested or absorbed. They may contain mercury or lead for example. When toxic wastes are disposed of on land, contaminated liquid may leach from the waste and pollute groundwater.

Treatment, Disposal, and Cleanup of Contaminated Sites

Reduction and cleanup of hazardous wastes can occur by producing less waste, converting the hazardous material to less hazardous or nonhazardous substances, and placing the toxic material into perpetual storage.

PRODUCE LESS WASTE

- **Recycle**—Improved technology seeks ways of collecting hazardous wastes and using them as raw materials for new products.
- **Reduce or eliminate toxicity**—Improved technology seeks substitutes for hazardous chemicals. For example, Puron replaced Freon.

CONVERSION TO LESS HAZARDOUS OR NONHAZARDOUS SUBSTANCES

Chemical, physical, and biological treatment

Bioremediation is the use of bacteria and enzymes to break down hazardous materials. Phytoremediation involves rhizofiltration (using sunflowers to absorb radioactive wastes), phytostabilization (e.g., using willow and poplar trees to absorb organic contaminants), phytodegradation (e.g., using poplars to absorb and break down contaminants), and phytoextraction (e.g., using Indian mustard and brake ferns to absorb inorganic metal contaminants). Pros: inexpensive, low energy use, little to no air pollution, easy to build. Cons: slow, effective only as far down as roots will reach, some toxic materials can evaporate through plants. Plants would be toxic and need to be properly disposed. Chemical methods involve use of cyclodextrin.

Incineration

Can release air pollutants and toxic ash (such as lead, mercury, and dioxins).

Thermal Treatment

Plasma arcs. Pros: small, mobile, no toxic ash. Cons: expensive and can release particulates, chlorine gas, toxic metals, and radioactive wastes.

PERPETUAL STORAGE

Arid Region Unsaturated Zone

The unsaturated zone is the subsurface between the land surface and underlying aquifers. It includes sites in the arid western United States that are being relied upon to isolate a significant portion of the nation's radioactive and other hazardous wastes for thousands of years.

Capping

Capping, often used in combination with other cleanup methods, covers buried wastes to prevent contaminants from spreading. Spreading or migration of buried wastes can be caused by rainwater or surface water moving through the site or by wind blowing dust off the site. The primary purpose of a cap is to minimize contact between rain or surface water and the buried waste. Caps: (a) minimize water movement through the wastes by using efficient drainage; (b) resist damage caused by settling; and (c) prevent standing water by funneling away as much water as the underlying filter or soils can handle. Capping is used when the underground contamination is so extensive that excavating and removing it isn't practical, or when removing wastes would be more dangerous to human health and the environment than leaving them in place. Wells are often used to monitor groundwater where a cap has been installed to detect any movement of the wastes.

Landfill

Pros: inexpensive. Cons: groundwater seepage and contamination.

Salt Formations

Toxic wastes are deposited in deep salt formations. The absence of flowing water within natural salt formations prevents dissolution and subsequent spreading of the waste products. Rooms and caverns in the salt can be sealed, thus isolating the waste from the biosphere.

Surface Impoundments

Excavated ponds, pits, or lagoons. Pros: low cost, low operating cost, built quickly, wastes can be retrieved and, if lined, can store wastes for long periods. Cons: groundwater contamination, VOC pollution, overflow if flooding occurs, earthquake issues, promotes waste production.

Underground Injection

Pros: low cost, wastes can be retrieved, simple technology. Cons: leaks, earthquake issues, groundwater contamination.

Waste Piles

Storage of toxic materials in drums, underground vaults, or above-ground buildings. Pros: easy to identify leaks. Cons: shipping of materials to facilities results in accidents.

CASE STUDY

The first synthesized chlorinated organic pesticide was DDT. It appeared to have low toxicity and was broad spectrum. It did not break down, so it did not have to be reapplied often. Crop production increased and mosquitoes decreased. In 1962, Rachel Carson published *Silent Spring*, which made the connection between DDT and nontarget organisms by direct and indirect toxicity. DDT persisted in the environment through bioaccumulation (an increase in concentration up the food chain) and biomagnification (the tendency for a compound to accumulate in tissues). DDT was found to decrease the eggshell thickness of various species of birds, nearly wiping out bald eagles and peregrine falcons. DDT was beginning to show up in native people of the Arctic, seals, and human breast milk. It was pulled off the U.S. market and is now being manufactured in Indonesia.

Bioaccumulation

Bioaccumulation is the increase in concentration of a pollutant from the environment in an organism or part of an organism. It involves the biological sequestering of substances that enter the organism through respiration, food intake, or epidermal (skin) contact with the substance. The level at which a given substance is bioaccumulated depends on:

• Rate of uptake
• Mode of uptake (gills, ingested along with food, contact with skin, etc.)
• Rate the substance is eliminated from the organism
• Transformation of the substance by metabolic processes
• Lipid (fat) content of the organism
• Environmental factors

Biomagnification

Biomagnification is the increase in concentration of a pollutant from one link in a food chain to another. In order for biomagnification to occur, the pollutant must be long-lived, mobile, soluble in fats, and biologically active. If a pollutant is short-lived, it will be broken down before it can become dangerous. If it is not mobile, it will stay in one place and be less likely to be taken up by many organisms. If the pollutant is soluble in water, it will be excreted by the organism. Pollutants that dissolve in fats, however, may be retained for a long time. It is traditional to measure the amount of pollutants in fatty tissues of organisms such as fish. In mammals, milk produced by females is often tested since the milk is high in fat and because the young are often more susceptible to damage from toxins.

Bioaccumulation vs. Biomagnification

For example, methylmercury is taken up by bacteria and phytoplankton. Small fish eat the bacteria and phytoplankton and *accumulate* the mercury. The small fish are in turn eaten by larger fish, which can become food for humans and animals resulting in the buildup (*biomagnification*) of large concentrations of mercury in human and animal tissue. As a general rule, the more fat-like a substance is, the more likely it is to bioaccumulate in organisms, such as fish.

ECONOMIC IMPACTS

A cost-benefit analysis is a technique for deciding whether to make a change. To use the technique, one adds up the value of the benefits of a course of action and subtracts the costs associated with it.

Costs are either one time or ongoing. Benefits are most often received over time. Time is factored into a cost-benefit analysis by calculating a payback period—the time for the benefits of a change to repay its costs.

In its simple form, a cost-benefit analysis is carried out using only financial costs and benefits. For example, a simple cost-benefit analysis of a new road would measure the cost of building the road and subtract this from the economic benefit of improving transport links. It would not measure either the cost of environmental damage or the benefit of quicker and easier travel to work. A more sophisticated approach is to attempt to put a financial value on intangible costs and benefits, which is highly subjective.

A cost-benefit analysis applies to three economic situations: First, it can help judge whether public services provided by the private sector are adequate. Second, it can be used when judging and assessing inefficiencies (market failures) in the private sector and their impact on the health, safety, and environmental needs of the country. Third, it helps in determining how to meet societal needs in a cost-effective manner in areas that only government can address. These include defense, preservation of scenic areas, environmental protection, and so on.

A cost-benefit analysis can be used for evaluating policy alternatives, shaping regulatory strategies, and evaluating specific regulations. A cost-benefit analysis requires:

1. Gathering all information and data about a public issue, including history and background

2. Defining the possible solutions to solving the issue

3. Brainstorming the possible environmental and societal consequences of the alternatives

4. Quantifying the benefits and the costs

5. Making decisions and balancing concerns

Framework of Cost-Benefit Analysis

Step	Description
Cost-benefit	Determine an action and levels of action that achieve the greatest net economic benefit. Exploring options and determining incremental levels of remediation provide the most benefit for the least cost.
Cost-effectiveness	Implementing a specific environmental, health, or safety objective at the least cost. Emphasis is on achieving the objective. Flexible regulatory guidelines are adapted to find the lowest cost to solve a problem.
Health or environmental protection standards	Reducing risk to the public whatever the cost.
Risk-benefit	Balancing health or environmental protection with the costs of providing the protection.
Technology	To achieve results that are predictable and certain.

External Costs

An external cost occurs when the product(s) or activities of one group has a negative impact on another group and when that impact is not fully accounted, or compensated for, in the price of the product. Example: A power station generates emissions of SO_2 causing damage to buildings and human health. In this case, the cost to the owners of the buildings and to those who suffer from respiratory issues is not taken into account in the cost of the electricity produced. In this example, the environmental costs are "external" because, although they are real costs to society, the owner of the power station is not taking them into account when setting the price of electricity.

The total cost of a good to society (called the social cost) includes the costs or production incurred by the industry as well as the external costs. One possibility for accounting for external costs in setting the true social cost would be via eco-taxes; i.e., taxing damaging fuels and technologies in proportion to their external costs. Another possible solution would be to subsidize cleaner technologies thus avoiding the social and environmental costs altogether. And finally, accounting for external cost estimates in the cost-benefit-analysis. In such an analysis, the costs to establish measures to reduce a certain environmental burden are compared with the benefits.

Ecotaxes (Green Taxes)

Ecotaxes are intended to promote ecologically sustainable activities by providing economic incentives and can complement or reduce the need for regulatory (command and control) approaches. Often, an ecotax may attempt to maintain overall tax revenue by proportionately reducing other taxes (green tax shift). Examples of ecotaxes would be:

- taxes on the use of fossil fuels by greenhouse gases produced.
- duties on imported goods produced by ecologically unsound methods.
- taxes on mineral, energy, and forestry products that were produced by ecologically unsound methods.
- fees for camping, hiking, fishing, and hunting.
- taxes on technologies and products, which are associated with substantial negative externalities.
- waste disposal taxes.
- taxes on effluents, pollution, and other hazardous wastes.

Cap and Trade

Cap and trade, also known as emissions trading, is a market approach used to control pollution by providing economic incentives for achieving reductions in the emissions of pollutants. With cap-and-trade policies, the government sets a limit or "cap" on the amount of a pollutant that can be emitted. Companies or other groups are then issued emission permits and are required to hold an equivalent number of allowances or credits, which represent the right to emit a specific amount of pollutants. The total amount of allowances and credits cannot exceed the cap, limiting total emissions to that level. Companies that need to increase their emission allowance must then buy credits from those who pollute less. The transfer of allowances is referred to as a "trade." In effect, the buyer is paying a charge for polluting, while the seller is being rewarded for having reduced emissions by more than was needed. Therefore, those who can reduce emissions more cheaply will do so, achieving pollution reduction at the lowest cost to society.

Sustainability

Sustainability deals with the continuity of the economic, social, and institutional aspects of human society while at the same time preserving biodiversity and the environment. Sustainable activities seek to provide the best outcomes for both human societies and natural ecosystems. Several issues are common to both interests:

- Consideration of risk, uncertainty, and irreversibility
- Ensuring appropriate valuation, appreciation, and restoration of nature
- Integration of environmental, social, and economic goals in policies and activities
- Equal opportunity and community participation
- Conservation of biodiversity and ecological integrity
- A commitment to best practice
- No net loss to either human or natural capital
- Continuous improvement
- The need for good governance

Unlimited economic and population growth puts many demands on natural resources. Its effects on pollution and the carrying capacity of Earth are factors that impact sustainability. These are all analyzed using life cycle assessments and ecological footprint analyses.

Ecological Footprint

The "ecological footprint" is a measure of human demand on the Earth's ecosystems. It is a standardized measure of demand for natural capital that may be contrasted with the planet's ecological capacity to regenerate and represents the amount of biologically productive land and sea area necessary to supply the resources a human population consumes, and to assimilate associated waste. Currently, humanity's total ecological footprint is estimated at 1.5 planet Earths—in other words, humanity uses ecological services 1.5 times as fast as Earth can renew them.

The "carbon footprint" is the amount of carbon being emitted by an activity or organization. The carbon component of the ecological footprint converts the amount of carbon dioxide being released into the amount of productive land and sea area required to sequester it and tells the demand on the Earth that results from burning fossil fuels. The carbon footprint is 54% of the ecological footprint and its most rapidly-growing component having increased 11-fold since 1961.

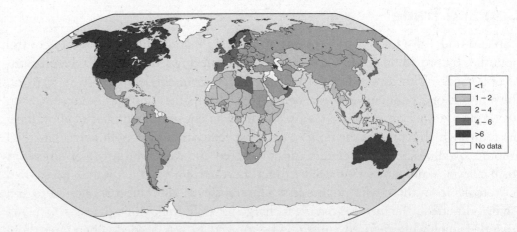

Figure 10.3 Ecological footprints of various countries measured in global hectares per person (gha)

QUICK REVIEW CHECKLIST

☐ **Hazards to Human Health**
- ☐ types of risks
- ☐ risk management strategies
- ☐ acute and chronic effects
- ☐ dose-response relationships
  - ☐ LD_{50}
  - ☐ EC_{50}
  - ☐ dose-response curves

☐ **Air Pollutants and Effect on Human Health**
- ☐ air toxics
- ☐ asbestos
- ☐ carbon monoxide
- ☐ indoor air pollutants
- ☐ lead
- ☐ nitrogen dioxide
- ☐ ozone
- ☐ PM_{10}
- ☐ sulfur dioxide
- ☐ smoking

☐ **Hazardous Chemicals in the Environment**
- ☐ corrosive
- ☐ discarded commercial products
- ☐ ignitable
- ☐ nonspecific sources
- ☐ reactive
- ☐ source specific
- ☐ toxic

☐ **Treatment, Disposal, and Cleanup of Contaminated Sites**
- ☐ conversion techniques
  - ☐ chemical, physical, or biological treatment
  - ☐ incineration
  - ☐ thermal treatment
  - ☐ perpetual storage
  - ☐ arid region unsaturated zone
  - ☐ landfill
  - ☐ salt formations
  - ☐ surface impoundments
  - ☐ underground injection
  - ☐ waste piles

☐ **Case Study—DDT**

QUICK REVIEW CHECKLIST (continued)

☐ **Relevant Laws**
  ☐ Federal Hazardous Substances Act (1960)
  ☐ Federal Environmental Pesticides Control Act (1972)
  ☐ Hazardous Materials Transportation Act (HAZMAT) (1975)
  ☐ Toxic Substances Control Act (1976)
  ☐ Comprehensive Environmental Response, Compensation, and Liability Act (CERCLA or Superfund) (1980)
  ☐ Nuclear Waste Policy Act (1982)

☐ **Biomagnification**

☐ **Cost-Benefit Analysis**

☐ **Externalities**

☐ **Marginal Costs**

☐ **Sustainability**

MULTIPLE-CHOICE QUESTIONS

1. About 70% of U.S. hazardous wastes come from

   (A) agricultural pesticides
   (B) smelting, mining, and metal manufacturing
   (C) nuclear power plants
   (D) chemical and petroleum industries
   (E) households

2. Which act established federal authority for emergency response and cleanup of hazardous substances that have been spilled, improperly disposed of, or released into the environment?

   (A) Resource Recovery Act
   (B) Resource Conservation and Recovery Act
   (C) Solid Waste Disposal Act
   (D) Superfund
   (E) Hazardous Materials Transportation Act

3. Effects produced from a long-term, low-level exposure are called

   (A) acute
   (B) chronic
   (C) pathological
   (D) symptomatic
   (E) synergistic

4. Which of the following techniques is NOT an example of bioremediation?

   (A) land farming
   (B) composting
   (C) rhizofiltration
   (D) phytoremediation
   (E) All are examples of bioremediation

5. The following dose-response curve shows that

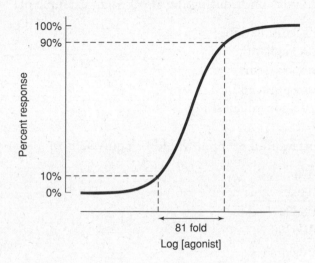

   (A) larger amounts of agonist produce a corresponding increase in response
   (B) 80% more agonist is required to achieve 80% more response
   (C) 81 times more agonist is needed to achieve a 90% response than a 10% response
   (D) an 80% high response is achieved with a tenfold increase in agonist
   (E) None of the above are true

6. A dose that is represented as LD_{50}

   (A) shows a response in 50% of the population
   (B) kills half of the study group
   (C) is a dose that has an acceptable risk level of 50%
   (D) is a dose that has a threshold response of 50%
   (E) is a dose that is administered to 50% of the population

7. Problem(s) associated with risk management include

   (A) people making a risk assessment vary in their conclusions of long-term versus short-term risks and benefits
   (B) some technologies benefit some groups and harm others
   (C) there is consideration of the cumulative impacts of various risks rather than consideration of each impact separately
   (D) there may be conflict of interest in those carrying out the risk assessment and review of the results
   (E) All of the above are true

8. Currently, the single-most significant threat to human health is

   (A) toxic chemicals
   (B) accidents
   (C) pathogenic organisms
   (D) pollution
   (E) nontransmissible diseases such as cancer and cardiovascular disease

9. The accumulation of DDT by peregrine falcons, brown pelicans, and other predatory birds during the 1960s is an example of

   (A) bioaccumulation
   (B) bioremediation
   (C) acute exposure
   (D) biomagnification
   (E) a case-controlled study

10. A concentration of 30 ppm would be equivalent to

    (A) 0.3%
    (B) 0.03%
    (C) 0.003%
    (D) 0.0003%
    (E) 0.00003%

11. Preserving the value of a resource for the future is a(n)

    (A) aesthetic value
    (B) cultural value
    (C) existence value
    (D) use value
    (E) option value

12. It costs a copper smelter $200 to reduce emissions by 1 ton and $250 for each additional ton. It costs an electric company $100 to reduce its emissions by 1 ton and $150 for each additional ton. What is the least expensive way of reducing total emissions by 2 tons?

    (A) Legislate that both firms must reduce emissions by 1 ton
    (B) Charge both firms $251 for every ton they emit
    (C) Allow each firm to buy a $151 permit to pollute
    (D) File an injunction to halt production until the firms reduce emissions by 2 tons
    (E) None of the above

13. In economic analysis, the optimum level of pollution

    (A) is always zero
    (B) is where the marginal benefits from further reduction equals the marginal cost of further reduction
    (C) occurs when demand crosses the private cost supply curve
    (D) should be determined by the private market without any government intervention
    (E) None of the above

14. Externalized costs of nuclear power include all of the following EXCEPT

    (A) disposing of nuclear wastes
    (B) government subsidies
    (C) costs associated with Three Mile Island
    (D) Price-Anderson Indemnity Act
    (E) All are external costs

15. Which of the following is NOT part of a cost-benefit analysis?

    (A) Judging whether public services provided by the private sector are adequate
    (B) Judging and assessing inefficiencies in the private sector and their impact on health, safety, and environmental need
    (C) Determining external costs to society
    (D) Meeting societal needs in a cost-effective manner
    (E) All are part of a cost-benefit analysis

16. A federal tax on a pack of cigarettes is an example of

    (A) full-cost pricing
    (B) an internal cost
    (C) an external cost
    (D) a marginal cost
    (E) a life cycle cost

17. The threshold dose-response model

    (A) cannot be used to determine how toxic a substance is
    (B) implies a risk associated with all doses
    (C) implies that no detectable harmful effects can occur below a certain dose
    (D) is similar to a linear dose-response model
    (E) All of the above are true

18. Approximately how many people aged 15–24 are newly infected world-wide with the AIDS virus each day?

    (A) 60
    (B) 600
    (C) 6,000
    (D) 60,000
    (E) 600,000

19. Which of the following is NOT part of risk assessment?

    (A) Determining the probability of a particular hazard
    (B) Determining types of hazards
    (C) Coming up with an estimate on the chances of how many people could be exposed to a particular risk
    (D) Informing the public about the chances of risks
    (E) All are part of risk assessment

20. Which factor listed below is generally considered to be the primary cause of reduced human life span?

    (A) AIDS
    (B) Infectious disease
    (C) Cancer
    (D) Poverty
    (E) Heart disease

FREE-RESPONSE QUESTION

By: Annaliese Berry, B.A.
California Institute of Technology
Williams College
A.P. Environmental Science Teacher
Marin Academy, San Rafael, California

TIP

For Free Response essays, be sure to practice making good arguments in favor of, or against, a particular position on an environmental issue. Make sure you understand that restating the question is a waste of time in a timed test, and doing so will never earn you any points.

Part (a): **2 points**
Part (b): **2 points**
Part (c): **4 points**
Part (d): **2 points**

Total: 10 points

In 1989, the oil tanker *Exxon Valdez* struck an offshore reef in Alaska and spilled more than 11 million gallons of crude oil into Prince William Sound. It was the largest spill in U.S. history. Although Exxon spent $2.2 billion in partially dispersing and removing the oil, many seabirds and mammals died as a result of this catastrophe.

Ironically, environmental accidents such as the *Exxon Valdez* oil spill actually stimulate our nation's economy by causing an increase in gross domestic product and gross national product. Using this information, answer the following questions.

(a) Explain what gross domestic product (GDP) or gross national product (GNP) is a measure of and give examples of a possible cause for an increase in the GDP or GNP following the *Exxon Valdez* oil spill.

(b) Give at least one criticism of GDP and GNP as progress indicators, and mention an alternative progress indicator. What does that alternative attempt to measure?

(c) Name and explain one internal cost and one external cost that might be associated with the oil industry. Include in your answer an explanation of what internal and external costs are.

(d) How would internalizing external costs (sometimes called full-cost analysis or true-cost analysis) affect the pricing and economic competitiveness of petroleum products?

MULTIPLE-CHOICE ANSWERS AND EXPLANATIONS

1. **(D)** In 1999, over 20,000 hazardous waste generators produced over 40 million tons of hazardous wastes.

2. **(D)** The Superfund program began in 1980 to locate, investigate, and clean up the most polluted sites nationwide.

3. **(B)** Chronic diseases are of long duration with slow progress. Examples would include congestive heart failure, Parkinson's disease, and cerebral palsy.

4. **(E)** Land farming, composting, rhizofiltration, and phytoremediation all involve the use of bacteria and enzymes to break down hazardous materials.

5. **(C)** "81 fold" means 81 more times. An agonist is a drug or other chemical that produces a reaction typical of a naturally occurring substance.

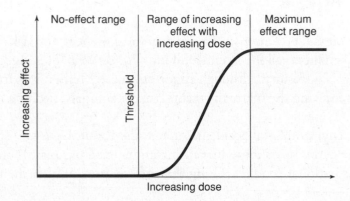

6. **(B)** The LD_{50} value is typically expressed in milligrams of material per kilogram of body weight. It indicates the quantity of material that will cause 50% of the subjects to perish.

7. **(E)** Risk assessment is the process of evaluating the likelihood of an adverse health effect. Risk assessment does NOT determine the level of allowable or acceptable risk—that is risk management.

8. **(E)** For most of human history, pathogenic (disease-causing) organisms were the greatest threat to health. Today cardiovascular, cancer, and other non-infectious diseases have become the major killers.

9. **(D)** Biomagnification is the increase in the concentration of toxic substances as one moves up the food chain. Animals at the top of the food chain receive the highest concentration of toxins and experience the worst effects.

10. **(C)** To change ppm to a percentage, move the decimal place four places to the left and add a % sign.

11. **(E)** An option value is the value that people place on having the option to enjoy something in the future.

12. **(C)** The copper smelter would pay $450 to emit 2 tons of pollutants. The electric company would pay $250 to emit 2 tons of pollutants. Clearly, $151 would be the least expensive option available.

13. **(B)** The marginal benefit of pollution control declines as the level of environmental quality goes up. At the same time, the marginal cost of pollution tends to increase. The optimal level of pollution from an economic standpoint is where the marginal benefit equals the marginal cost. As long as the marginal benefit exceeds the marginal cost, there is economic incentive for cleaning up.

14. **(E)** External costs are the costs that are borne by people other than the producer of a good.

15. **(C)** Cost-benefit analysis is a technique for deciding whether to make a change and requires adding up the value of the benefits of a course of action and subtracting the costs associated with it.

16. **(A)** Full-cost pricing accounts for the cost of a good when its internal costs and its estimated short- and long-term external costs are included in its market price. Washington State increased its cigarette tax to $1.425 a pack on Jan. 1, 2002, making it the nation's highest. Virginia has the lowest state sales tax of 2.5 cents per pack. There is a 39-cent per pack federal excise tax.

17. **(C)** The threshold dose-response model has long been recognized as the most important tool in understanding the risk assessment processes used by regulatory and public health agencies worldwide.

18. **(C)** An estimated 10 million people aged 15–24 are living with HIV/AIDS. Half of all new infections, more than 6,000 each day, occur among the young. However, in countries where HIV transmissions have been reduced, the greatest reductions have been among young people.

19. **(D)** Risk assessment is the process of quantifying the probability of a harmful effect. In most countries, the approval of certain pesticides, industrial chemicals, power plants, and so on is not allowed unless it can be demonstrated through risk assessment that they do not increase the risk, death, or illness above a certain level. Educating the public is not part of risk assessment. That is left up to the media, public interest groups, and special interest groups.

20. **(D)** Poverty is the fundamental issue that affects human life span. People who are without financial resources suffer from malnutrition, exposure, and disease. Those with financial resources are able to obtain proper nutrition, have shelter, and are able to seek medical treatment.

FREE-RESPONSE ANSWER

(a) Gross domestic product (GDP) is a measure of the market value of goods and services transacted within a country during a year, so it increases any time products or services are paid for. This means that the additional expense that Exxon (now Exxon-Mobil) incurred by controlling the spill ($2.2 billion) increased the GDP. For example, money was spent to hire boats with skimmers in an attempt to capture and contain floating oil.

Notable points:

Explanation of GDP or GNP.

Example of how cleaning up the spill or replacing lost equipment/oil causes goods or services to be transacted.

GDP and GNP are economic measures, not direct quality of life measures.

Example of alternative indicator. Examples include the NNP, ISEW, GPI, or NEW. A general alternative is to take into account any standard of living data (literacy rate, percent of population below poverty line, and so on).

Explanation of what example indicator measures.

Naming of an internal cost. Costs may be expenses related to cleanup, ship or oil replacement, cost of labor, and so on.

Explanation that internal costs are costs paid for by the organization producing the product.

Naming of an external cost such as water pollution left behind after a spill, air or water pollution from operating tankers, or an increase of CO₂ in the atmosphere when the petroleum products are consumed.

Explanation that external costs are harmful effects of the product that are not paid for by the organization.

Price of product would increase as external costs are paid for by the producer and passed along to the consumer.

Petroleum products would become less economically competitive.

(b) GDP and GNP have been criticized as progress indicators because economic growth is only indirectly related to quality of life. The GDP and GNP are often stimulated by events that directly lower quality of life, such as the *Exxon Valdez* oil spill, the Oklahoma City bombing, Hurricane Katrina in New Orleans, or the events of 9/11 in New York City. Alternative progress indicators attempt to capture quality of life or standard of living, although these are often harder to define and measure than the flow of dollars. One such indicator is the net national product (NNP), which accounts for depletion and destruction of natural resources along with changes in GNP.

(c) An internal cost is a cost that is paid for by the organization producing a product. This cost is typically passed along to the consumers when they purchase the product. One internal cost of the petroleum industry could be the cost of extracting petroleum using oil drills and platforms.

(d) Internalizing external costs would mean that organizations producing a product would pay for any harmful effects (the external costs) of their products. This would allow the market price of a product to reflect the full cost of producing and cleaning up the product. It would also increase the price of any product that has external costs. A product with high external costs would become substantially more expensive and less economically competitive when this happened. In the energy industry, sources of power such as oil and gas have relative high external costs and could become more expensive than lower-impact sources of energy such as wind power.

UNIT VII: GLOBAL CHANGE (10–15%)

Areas on Which You Will Be Tested

A. **Stratospheric Ozone**—formation of stratospheric ozone, ultra-violet radiation, causes of ozone depletion, effects of ozone depletion, strategies for reducing ozone depletion, and relevant laws and treaties.

B. **Global Warming**—greenhouse gases and the greenhouse effect, impacts and consequences of global warming, reducing climate change, and relevant laws and treaties.

C. **Loss of Biodiversity**
1. Habitat loss—overuse, pollution, introduced species, and endangered and extinct species.
2. Maintenance through conservation.
3. Relevant laws and treaties.

Stratospheric Ozone and Global Warming

CHAPTER 11

STRATOSPHERIC OZONE

The stratosphere contains approximately 97% of the ozone in the atmosphere, and most of it lies between 9 and 25 miles (15 to 40 km) above Earth's surface. Most ozone is formed over the tropics. However, slow circulation currents carry the majority of it to the poles, resulting in the thickest layers over the poles and the thinnest layer above the tropics. It also varies somewhat due to season, being somewhat thicker in the spring and thinner during the autumn. The increase in temperature with height in the stratosphere occurs because of absorption of ultraviolet radiation by ozone.

Ozone is formed in the stratosphere by the reaction of ultraviolet radiation striking an oxygen molecule. This results in the oxygen molecule splitting apart and forming atomic oxygen: $O_2 + hv \rightarrow O + O$. Atomic oxygen can now react with molecular oxygen to form ozone: $O + O_2 \rightarrow O_3$. The reverse reaction also occurs when ultraviolet radiation strikes an ozone molecule, causing it to form atomic oxygen and molecular oxygen: $O_3 + hv \rightarrow O + O_2$. Atomic oxygen can also react with ozone to produce oxygen gas: $O + O_3 \rightarrow O_2 + O_2$. Generally, these reactions balance each other so that the concentration of ozone in the stratosphere remains fairly constant, which keeps the amount of UV radiation reaching Earth also constant. Various forms of life on Earth have evolved over millions of years with fairly constant amounts of UV radiation.

Ultraviolet Radiation

The sun emits a wide variety of electromagnetic radiation, including infrared, visible, and ultraviolet. Ultraviolet radiation can be subdivided into three forms: UVA, UVB, and UVC.

UVA

320 to 400 nm wavelength. Closest to blue light in the visible spectrum. The form that usually causes skin tanning. UVA radiation is 1,000 times less effective than UVB in producing skin redness, but more of it reaches Earth's surface than UVB. Birds, reptiles, and bees can see UVA. Many fruits, flowers, and seeds also stand out more strongly from the background in ultraviolet wavelengths. Many birds have patterns in their plumage that are not visible in the normal spectrum (white light) but become visible in ultraviolet. Urine of some animals is also visible only in the UVA spectrum.

UVB

290 to 320 nm wavelength. Causes blistering sunburns and is associated with skin cancer.

UVC

10 to 290 nm. Found only in the stratosphere and largely responsible for the formation of ozone.

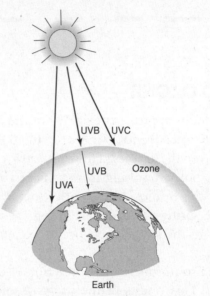

Figure 11.1 Ultraviolet radiation reaching Earth

Causes of Ozone Depletion

Thinning of the ozone layer was first discovered in 1985. It occurs seasonally and is due to the presence of human-made compounds containing halogens (chlorine, bromine, fluorine, or iodine). Measurements indicate that the ozone over the Antarctic has decreased as much as 60% since the late 1970s with an average net loss of about 3% per year worldwide.

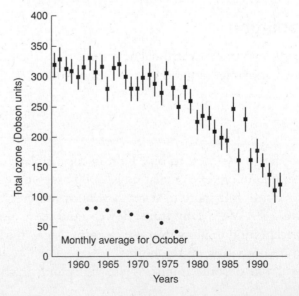

Figure 11.2 Ozone concentration over Antarctica (1955–1995)

The main culprits in the depletion of the ozone layer are compounds known as CFCs (chlorofluorocarbons). First manufactured during the 1920s, they are used as refrigerants (for example, Freon), aerosol propellants, electrical part cleaning solvents, and in the manufacture of foam products and insulation. By 1974, nearly 1 million tons of CFC gases were produced each year, and the chemicals were generating $8 billion worth of business. The largest single source of CFCs to the atmosphere is leakage from air conditioners. The average residence time of CFCs in the environment is 200 years.

When a CFC molecule enters the stratosphere, ultraviolet radiation causes it to decompose and produce atomic chlorine:

This atomic chlorine then reacts with the ozone in the stratosphere to produce chlorine monoxide (ClO):

$$Cl + O_3 \rightarrow ClO + O_2$$

The chlorine monoxide then reacts with more ozone to produce even more atomic chlorine in what becomes essentially a chain reaction:

$$ClO + O_3 \rightarrow Cl + 2O_2$$

Thus, one chlorine atom released from a CFC can ultimately destroy over 100,000 ozone molecules. Much of the destruction of the ozone layer that is occurring now is the result of CFCs that were produced many years ago since a CFC molecule takes 8 years to reach the stratosphere and the residence time in the stratosphere for a CFC molecule is over 100 years.

Bromine, found in much smaller quantities than chlorine, is about 50 times more effective than chlorine in its effect on stratospheric ozone depletion and is responsible for about 20% of the problem. Bromine is found in halons, which are used in fire extinguishers. Methyl bromide is used in fumigation and agriculture. It is naturally released from phytoplankton and biomass burning.

Effects of Ozone Depletion

During the onset of the 1998 Antarctic spring, a hole three times the size of Australia (over 3,500 miles [5,600 km] in diameter) developed in the ozone layer over the South Pole. Stratospheric ozone protects life from harmful ultraviolet radiation. Harmful effects of increased UV radiation include:

- Increases in skin cancer
- Increases in sunburns and damage to the skin
- Increases in cataracts of the eye
- Reduction in crop production
- Deleterious effects to animals (they don't wear sunglasses or sunscreen)
- Reduction in the growth of phytoplankton and the cumulative effect on food webs
- Increases in mutations since UV radiation causes changes in DNA structure

- Cooling of the stratosphere
- Reduction in the body's immune system
- Climatic change

Strategies for Reducing Ozone Depletion

Although most developed countries have phased out ozone-destroying chemicals, they are still legal in developing nations. There are several alternatives to CFC use. First, HCFC replaces chlorine with hydrogen. Unfortunately, it is still capable of destroying ozone albeit less effectively because it breaks down more readily in the troposphere. Second, alternatives to halons can be used in fire extinguishers. Third, helium, ammonia, propane, or butane can be used as a coolant. Helium-cooled refrigerators use 50% less electricity than those using CFCs or HCFCs. Individuals can use pump sprays instead of aerosol spray cans when possible, comply with disposal requirements of the Clean Air Act for old refrigerators and air conditioners, read labels and, when choices are available, use ozone friendly products, and support legislation that reduces ozone-destroying products.

> ### RELEVANT TREATY
>
> **Montreal Protocol (1987):** Designed to protect the stratospheric ozone layer. The treaty was originally signed in 1987 and substantially amended in 1990 and 1992. The Montreal Protocol stipulated that the production and consumption of compounds that deplete ozone in the stratosphere—chlorofluorocarbons (CFCs), halons, carbon tetrachloride, and methyl chloroform—were to be phased out by 2000 (2005 for methyl chloroform).

GLOBAL WARMING

When sunlight strikes Earth's surface, some of it is reflected back toward space as infrared radiation (heat). Greenhouse gases absorb this infrared radiation and trap the heat in the atmosphere.

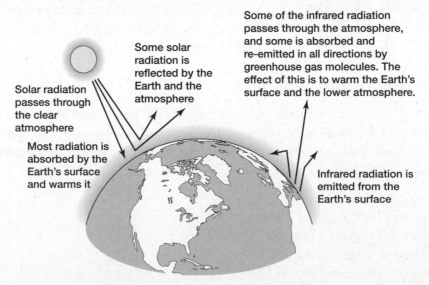

Figure 11.3 The greenhouse effect

Greenhouse Gases

Greenhouse Gas	Average Time in Troposphere (Years)	Relative Warming Potential ($CO_2 = 1$)	Source
Carbon dioxide (CO_2)	100	1	Burning oil, coal, plants, deforestation, cellular respiration
Carbon tetrachloride (CCl_4)	45	1,500	Cleaning solvent
Chlorofluorocarbons (CFCs)	15 (100 in stratosphere)	1,000–8,000	Air conditioners, refrigerators, foam products, insulation
Halons	65	6,000	Fire extinguishers
Hydrochlorofluorocarbons (HCFCs)	10–400	500–2,000	Air conditioners, refrigerators, foam products, insulation
Hydrofluorocarbons (HFCs)	15–400	150–13,000	Air conditioners, refrigerators, foam products, insulation
Methane (CH_4)	15	25	Rice cultivation, enteric fermentation, production of coal, natural gas leaks
Nitrous oxide (N_2O)	115	300	Burning fossil fuels, fertilizers, livestock wastes, plastic manufacturing
Sulfur hexafluoride (SF_6)	3,200	24,000	Electrical industry as a replacement for PCBs
Tropospheric Ozone (O_3)	Variable	3,000	Combustion of fossil fuels

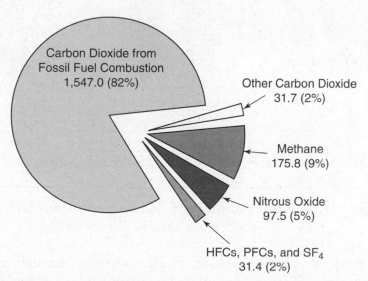

Figure 11.4 U.S. anthropogenic greenhouse gas emissions (2001)
(Million Metric Tons of Carbon Equivalent)

Levels of several important greenhouse gases have increased by about 25% since large-scale industrialization began around 150 years ago. During the past 20 years, about three-quarters of human-made carbon dioxide emissions were from burning fossil fuels. In the United States, greenhouse gas emissions come mostly from energy use. These are driven largely by economic growth, fuel used for electricity generation, and weather patterns affecting heating and cooling needs. Energy-related carbon dioxide emissions, resulting from petroleum and natural gas, represent 82% of total U.S. anthropogenic (caused by humans) greenhouse gas emissions.

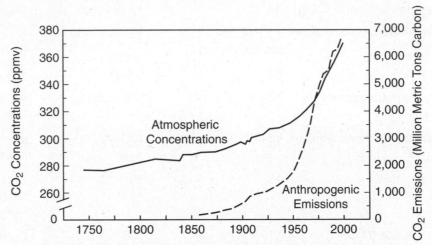

Figure 11.5 Trends in atmospheric CO_2 concentrations
Source: Oak Ridge National Laboratory

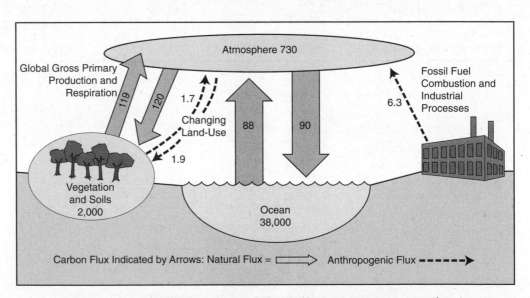

Figure 11.6 Global carbon cycle (billion metric tons carbon)
Source: Intergovernmental Panel on Climate Change

Concentrations of carbon dioxide in the atmosphere are naturally regulated by numerous processes collectively known as the carbon cycle. The movement (flux) of carbon between the atmosphere and the land and oceans is dominated by natural processes, such as plant photosynthesis. Although these natural processes can absorb some of the net 6.1 billion metric tons of anthropogenic carbon dioxide emissions produced each year (measured in carbon equivalent terms), an estimated 3.2 billion metric tons is added to the atmosphere annually. Earth's positive imbalance between emissions and absorption results in the continuing growth in greenhouse gases in the atmosphere.

Impacts and Consequences of Global Warming

Global warming affects the weather, the economy, and numerous other aspects of life.

ACIDIFICATION

The oceans currently absorb 1 metric ton of carbon dioxide per person per year. Increased carbon dioxide absorption of oceans will lower the pH of seawater. This adversely affects corals, plankton, organisms with shells, and reproduction rates as eggs and sperm are exposed to higher levels of acid.

CHANGES IN TROPOSPHERIC WEATHER PATTERNS

Air temperatures today average 5°F to 9°F (3°C to 5°C) warmer than they were before the Industrial Revolution. Historic increases in air temperature averaged less than 2°F (1°C) per 1,000 years. Higher temperatures result in higher amounts of rainfall due to higher rates of evaporation. Worldwide, hurricanes of category 4 or 5 have risen from 20% of all hurricanes in the 1970s to 35% in the 1990s. Precipitation due to hurricanes in the United States has increased 7% during the 20th century. More rainfall increases erosion, which then leads to higher rates of desertification due to deforestation. This then leads to losses in biodiversity as some species are forced out of their habitat. El Niño and La Niña patterns and their frequencies have also changed.

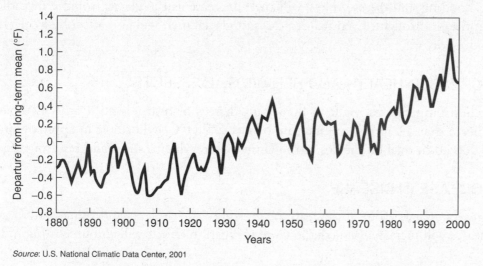

Source: U.S. National Climatic Data Center, 2001

Figure 11.7 Global temperature changes (1880–2000)

DISPLACEMENT OF PEOPLE

The United Nations estimates that by the year 2050, 150 million people will need to be relocated worldwide. This will occur due to the effects of coastal flooding, shoreline erosion, and agricultural disruption.

ECOLOGICAL PRODUCTIVITY

Satellite photos have shown that productivity in the Northern Hemisphere has increased since 1982. However, biomass increases due to warmer temperatures reaches a certain point—the point where limiting factors of water and nutrients curb future productivity increases. In the tropics, plants increase productivity more so than trees (which are carbon sinks). With higher percentages of plants due to increased temperatures and carbon dioxide concentrations, the rates of decomposition increase because plants are shorter lived. As a result, more carbon enters the carbon cycle.

FOREST FIRES

Boreal forest fires in North America used to average 2.5 million acres (10,000 sq km). They now average 7 million acres (28,000 sq km). Forest management practices may also be contributing to the increase.

GLACIER MELTING

Total surface area of glaciers worldwide has decreased by 50% since the end of the 19th century. Temperatures of the Antarctic Southern Ocean rose 0.31°F (0.17°C) between the 1950s and the 1980s. Glacier melting causes landslides, flash floods, glacial lake overflow, and increased variation in water flows into rivers. Hindu Kush and Himalayan glacier melts are reliable water sources for many people in China, India, and much of Asia. Global warming initially increases water flow, causing flooding and disease. Flow will then decrease as the glacier volume dwindles, resulting in drought. Eventual decreases in glacial melt will also affect hydroelectric production.

INCREASED HEALTH AND BEHAVIORAL EFFECTS

Higher temperatures result in higher incidences of heat-related deaths. Estimates indicate that a temperature increase of just 2°F (1°C) will result in approximately 25,000 additional homicides in the United States due to stress and resulting rage.

INCREASE IN DISEASE

Rates of malaria (due to increase in mosquitoes), cholera, and other waterborne diseases will increase. Remediation and mitigation efforts will end up costing more.

INCREASED PROPERTY LOSS

Weather-related disasters have increased 3-fold since the 1960s. Insurance payouts have increased 15-fold (adjusted for inflation) during this same time period. Much of this can be attributed to people moving to vulnerable coastal areas.

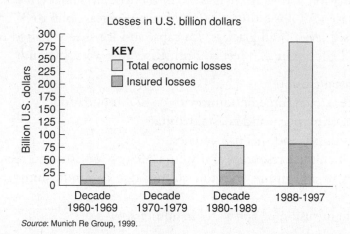

Losses in U.S. billion dollars

KEY
- Total economic losses
- Insured losses

Source: Munich Re Group, 1999.

Figure 11.8 Weather and flood catastrophes since 1960

LOSS IN ECONOMIC DEVELOPMENT

Money that was earmarked for education, improved health care, reduced hunger, and improved sanitation and fresh water supplies, will instead be spent on mitigating the effects of global warming.

LOSS OF BIODIVERSITY

Arctic fauna will be most affected. The food webs of polar bears that depend on ice flows, birds, and marine mammals will be negatively impacted. Many refugee species currently have shifted their ranges toward the poles, averaging 4 miles (6 km) per decade. Bird migrations are averaging over two days earlier per decade. Grasses have become established in Antarctica for the first time. Many species of fish and krill that require cooler waters will be negatively impacted. Decreased glacier melt will impact migratory fish, such as salmon, that need sufficient river flow.

RELEASES OF METHANE FROM HYDRATES
IN COASTAL SEDIMENTS

Methane hydrate (methane clathrate) is a form of water ice that contains methane within its crystal structure. Extremely large deposits of this resource have been discovered in ocean sediments.

RELEASES OF METHANE FROM THAWING PERMAFROST REGIONS

Thawing of permafrost would increase bacterial levels in the soil and eventually lead to higher releases of methane. Estimates of melting of permafrost peat bogs in Siberia could release as much as 70,000 million metric tons of methane (a greenhouse gas) within the next few decades.

RISE IN SEA LEVEL

Sea levels have risen 400 feet (120 m) since the peak of the last ice age (18,000 years ago). From 3,000 years ago to the start of the Industrial Revolution, the rate of sea level rise averaged 0.1 to 0.2 mm per year. Since 1900, sea level has risen about 3 mm per year (over a 10-fold increase). An increase in global temperatures of 3°F to 8°F (1.5°C to 4.5°C) is estimated to lead to an increase of 6 to 37 inches (15 cm to 95 cm) in sea level. If all glaciers, ice caps, and ice sheets melted on Earth, the sea level would rise about 225 feet (69 m). Rises in sea level would:

- Increase coastal erosion
- Create higher storm surge flooding with coastal inundation
- Increase loss of property and coastal habitats
- Cause losses in fish and shellfish catches
- Cause loss of cultural resources and values such as tourism and recreation
- Cause losses in agriculture and aquaculture due to diminishing soil and water quality
- Result in the intrusion of salt water in aquifers.

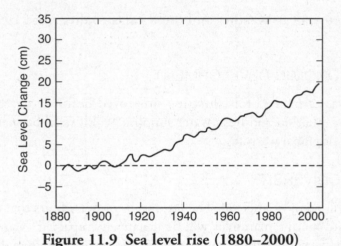

Figure 11.9 Sea level rise (1880–2000)

SLOWING OR SHUTDOWN OF THERMOHALINE CIRCULATION

Melting of the glaciers in Greenland would shift the salt water–freshwater balance in the North Atlantic. This would result in a decrease of heavier saline waters sinking than in traditional ocean circulation patterns. This would have significant effects on the fishing industry. Localized cooling in the North Atlantic brought about through the reduction of thermohaline circulation currents (North Atlantic drift) could result in much colder temperatures in Great Britain and Scandinavia.

Reducing Climate Change

World carbon dioxide emissions are expected to increase by 2% annually between 2001 and 2025. Much of the increase in these emissions is expected to occur in the developing world, such as China and India, where emerging economies fuel economic development with fossil energy. The United States was surpassed by China in 2007 in being the largest emitter of greenhouse gases. Developing countries' emissions are expected to grow by 3% annually between 2001 and 2025 and surpass emissions of industrialized countries by 2018.

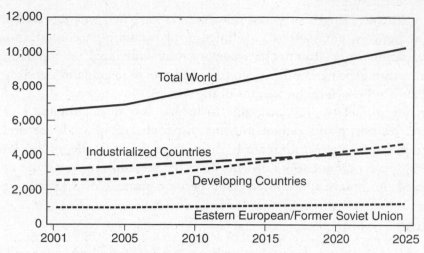

Figure 11.10 World carbon dioxide emissions by region, 2001–2025 (million metric tons of carbon equivalent)
Source: Energy Information Administration

The United States produces about 25% of global carbon dioxide emissions from burning fossil fuels. This is due primarily to a robust economy and 85% of the U.S. energy needs being met through burning fossil fuels.

To stabilize the current global warming crisis would require:

1. A decrease in methane emissions by 8%

2. A decrease in nitrous oxide emissions by 50%

3. A decrease in carbon dioxide emissions by up to 80%

Six methods can be used to reduce dramatic climatic changes brought on by global warming.

1. Increase the efficiency of cars.

2. Move to more renewable sources of energy (wind, solar, geothermal, etc.).

3. Find chemical substitutes that do not impact global warming.

4. Slow down the rate of deforestation and encourage reforestation.

5. Reduce dependence on inorganic, nitrogen-based fertilizers and utilize conservation tillage techniques.

6. Support treaties and protocols that require reductions in greenhouse gas emissions.

LOSS OF BIODIVERSITY

Information on habitat destruction needs to be compared with past habitat conditions. Long-term history of habitat conditions can be obtained from fossil records, ice core samples, tree ring analysis, and the analysis of pollen. Organisms can cope with habitat destruction by migration, adaptation, and acclimatization. Migration depends on the magnitude of degradation, the rate of degradation, the organism's ability to migrate, access routes or corridors, and the proximity and availability of suitable new habitats. Adaptation is the ability to survive due to changing environmental conditions. Adaptation depends on the magnitude and rate of degradation, birth rate, length of generation, population size, genetic variability, and gene flow between populations as a function of variation. Acclimatization is the ability to adjust to environmental changes on an individual or population level. Acclimatization depends on the magnitude and rate of degradation and the physiological and/or behavioral limitations of the species.

Plants are initially more susceptible to habitat loss than animals. This occurs for several reasons: plants cannot migrate; dispersal rates of seeds are slow events (for example, spruce trees can increase their range about 1 mile, or 1.6 km, every 100 years); plants cannot seek nutrients or water; seedlings must survive and grow in degraded conditions; and stressed plants become prone to disease and infestation.

Overuse

Worldwide, more than 3 million fishing boats remove 70 million to 90 million tons of fish and shellfish from the oceans each year. Of the world's marine fish stocks, 70% are fully exploited, overfished, or depleted. Twenty-million metric tons of bycatch (fish and shellfish not sought after) are destroyed each year and represents one-fourth of all catch taken from the sea. Around 80,000 whales, dolphins, seals, and other marine mammals perish as bycatch each year. Shrimp trawlers throw back 5.2 pounds (2.4 kg) of marine life that will eventually die for every pound (0.5 kg) of shrimp they catch. Overall, this amounts to almost 3 billion pounds (1.4 billion kg) of fish destroyed annually by the shrimp industry.

Pollution

Pollution results from human activities such as urban sprawl, transportation, and industry. More significantly in terms of wildlife and biodiversity, human activity expands the effects of pollution into areas where environmental quality may not yet have been severely compromised. Industrial and transportation sources release pollutants into the air, water, and soil that are toxic to most species and degrade their habitats. In addition, light and noise associated with human activity often have negative effects on wildlife behavior and natural cycles.

Pollutants have devastating effects on wildlife when they are released into the soil and water. They affect an animal's endocrine, reproductive, and immune systems. Wildlife can be exposed to pollutants through skin absorption, the animals or plants they eat, and the water they drink. Amphibians and other aquatic species are especially vulnerable to pollutants in streams, rivers, and lakes.

Plants or wildlife (especially species with thin, moist skin such as amphibians) can inhale or absorb air pollutants, causing health problems. Air pollution also contaminates water and soil through acid rain. Sulfur dioxide, a main component of acid rain, can damage plant communities by impairing photosynthesis processes. Plants weakened by acid rain may become vulnerable to root rot, insect damage, and disease, causing serious damage to plant communities and forest ecosystems.

Noise can affect wildlife behavior and physiology. Noise can also cause chronic stress, increased heart rates, metabolism and hormone imbalance, and nervous system stimulation. These responses may cause energy loss, food intake reduction, an avoidance and abandonment of habitat, and even injury or death. Noise will often flush birds from their nests, causing broken eggs, injured young, or exposure to predators. Finally, birdsongs and calls that are used to mate and establish territories may be disrupted by excessive human noise pollution.

Developed areas produce artificial light that causes problems for wildlife that depend on the light from the moon and stars to navigate. Streetlights, neon signs, parking lots, and late-night golf driving ranges may potentially keep wildlife from moving, nesting, or foraging near urban areas.

Introduced Species

Invasive or exotic species are animals and plants that are transported to any area where they do not naturally live. The spread of nonnative species has emerged in recent years as one of the most serious threats to biodiversity. It has undermined the ecological integrity of many native habitats and pushed some rare species to the edge of extinction. Some introduced species simply outcompete native plants and animals for space, food, or water. Others may also fundamentally alter natural disturbance regimes and other ecological processes, making it difficult or impossible for native species to survive. About 15% of the estimated 6,000 nonnative plant and animal species in the United States cause severe economic or ecological impacts. Exotic species have been implicated in the decline of approximately 40% of the species listed for protection under the federal Endangered Species Act.

A major marine source is marine ballast. Every 60 seconds, ships discharge 40,000 gallons (150,000 L) of ballast water that contains foreign plant and animal species into U.S. harbors. Often these plants and animals grow at uncontrolled rates because they have no natural predators in these new locations. These introduced species often become pests and crowd out native plants and animals. Over 250 invasive species of plants and animals are found in San Francisco Bay alone. Common examples of invasive species include zebra mussels, tumbleweed, kudzu, the mongoose, and gypsy moths.

Endangered and Extinct Species

Since 1500 C.E., 816 species have become extinct, 103 of them since 1800—a rate 50 times greater than the natural background rate. In the next 25 years, extinction rates are expected to rise as high as 25%. According to the Nature Conservancy, about one-third of all U.S. plant and animal species are at risk of becoming extinct. Mammals listed as "endangered" rose from 484 in 1996 to 520 in 2000, with primates increasing from 13 to 19. Endangered birds increased from 403 species to 503 species. Endangered freshwater fish more than doubled—from 10 species to 24 species in four years. One-fourth of all mammals and reptiles, one-fifth of all amphibians, one-eighth of all birds, and one-sixth of all conifers are in some manner endangered or are threatened to the point of extinction.

Maintenance Through Conservation

Maintaining and protecting wildlife consists of three major approaches:

- **Species approach**—Protecting *endangered species* through legislation.
- **Ecosystem approach**—Preserving balanced *ecosystems*.
- **Wildlife management approach**—Managing *game species* for sustained yield through international treaties to protect migrating species, improving wildlife habitats, regulating hunting and fishing, creating harvest quotas, and developing population management plans.

Biodiversity can be protected by:

1. Properly designing and updating laws that legally protect endangered and threatened species (such as the Endangered Species Act)

2. Protecting the habitats of endangered species through private or governmental land trusts

3. Reintroducing species into suitable habitats

4. Managing habitats and monitoring land use

5. Establishing breeding programs for endangered or threatened species

6. Creating and expanding wildlife sanctuaries

7. Restoring compromised ecosystems

8. Reducing nonnative and invasive species

RELEVANT LAWS AND TREATIES

Multiple-Use Act (1960): Directed that the national forests be managed for timber, watershed, range, outdoor recreation, wildlife, and fish purposes.

The Convention on International Trade in Endangered Species of Wild Fauna and Flora (CITES) (1963): An international agreement between governments to ensure that international trade in wild animals and plants did not threaten their survival.

Marine Mammal Protection Act (1972): Established federal responsibility to conserve marine mammals.

Endangered Species Act (1973): Provided a program for the conservation of threatened and endangered plants and animals and the habitats in which they are found.

QUICK REVIEW CHECKLIST

- ☐ **Stratospheric Ozone**
  - ☐ ultraviolet radiation
    - ☐ UVA
    - ☐ UVB
    - ☐ UVC
  - ☐ causes of ozone depletion
  - ☐ effects of ozone depletion
  - ☐ strategies for reducing ozone depletion
  - ☐ relevant laws and treaties
    - ☐ Montreal Protocol (1987)
    - ☐ London (1990)
    - ☐ Copenhagen (1992)

QUICK REVIEW CHECKLIST (continued)

- [ ] **Global Warming**
  - [ ] greenhouse gases
    - [ ] carbon dioxide
    - [ ] methane
    - [ ] nitrous oxide
    - [ ] HFC, CFC, HCFC, SF_6
  - [ ] impacts and consequences of global warming
    - [ ] acidification
    - [ ] changes in tropospheric weather patterns
    - [ ] displacement of people
    - [ ] ecological productivity
    - [ ] forest fires
    - [ ] glacier melting
    - [ ] health and behavioral effects
    - [ ] disease
    - [ ] property loss
    - [ ] economic development
    - [ ] biodiversity
    - [ ] methane hydrates in coastal sediments
    - [ ] thawing of permafrost and release of methane
    - [ ] rise in sea level
    - [ ] effect on thermohaline circulation

- [ ] **Reducing Climate Change**

- [ ] **Loss of Biodiversity**

- [ ] **Overuse**

- [ ] **Pollution**

- [ ] **Introduced Species**

- [ ] **Endangered and Extinct Species**

- [ ] **Conservation**
  - [ ] species approach
  - [ ] ecosystem approach
  - [ ] wildlife management approach
  - [ ] specific measures to protect biodiversity

- [ ] **Relevant Laws and Treaties**
  - [ ] Multiple-use Act (1960)
  - [ ] Convention on International Trade in Endangered Species of Wild Fauna and Flora (CITES)—1963
  - [ ] Marine Mammal Protection Act (1972)
  - [ ] Endangered Species Act (1973)

MULTIPLE-CHOICE QUESTIONS

1. Which of the following represents the greatest contribution of methane emission to the atmosphere?

    (A) Enteric fermentation or flatulence from animals
    (B) Coal mining
    (C) Landfills
    (D) Rice cultivation
    (E) Burning of biomass

2. Ozone depletion reactions that occur in the stratosphere are facilitated by

    (A) NO_2^-
    (B) NH_3
    (C) NH_4^+
    (D) N_2O
    (E) NO_3^-

3. In addition to absorbing harmful solar rays, how do ozone molecules help to stabilize the upper atmosphere?

    (A) They release heat to the surroundings.
    (B) They create a buoyant lid on the atmosphere.
    (C) They create a warm layer of atmosphere that keeps the lower atmosphere from mixing with space.
    (D) All of the above are correct.
    (E) None of the above are correct.

4. The bromine released from methyl bromide is about 40 times more damaging to the ozone layer than chlorine on a molecule of chlorofluoro-carbon. Under the Montreal Protocol, methyl bromide in developing countries will be phased out in 2005. What is methyl bromide used for?

    (A) Sterilizing soil in fields and greenhouses
    (B) Killing pests on fruits, vegetables, and grain before export
    (C) Fumigating soils
    (D) Killing termites in buildings
    (E) All of the above

5. Rising sea levels due to global warming would be responsible for all of the following EXCEPT

    (A) destruction of coastal wetlands
    (B) beach erosion
    (C) increased damage due to storms and floods
    (D) increased salinity of estuaries and aquifers
    (E) All of the above

6. True statements about global warming include which of the following?

    I. Average carbon dioxide levels in the atmosphere today are the highest they have ever been in Earth's history.

    II. Average global air temperatures today are the highest they have ever been in Earth's history.

    III. The increases in average global air temperatures are equally distributed across the Earth's surface.

    (A) I only
    (B) II only
    (C) I and II
    (D) III only
    (E) All choices are false

7. One of the reasons the vortex winds in the Antarctic are important in the formation of the hole in the ozone layer is

    (A) they prevent warm, ozone-rich air from mixing with cold, ozone-depleted air
    (B) they quickly mix ozone-depleted air with ozone-rich air
    (C) they harbor vast quantities of N_2O
    (D) they bring in moist, warm air, which accelerates the ozone-forming process
    (E) None of the above

8. Although the Montreal Protocol curtailed production of ozone-depleting substances, the peak concentration of chemicals in the stratosphere is only now being reached. When do scientists expect a recovery in the ozone levels in the stratosphere to occur?

    (A) Immediately
    (B) Within 5 years
    (C) Within 20 years
    (D) Within 50 years
    (E) In about 100 years

9. Over the past 1 million years of Earth's history, the average trend in Earth's temperature would best be described as:

    (A) very large fluctuations, from periods of extreme heat to periods of extreme cold
    (B) very little change, if any, fairly uniform and constant temperature
    (C) a general cooling trend
    (D) a general warming trend
    (E) a number of fluctuations varying by a few degrees

10. The agency responsible for the identification and listing of endangered species is the

   (A) Forest Service
   (B) Agriculture Department
   (C) National Park Service
   (D) Fish and Wildlife Service
   (E) Endangered Species Department of the Interior

11. A treaty that controls international trades in endangered species is known as (the)
   (A) Endangered Species Act
   (B) CITES
   (C) International Treaty on Endangered Species
   (D) Lacey Act
   (E) Federal Preserve System

12. Managing game species for sustained yields would be consistent with what conservation approach?

   (A) Wildlife management approach
   (B) Species approach
   (C) Ecosystem approach
   (D) Sustainable yield approach
   (E) Holistic approach

13. A certain insect was causing extensive damage to local crops. The farmers, who were environmentally conscious and did not want to use pesticides, decided to introduce another insect into their fields that studies have shown would prey on the pest insect. Before any action is taken, the farmers should consider (the)

   (A) principle of natural balance
   (B) principle of unforeseen events
   (C) Murphy's law
   (D) law of supply and demand
   (E) precautionary principle

14. Which of the following is the biggest threat to wildlife preserves?

   (A) Hunters
   (B) Poachers
   (C) Global warming
   (D) Invasive species
   (E) Tourists

15. A certain species of plant is placed on the threatened species list. Several years later it is placed on the endangered species list. This is an example of

    (A) a negative-negative feedback loop
    (B) a positive-negative feedback loop
    (C) a negative-positive feedback loop
    (D) a negative-feedback loop
    (E) a positive-feedback loop

16. The specific form of radiation largely responsible for the formation of ozone in the stratosphere is
    (A) UVA
    (B) UVB
    (C) UVC
    (D) infrared
    (E) gamma rays

17. Which of the following is NOT an effect caused by increased levels of ultraviolet radiation reaching Earth?

    (A) Warming of the stratosphere
    (B) Cooling of the stratosphere
    (C) Increased genetic damage
    (D) Reduction in immunity
    (E) Reduction in plant productivity

18. Which of the following gases is NOT considered a greenhouse gas, responsible for global warming?

    (A) SO_2
    (B) N_2O
    (C) H_2O
    (D) SF_6
    (E) CH_4

19. Which of the following effects would result from the shutdown or slowdown of the thermohaline circulation pattern?

    (A) A warmer Scandinavia and Great Britain
    (B) Freshwater fish moving into the open ocean
    (C) A colder Scandinavia and Great Britain
    (D) A saltier ocean
    (E) A drier Scandinavia and Great Britain

20. The largest number of species are being exterminated per year in

    (A) grasslands
    (B) deserts
    (C) forests
    (D) tropical rain forests
    (E) the tundra

FREE-RESPONSE QUESTION

Part (a): **3 points**
Part (b): **4 points**
Part (c): **3 points**

Total: **10 points**

The following charts show global temperature changes and carbon dioxide emissions.

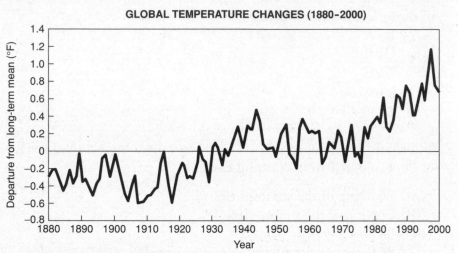

Source: U.S. National Climatic Data Center, 2001

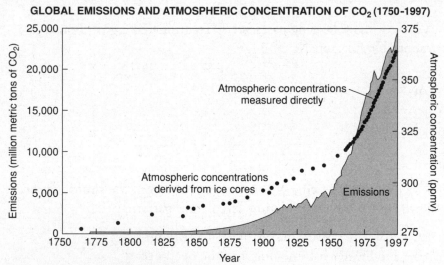

Source: Carbon Dioxide Information Analysis Center, 2001

(a)   Describe global warming.
   (i)   What role do naturally occurring atmospheric gases play in global warming?
   (ii)   Describe how a certain degree of global warming is necessary for life to exist on Earth.
   (iii)   How have atmospheric concentrations of carbon dioxide (CO_2) gas changed since the Industrial Revolution?

(b) What factors may be increasing global temperatures?

    (i)   Compare natural contributions of CO_2 to the atmosphere vs. release of CO_2 through human activities.

    (ii)   Name and describe TWO other gases besides CO_2 that contribute to global warming?

(c) What are the environmental consequences of global warming?

    (i)   How have global mean temperatures changed in the last 200 years?

    (ii)   How have the changes in global mean temperatures in (i) affected weather patterns?

    (iii)   List TWO consequences to the environment of increases in global air temperature.

MULTIPLE-CHOICE ANSWERS AND EXPLANATIONS

1. **(C)** Since 1800, there has been a 150% increase in methane concentrations in the atmosphere compared with a 30% increase in carbon dioxide. Methane is about 30 times more effective in contributing to global warming than carbon dioxide.

2. **(D)** Nitrous oxide is responsible for about 7% of the anthropogenic greenhouse gases and stays in the atmosphere for about 120 years. The concentration of nitrous oxide in the atmosphere has increased about 15% since the Industrial Revolution. It has an ozone-depleting potential comparable with many HCFCs. The main source of nitrous oxide are inorganic fertilizers and industrial processes. Natural sources of nitrous oxide arise from biological processes within the soil and oceans.

3. **(D)** Solar energy that is absorbed by ozone molecules and is partly turned into heat creates a warm region in the stratosphere. This creates a stable air mass that resists sinking and mixing with the lower atmosphere, effectively forming a barrier. In the ozone layer, temperature increases with height, creating a stable and buoyant air mass that keeps an effective lid on the lower atmosphere.

4. **(E)** No explanation needed.

5. **(E)** No explanation needed.

6. **(E)** Earth's primitive forests first appeared around 300 million years ago during the Carboniferous Period. Before then, the atmosphere held far more CO_2 but concentrations declined throughout the Carboniferous Period as plants flourished and absorbed more and more CO_2 from the atmosphere through photosynthesis, creating carbon sinks (coal, oil, etc.). During the Carboniferous Period, the atmosphere became greatly depleted of CO_2 (declining from about 2500 ppm to 350 ppm) so that by the end of the Carboniferous Period the atmosphere was less favorable to plant life and plant growth slowed. Today, CO_2 concentrations are approximately 380 ppm (0.038% of our atmosphere). The Arctic is feeling the effects of increases in average global air temperatures the most. Average temperatures in Alaska, western Canada, and eastern Russia are rising at twice the rate of the global average.

7. **(A)** Ozone depletion follows an annual cycle that corresponds to the amount of light that reaches the Antarctic. The cycle begins every year around June when vortex winds develop in the Antarctic. Cold temperatures pro-

duced by these winds create polar stratospheric clouds that capture floating chlorofluorocarbons (CFCs) and other ozone-depleting compounds. For the next two months, a reaction occurs on the cloud surface that frees the chlorine in the CFCs but keeps it contained within the vortex. In September, sunlight returns to the Antarctic and triggers a chemical reaction, causing chlorine to convert ozone to oxygen gas. November brings a breakdown in the vortex and allows the ozone-rich air to combine with the thinning ozone. Wind currents carry this mixture over the southern hemisphere.

8. **(E)** Concentrations of ozone-depleting chlorofluorocarbons (CFCs) have leveled off in the stratosphere and have actually begun to decline in the lower atmosphere. However, the largest Antarctic ozone hole ever recorded occurred in 2000. A CFC molecule can take about 8 years after being released at ground level to reach the stratosphere. Decades may pass until it is converted by sunlight into a form that depletes ozone.

9. **(E)** The mean surface temperature of the planet has been warming at a rate of 0.2°C per decade for the past 30 years. The global mean temperature is now within 1°C of the maximum for the past million years. The last time Earth was this warm was about 3 million years ago when sea level was about 25 meters higher than today. If carbon dioxide emissions are not curbed, global temperatures are likely to rise 2 to 3°C by 2100.

10. **(D)** The Fish and Wildlife Service, in the Department of the Interior, and the National Oceanic and Atmospheric Administration (NOAA) in the Department of Commerce, share responsibility for administration of the Endangered Species Act. An endangered species is one that is in danger of extinction throughout all or a significant portion of its range. A threatened species is one that is likely to become endangered in the foreseeable future.

11. **(B)** CITES (the Convention on International Trade in Endangered Species) is an international agreement between governments. Its aim is to ensure that international trade in specimens of wild animals and plants does not threaten their survival.

12. **(A)** The strength of the traditional wildlife management approach is that it explicitly uses and enhances natural processes to perpetuate populations.

13. **(E)** The precautionary principle states that if the consequences of an action are unknown but are judged to have some potential for major or irreversible negative consequences, that action should be avoided. The concept includes risk prevention, cost effectiveness, ethical responsibilities toward maintaining the integrity of natural biological systems, and risk assessment. It is not the risk that must be avoided but the potential risk that must be prevented.

14. **(C)** All choices but (C) can be dealt with on a local level through specific enforcement and management practices. Global warming is an issue that affects all ecosystems and must be solved on a global scale, requiring international cooperation.

15. **(E)** A positive-feedback loop occurs in a situation in which a change in a certain direction (toward extinction in this case) causes the system to change in the same direction.

16. **(C)** Ozone is produced by oxygen and sunlight in the UVC wavelength range (<240 nm). In the atmosphere, this reaction works only at higher altitudes where there is adequate high-energy UV penetration (>10 km). The atomic oxygen so formed can combine with oxygen gas (O_2) to form ozone (O_3).

17. **(A)** The stratosphere has been cooling over the past three decades. The stratosphere contains the ozone layer, which absorbs sunlight and heats the stratosphere. This long-term cooling trend is generally accepted to result from the loss of the ozone layer as a result of human-made influences. However, the cooling trend is not uniform like ozone loss but, rather, broken into a series of jumps or discontinuities most likely associated with major volcanic eruptions that inject aerosols into the stratosphere. The aerosols also absorb sunlight and heat from the stratosphere, thus temporarily offsetting the cooling trend from ozone loss.

18. **(A)** Sulfur dioxide (SO_2) contributes to acid rain, not global warming.

19. **(C)** There is speculation that global warming could melt glaciers in Greenland, increasing the amount of freshwater in the North Sea. This disruption in balance between salt water and freshwater could theoretically slow down or shut down thermohaline circulation that is responsible for the North Atlantic Drift (a section of the Gulf Stream) that currently stabilizes temperatures in Great Britain and Scandinavia.

20. **(D)** Deforestation in the tropics in order to convert the land to agricultural purposes and cattle grazing is occurring at unprecedented levels. The tropics contain the highest biodiversity anywhere on Earth.

FREE-RESPONSE ANSWER

(a) Energy from the sun drives Earth's weather and climate and heats Earth's surface. In turn, Earth radiates energy back into space. Natural atmospheric greenhouse gases such as water vapor, carbon dioxide, methane, and nitrous oxide trap some of the outgoing energy similar to the glass panels of a greenhouse. Without the greenhouse effect, much of the heat reaching Earth would escape into space. Life as we know it, could not exist. However, Earth's climate is predicted to change due to human activities that have caused these natural gas concentrations to increase dramatically. According to the graph, since the beginning of the Industrial Revolution, atmospheric concentrations of carbon dioxide have increased nearly 30%. In addition, methane concentrations have more than doubled, and nitrous oxide concentrations have increased about 15%. Not all gases contribute to global warming equally. As a reference point, if the global warming potential of carbon dioxide gas is given as a 1, then methane is 20 times more effective in trapping heat energy, and nitrous oxide is over 300 times more effective.

Notable points:

Connection with sun as energy driver.

Energy hitting Earth from the sun is absorbed and reradiated back into space.

Some gases prevent energy from radiating back into space.

Not all gases trap heat equally.

Greenhouse effect is necessary for life on Earth.

Both natural and human-made gases cause the greenhouse effect.

Carbon dioxide is the gas most responsible for the greenhouse effect.

Increase in production of CO_2 is primary factor in greenhouse effect.

How CO_2 is produced.

Comparison of CO_2 levels today as compared with years past.

Mentioning one other greenhouse gas such as methane, water vapor, sulfur hexafluoride, or nitrous oxide.

(b) Scientists generally believe that the combustion of fossil fuels and other human activities are the primary reason for the increased concentration of carbon dioxide. Plant respiration and the decomposition of organic matter release more than 10 times the carbon dioxide released through human activities. However, these releases have generally been in balance during the centuries leading up to the Industrial Revolution, with carbon dioxide being absorbed by plants both terrestrial and oceanic. What has changed in the last few hundred years is the additional release of carbon dioxide by human activities. Fossil fuels burned to run cars and trucks, heat homes and businesses, and power factories are responsible for about 98% of U.S. carbon dioxide emissions, 24% of methane emissions, and 18% of nitrous oxide emissions. Increased agriculture, deforestation, landfills, industrial production, and mining also contribute a significant share of emissions.

In 1997, the United States emitted about one-fifth of total greenhouse gases. By 2100, in the absence of emissions control policies, carbon dioxide concentrations are projected to be 30%–150% higher than today's levels. Water vapor is the most abundant greenhouse gas. It occurs naturally and makes up about two-thirds of the natural greenhouse effect. Other gases that accelerate the greenhouse effect include nitrous oxide, hydrofluorocarbons, perfluorocarbons, and sulfur hexafluoride. Since preindustrial times, atmospheric concentrations of nitrous oxide have increased by 17%, carbon dioxide by 31%, and methane by 151%. Scientists have confirmed that this is primarily due to human activity.

—Decreased snow cover

—Floating ice

—Rise in sea level

—Increased precipitation due to warmer temperatures

—Increased surface temperatures

—Decreased soil moisture in some areas

—Increase in storm intensities

—Increase in rates of disease

—Increase in heatstroke and other heat-associated effects

—Increased flooding in some areas

—Increased droughts in some areas

(c) According to the graph, global mean surface temperatures have increased 0.5°F to 1.0°F since the late 1800s. The 20th century's 10 warmest years all occurred in the last 15 years. The snow cover in the Northern Hemisphere and floating ice in the Arctic Ocean have decreased. Globally, sea level has risen 4 to 8 inches in the last 100 years. Worldwide precipitation over land has increased by about 1% (warmer temperatures usually cause greater amounts of water to evaporate). The frequency of extreme rainfall events has increased throughout much of the United States. Increasing concentrations of greenhouse gases are likely to accelerate the rate of climatic change. Scientists expect that the average global surface temperature could rise 2°F to 10°F in the next century, with significant regional variation. Evaporation will increase as the climate warms, which will increase average global precipitation. Soil moisture is likely to decline in many regions, and intense rainstorms are likely to become more frequent. Sea levels are likely to rise 2 feet along most of the U.S. coasts. A few degrees of warming will increase the chances of more frequent and severe heat waves, which can cause more heat-related death and illness. Greater heat results in increased air pollution as well as damaged crops and depleted water resources. Warming is likely to allow tropical diseases, such as malaria, to spread northward in some areas of the world. It will also intensify Earth's hydrological cycle. Both evaporation and precipitation will increase. Some areas will receive more rain, while other areas will be drier. At the same time, extreme events like floods and droughts are likely to become more frequent. Warming will cause glaciers to melt and oceans to expand. Projections state that sea levels will rise between 4 inches and 3 feet over the next century, threatening low-lying coastal areas.

PRACTICE EXAMS

Practice Exam 1

With Answers and Analysis

SECTION I (MULTIPLE-CHOICE QUESTIONS)

Time: 90 minutes

100 questions

60% of total grade

No calculators allowed

This section consists of 100 multiple-choice questions. Mark your answers carefully on the answer sheet.

General Instructions

Do not open this booklet until you are told to do so by the proctor.

Be sure to write your answers for Section I on the separate answer sheet. Use the test booklet for your scratch work or notes. Remember, though, that no credit will be given for work, notes, or answers written only in the test booklet. Once you have selected an answer, thoroughly blacken the corresponding circle on the answer sheet. To change an answer, erase your previous mark completely, and then record your new answer. Mark only one answer for each question.

Example	Sample Answer
The Pacific is	Ⓐ Ⓑ ● Ⓓ Ⓔ

    (A) a river
    (B) a lake
    (C) an ocean
    (D) a sea
    (E) a gulf

There is no penalty for wrong answers on the multiple-choice section, so you should answer all multiple-choice questions. Even if you have no idea of the correct answer, you should try to eliminate any obvious incorrect choices, and then guess.

Because it is not expected that all test takers will complete this section, do not spend too much time on difficult questions. First answer the questions you can answer readily. Then, if you have time, return to the difficult questions later. Do not get stuck on one question. Work quickly but accurately. Use your time effectively.

Answer Sheet
PRACTICE EXAM 1

1 Ⓐ Ⓑ Ⓒ Ⓓ Ⓔ   26 Ⓐ Ⓑ Ⓒ Ⓓ Ⓔ   51 Ⓐ Ⓑ Ⓒ Ⓓ Ⓔ   76 Ⓐ Ⓑ Ⓒ Ⓓ Ⓔ
2 Ⓐ Ⓑ Ⓒ Ⓓ Ⓔ   27 Ⓐ Ⓑ Ⓒ Ⓓ Ⓔ   52 Ⓐ Ⓑ Ⓒ Ⓓ Ⓔ   77 Ⓐ Ⓑ Ⓒ Ⓓ Ⓔ
3 Ⓐ Ⓑ Ⓒ Ⓓ Ⓔ   28 Ⓐ Ⓑ Ⓒ Ⓓ Ⓔ   53 Ⓐ Ⓑ Ⓒ Ⓓ Ⓔ   78 Ⓐ Ⓑ Ⓒ Ⓓ Ⓔ
4 Ⓐ Ⓑ Ⓒ Ⓓ Ⓔ   29 Ⓐ Ⓑ Ⓒ Ⓓ Ⓔ   54 Ⓐ Ⓑ Ⓒ Ⓓ Ⓔ   79 Ⓐ Ⓑ Ⓒ Ⓓ Ⓔ
5 Ⓐ Ⓑ Ⓒ Ⓓ Ⓔ   30 Ⓐ Ⓑ Ⓒ Ⓓ Ⓔ   55 Ⓐ Ⓑ Ⓒ Ⓓ Ⓔ   80 Ⓐ Ⓑ Ⓒ Ⓓ Ⓔ
6 Ⓐ Ⓑ Ⓒ Ⓓ Ⓔ   31 Ⓐ Ⓑ Ⓒ Ⓓ Ⓔ   56 Ⓐ Ⓑ Ⓒ Ⓓ Ⓔ   81 Ⓐ Ⓑ Ⓒ Ⓓ Ⓔ
7 Ⓐ Ⓑ Ⓒ Ⓓ Ⓔ   32 Ⓐ Ⓑ Ⓒ Ⓓ Ⓔ   57 Ⓐ Ⓑ Ⓒ Ⓓ Ⓔ   82 Ⓐ Ⓑ Ⓒ Ⓓ Ⓔ
8 Ⓐ Ⓑ Ⓒ Ⓓ Ⓔ   33 Ⓐ Ⓑ Ⓒ Ⓓ Ⓔ   58 Ⓐ Ⓑ Ⓒ Ⓓ Ⓔ   83 Ⓐ Ⓑ Ⓒ Ⓓ Ⓔ
9 Ⓐ Ⓑ Ⓒ Ⓓ Ⓔ   34 Ⓐ Ⓑ Ⓒ Ⓓ Ⓔ   59 Ⓐ Ⓑ Ⓒ Ⓓ Ⓔ   84 Ⓐ Ⓑ Ⓒ Ⓓ Ⓔ
10 Ⓐ Ⓑ Ⓒ Ⓓ Ⓔ   35 Ⓐ Ⓑ Ⓒ Ⓓ Ⓔ   60 Ⓐ Ⓑ Ⓒ Ⓓ Ⓔ   85 Ⓐ Ⓑ Ⓒ Ⓓ Ⓔ
11 Ⓐ Ⓑ Ⓒ Ⓓ Ⓔ   36 Ⓐ Ⓑ Ⓒ Ⓓ Ⓔ   61 Ⓐ Ⓑ Ⓒ Ⓓ Ⓔ   86 Ⓐ Ⓑ Ⓒ Ⓓ Ⓔ
12 Ⓐ Ⓑ Ⓒ Ⓓ Ⓔ   37 Ⓐ Ⓑ Ⓒ Ⓓ Ⓔ   62 Ⓐ Ⓑ Ⓒ Ⓓ Ⓔ   87 Ⓐ Ⓑ Ⓒ Ⓓ Ⓔ
13 Ⓐ Ⓑ Ⓒ Ⓓ Ⓔ   38 Ⓐ Ⓑ Ⓒ Ⓓ Ⓔ   63 Ⓐ Ⓑ Ⓒ Ⓓ Ⓔ   88 Ⓐ Ⓑ Ⓒ Ⓓ Ⓔ
14 Ⓐ Ⓑ Ⓒ Ⓓ Ⓔ   39 Ⓐ Ⓑ Ⓒ Ⓓ Ⓔ   64 Ⓐ Ⓑ Ⓒ Ⓓ Ⓔ   89 Ⓐ Ⓑ Ⓒ Ⓓ Ⓔ
15 Ⓐ Ⓑ Ⓒ Ⓓ Ⓔ   40 Ⓐ Ⓑ Ⓒ Ⓓ Ⓔ   65 Ⓐ Ⓑ Ⓒ Ⓓ Ⓔ   90 Ⓐ Ⓑ Ⓒ Ⓓ Ⓔ
16 Ⓐ Ⓑ Ⓒ Ⓓ Ⓔ   41 Ⓐ Ⓑ Ⓒ Ⓓ Ⓔ   66 Ⓐ Ⓑ Ⓒ Ⓓ Ⓔ   91 Ⓐ Ⓑ Ⓒ Ⓓ Ⓔ
17 Ⓐ Ⓑ Ⓒ Ⓓ Ⓔ   42 Ⓐ Ⓑ Ⓒ Ⓓ Ⓔ   67 Ⓐ Ⓑ Ⓒ Ⓓ Ⓔ   92 Ⓐ Ⓑ Ⓒ Ⓓ Ⓔ
18 Ⓐ Ⓑ Ⓒ Ⓓ Ⓔ   43 Ⓐ Ⓑ Ⓒ Ⓓ Ⓔ   68 Ⓐ Ⓑ Ⓒ Ⓓ Ⓔ   93 Ⓐ Ⓑ Ⓒ Ⓓ Ⓔ
19 Ⓐ Ⓑ Ⓒ Ⓓ Ⓔ   44 Ⓐ Ⓑ Ⓒ Ⓓ Ⓔ   69 Ⓐ Ⓑ Ⓒ Ⓓ Ⓔ   94 Ⓐ Ⓑ Ⓒ Ⓓ Ⓔ
20 Ⓐ Ⓑ Ⓒ Ⓓ Ⓔ   45 Ⓐ Ⓑ Ⓒ Ⓓ Ⓔ   70 Ⓐ Ⓑ Ⓒ Ⓓ Ⓔ   95 Ⓐ Ⓑ Ⓒ Ⓓ Ⓔ
21 Ⓐ Ⓑ Ⓒ Ⓓ Ⓔ   46 Ⓐ Ⓑ Ⓒ Ⓓ Ⓔ   71 Ⓐ Ⓑ Ⓒ Ⓓ Ⓔ   96 Ⓐ Ⓑ Ⓒ Ⓓ Ⓔ
22 Ⓐ Ⓑ Ⓒ Ⓓ Ⓔ   47 Ⓐ Ⓑ Ⓒ Ⓓ Ⓔ   72 Ⓐ Ⓑ Ⓒ Ⓓ Ⓔ   97 Ⓐ Ⓑ Ⓒ Ⓓ Ⓔ
23 Ⓐ Ⓑ Ⓒ Ⓓ Ⓔ   48 Ⓐ Ⓑ Ⓒ Ⓓ Ⓔ   73 Ⓐ Ⓑ Ⓒ Ⓓ Ⓔ   98 Ⓐ Ⓑ Ⓒ Ⓓ Ⓔ
24 Ⓐ Ⓑ Ⓒ Ⓓ Ⓔ   49 Ⓐ Ⓑ Ⓒ Ⓓ Ⓔ   74 Ⓐ Ⓑ Ⓒ Ⓓ Ⓔ   99 Ⓐ Ⓑ Ⓒ Ⓓ Ⓔ
25 Ⓐ Ⓑ Ⓒ Ⓓ Ⓔ   50 Ⓐ Ⓑ Ⓒ Ⓓ Ⓔ   75 Ⓐ Ⓑ Ⓒ Ⓓ Ⓔ   100 Ⓐ Ⓑ Ⓒ Ⓓ Ⓔ

Directions: For each question or statement, select the one lettered choice that is the best answer and fill in the corresponding circle on the answer sheet.

1. Which of the following would be most likely to increase competition among the members of a squirrel population in a given area?

   (A) An epidemic of rabies within the squirrel population
   (B) An increase in the number of hawk predators
   (C) An increase in the reproduction of squirrels
   (D) An increase in temperature
   (E) An increase in the food supply

2. Approximately how many years ago did life first appear on Earth?

   (A) 1 million
   (B) 500 million
   (C) 1 billion
   (D) 3.5 billion
   (E) 5 billion

3. Which one of the following statements is FALSE?

   (A) The greenhouse effect is a natural process that makes life on Earth possible with 98% of total global greenhouse gas emissions being from natural sources (mostly water vapor) and 2% from human-made sources.
   (B) The United States is the number one contributor to global warming.
   (C) The Kyoto Protocol would have allowed the United States to increase its greenhouse gas emissions—primarily carbon dioxide (CO_2), methane (CH_4), and nitrous oxide (N_2O)—by only 2% per year based on 1990 levels.
   (D) The effects of global warming on weather patterns may lead to adverse human health impacts.
   (E) Global warming may increase the incidence of many infectious diseases.

4. Which of the following would be an external cost?

   (A) The cost of steel in making a refrigerator
   (B) The cost of running a refrigerator for one month
   (C) The cost of labor in producing refrigerators
   (D) The taxes paid by consumers in purchasing refrigerators
   (E) The costs associated with health care when the refrigerator leaks refrigerant into the atmosphere

5. In which stage of the nitrogen cycle do soil bacteria convert ammonium ions (NH_4^+) into nitrate ions (NO_3^-); a form of nitrogen that can be used by plants?

   (A) Nitrogen fixation
   (B) Nitrification
   (C) Assimilation
   (D) Ammonification
   (E) Denitrification

6. Which of the following is NOT an example of environmental mitigation?

   (A) Promoting sound land use planning, based on known hazards
   (B) Relocating or elevating structures out of the floodplain
   (C) Constructing living snow fences
   (D) Organizing a beach cleanup
   (E) Developing, adopting, and enforcing effective building codes and standards

7. Which of the following statements regarding an El Niño is FALSE?

   (A) Depression of the thermocline occurs, which cuts off cold water upwelling.
   (B) A change in atmospheric pressures occurs and is associated with changing ocean water temperatures.
   (C) El Niño affects weather patterns globally.
   (D) An increase in greenhouse gases may increase the incidents of El Niños.
   (E) Northeast and southeast trade winds increase.

8. Most of the Earth's freshwater supply is found in

   (A) lakes
   (B) ice caps and glaciers
   (C) aquifers
   (D) estuaries
   (E) rivers

9. Most municipal solid wastes in the United States consist of

   (A) yard wastes
   (B) food wastes
   (C) plastic
   (D) paper
   (E) glass

10. Rising sea levels due to global warming would be responsible for all of the following EXCEPT

    (A) destruction of coastal wetlands
    (B) beach erosion
    (C) increases due to storm and flood damage
    (D) increased salinity of estuaries and aquifers
    (E) all would be the result of rising sea level

11. Which of the following characteristics are NOT typical of a ground fire?

    I. Fire smolders and/or creeps slowly through the litter and humus layers, consuming all or most of the organic cover, and exposing mineral soil or underlying rock.
    II. Burns the upper litter layer and small branches that lie on or near the ground. Usually move rapidly through an area, and do not consume all the organic layer.
    III. Release considerable amounts of nutrients from the burned fuels, destroy many small organisms and fungi that live in the humus and organic layers, consume seed stored in the litter, and kill roots in all but deep soil layers.

    (A) I only
    (B) II only
    (C) III only
    (D) I and II
    (E) I and III

12. Place the following economic activities in order, starting with those activities closest to natural resources and ending with those furthest away.

    I. Use raw materials to produce or manufacture something new and more valuable.
    II. Professions that process, administer, and disseminate information (e.g., computer engineers, professors, and lawyers).
    III. Agriculture, fishing, hunting, herding, forestry, and mining.
    IV. Include all activities that amount to doing service for others (e.g., doctors, teachers, and secretaries).

    (A) I, II, III, IV
    (B) I, III, IV, II
    (C) III, I, II, IV
    (D) III, I, IV, II
    (E) II, IV, I, III

13. Most of the municipal trash floating in the ocean is composed of

    (A) plastic
    (B) paper
    (C) wood
    (D) metal cans
    (E) yard wastes

14. Most of Earth's mass is found in this region, which is composed of iron, magnesium, aluminum, and silicon-oxygen compounds. At over 1800°F (1000°C), most of this region is solid but the upper third is more plasticlike in nature. Which region is being described?

    (A) Troposphere
    (B) Lithosphere
    (C) Crust
    (D) Mantle
    (E) Core

15. Which of the following statements regarding coral reefs is FALSE?

    (A) Modern reefs can be as much as 2.5 million years old.
    (B) Coral reefs capture about half of all the calcium flowing into the ocean every year, fixing it into calcium carbonate rock at very high rates.
    (C) Coral reefs store large amounts organic carbon and are very effective sinks for carbon dioxide from the atmosphere.
    (D) Coral reefs are among the most biologically diverse ecosystems on the planet.
    (E) Coral reefs are among the most endangered ecosystems on Earth.

16. Which of the following strategies to control pollution would incur the greatest governmental cost?

    (A) Green taxes
    (B) Government subsidies for reducing pollution
    (C) Regulation
    (D) Charging a user fee
    (E) Tradable pollution rights

17. In terms of annual production, which two crops listed below had the greatest success during The Green Revolution (1967–1978)?

    (A) maize and corn
    (B) rice and corn
    (C) wheat and rice
    (D) wheat and maize
    (E) soybean and rice

18. In 1989, the *Exxon Valdez* spilled 10.8 million gallons of crude oil into Prince William Sound in Alaska. What happened to most of the oil?

    (A) It was cleaned up by Exxon.
    (B) It eventually evaporated into the air.
    (C) It sank into the ground.
    (D) It biodegraded and photolyzed.
    (E) It dispersed into the water column.

19. Which of the following contributes LEAST to speciation?

   (A) Sexual reproduction
   (B) Asexual reproduction
   (C) Selection
   (D) Variation
   (E) Isolation

20. Which Act's primary goal is to protect human health and the environment from the potential hazards of waste disposal and calls for conservation of energy and natural resources, reduction in waste generated, and environmentally sound waste management practices?

   (A) RCRA
   (B) FIFRA
   (C) CERCLA
   (D) OSHA
   (E) FEMA

21. One of the earliest sites that utilized Superfund resources was

   (A) Bhopal, India
   (B) Love Canal, New York
   (C) Prince William Sound, Alaska
   (D) Chernobyl, Ukraine
   (E) Donora, Pennsylvania

22. Which growth curve best represents the effects of environmental resistance acting on a sustainable population?

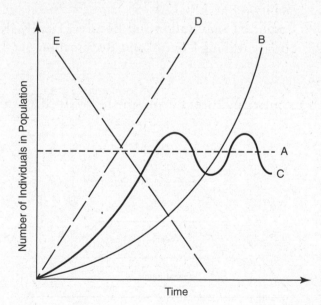

   (A) A
   (B) B
   (C) C
   (D) D
   (E) E

23. The location where two tectonic plates slide apart from each other with the space that was created being filled with molten magma from below, such as the Mid-Atlantic Ridge, the East Pacific Rise, and the East African Great Rift Valley, are known as

    (A) divergent boundaries
    (B) convergent boundaries
    (C) transform boundaries
    (D) tectonic boundaries
    (E) lithospheric boundaries

24. Radioactive materials, lead, arsenic, mercury, nickel, and benzene are all pollutants released from what type of electricity generating plant?

    (A) Hydroelectric
    (B) Solar
    (C) Coal-burning
    (D) Nuclear
    (E) Natural gas

25. On Michigan's Keweenaw Peninsula, copper mills discharged an estimated 200 million tons of copper-contaminated waste directly into Torch Lake, reducing its volume by 20% and leaving a toxic threat to fish and anyone who eats the fish. Which law listed below would address the damage that has been done at Torch Lake?

    (A) 1872 Mining Act
    (B) 1920 Mineral Leasing Act
    (C) 1980 Comprehensive Environmental Response, Compensation, and Liability Act (CERCLA)
    (D) 1976 Resource Conservation and Recovery Act (RCRA)
    (E) 1977 Surface Mining Control and Reclamation Act

Questions 26–27
Base your answers to questions 26 and 27 on the following diagram.

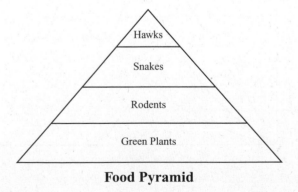

Food Pyramid

26. The greatest amount of energy present in this pyramid is found at the level of the

    (A) hawks
    (B) snakes
    (C) rodents
    (D) green plants
    (E) decomposers

27. The pyramid implies that in order to live and grow, 1,000 kilograms of snakes would require

    (A) less than 1,000 kilograms of green plants
    (B) 1,000 kilograms of rodents
    (C) more than 1,000 kilograms of rodents
    (D) no rodents
    (E) less than 1,000 kilograms of hawks

28. The type of planting in which an agricultural crop is grown simultane-ously with a long-term tree crop to provide annual income while the tree crop matures is known as

    (A) crop rotation
    (B) alley cropping
    (C) monocropping
    (D sequential cropping
    (E) intercropping

29. If a city with a population of 100,000 experiences 4,000 births, 3,000 deaths, 500 immigrants, and 200 emigrants within the course of one year, what is the net annual percentage growth rate?

    (A) 0.3%
    (B) 1.3%
    (C) 13%
    (D) 101.3%
    (E) 130%

30. The annual productivity of any ecosystem is greater than the annual increase in biomass of the herbivores in the ecosystem because

    (A) plants convert energy input into biomass more efficiently than animals
    (B) there are always more animals than plants in any ecosystem
    (C) plants have a greater longevity than animals
    (D) during each energy transformation, some energy is lost
    (E) animals convert energy input into biomass more efficiently than plants

Questions 31–33

Following is a diagram showing the relationships that exist in an arid ecosystem. Base your answers to questions 31 through 33 on the diagram and your knowledge of environmental science.

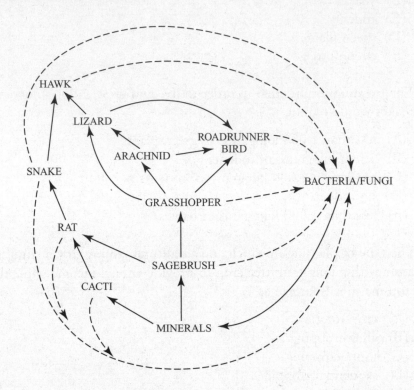

31. The snake is acting as a

    (A) producer
    (B) primary consumer
    (C) secondary consumer
    (D) autotroph
    (E) secondary producer

32. Between which two organisms would there MOST LIKELY be the greatest competition?

    (A) Rat and snake
    (B) Lizard and arachnid
    (C) Rat and roadrunner
    (D) Grasshopper and bacteria/fungi
    (E) Arachnid and roadrunner

33. In the diagram, which statement correctly describes the role of bacteria/fungi?

    (A) The bacteria/fungi convert radiant energy into chemical energy.

    (B) The bacteria/fungi directly provide a source of nutrition for animals.

    (C) The bacteria/fungi are saprophytic agents restoring inorganic material to the environment.

    (D) The bacteria/fungi convert atmospheric nitrogen into minerals and are found in the nodules of cacti.

    (E) The bacteria/fungi consume live plants and animals.

34. When resources are scarce,

    (A) prices go down

    (B) recycling is not profitable

    (C) investment and potential profits decrease

    (D) there is an impetus for conservation

    (E) there is little competition to discover new or substitute products

35. Examine the three age-structure diagrams below:

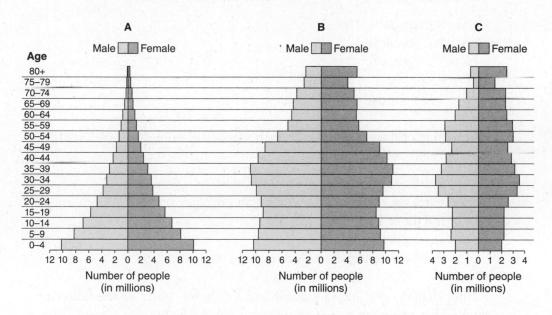

Which of the age structure diagrams above would be typical for a country experiencing a rapid rate of population growth?

    (A) A

    (B) B

    (C) C

    (D) B and C

    (E) None of the above

36. The annual amount of incoming solar energy varies considerably from tropical to polar latitudes with the polar regions radiating away more thermal energy than they absorb from the Sun in the course of a year. The primary reason that the polar regions are not becoming progressively colder each year is because

    (A) large areas of upwelling in the polar seas keep heat circulating in the region
    (B) as ice melts, it releases its heat energy to the surroundings
    (C) the extensive cloud cover in the polar regions acts as a thermal blanket keeping whatever heat there is from escaping
    (D) underground "hot spots" in the polar seas keep the ocean temperatures in the area fairly stable
    (E) global heat energy is constantly being circulated through both ocean and air currents

37. Approximately five pounds of carbon is released into the atmosphere in the form of carbon dioxide for every gallon of gasoline that is burned. Two cars are making a 500 mile trip. Car A gets 15 miles to the gallon and car B gets 30 miles to the gallon. Approximately how much more carbon (in the form of CO_2) will car A produce than car B?

    (A) 5 pounds
    (B) 16.7 pounds
    (C) 33.3 pounds
    (D) 83 pounds
    (E) 100 pounds

38. Which factor does NOT significantly affect the amount of solar energy reaching the surface of Earth?

    (A) Earth's rotation once every 24 hours
    (B) Earth's revolution around the Sun once per year
    (C) the tilt of Earth's axis (23.5°)
    (D) atmospheric conditions
    (E) the distance between Earth and the Sun

39. Other than the melting of glaciers and ice sheets, which of the following factors has made the largest contribution to the global rise in sea level over the past 100 years?

    (A) Urbanization and its impact on estuaries
    (B) Melting of sea ice
    (C) Increased river runoff
    (D) Warming of ocean surface waters
    (E) Increase in rainfall

40. A chain is dragged across the seafloor, pulling a huge net behind it. A single pass of this net can remove up to a quarter of seafloor life. Repeated passes can remove nearly all seafloor life, including sessile animals and plants plus many species of fish and marine invertebrates. This type of fishing is known as

    (A) bottom trawling or dredging
    (B) chain-fishing
    (C) scraping
    (D) scoop netting
    (E) drag-lining

41. Which of the following items should NOT be placed in a compost pile?

    I.   Chemically treated wood products
    II. Pet feces
    III. Manure
    IV. Pernicious weeds
    V.   Bones

    (A) II
    (B) III
    (C) IV
    (D) I, II, IV, V
    (E) I, II, III, IV, V

42. Two-thirds of Iceland's energy sources come from clean, renewable hydro-electric and geothermal sources. Research in Iceland is currently underway in developing hydrogen fuel cells. Iceland is an example of a country that is practicing

    (A) sustainability
    (B) remediation
    (C) conservation
    (D) preservation
    (E) mitigation

43. An electrical power plant that uses X joules of energy derived from natural gas can generate a maximum of Y joules of electrical energy within the same amount of time. Which of the following is always true?

    (A) X = Y due to the Law of Conservation of Energy.
    (B) X < Y due to the First Law of Thermodynamics.
    (C) X > Y due to the Second Law of Thermodynamics.
    (D) X < Y due to the Second Law of Thermodynamics.
    (E) Depending upon the efficiency of the power plant, X could be equal to Y.

44. Which of the following statements are TRUE?

   I.   The annual fluctuation of air temperature on land masses influenced by polar cells is greater than the change in temperature occurring in a 24-hour cycle.

   II. In land masses influenced by polar cells, precipitation rather than temperature is the critical factor in plant distribution and soil development.

   III. Land masses influenced primarily by Ferrel cells have defined seasons with strong annual cycles of temperature and precipitation.

   IV. Land masses in equatorial regions primarily influenced by Hadley cells are characterized by low humidity and little precipitation.

   V.   Subtropical land masses primarily influenced by Hadley cells are characterized by high relative humidity and tropical forests.

   (A) I and III
   (B) I, II, and IV
   (C) II and III
   (D) III and V
   (E) I, II, IV, and V

45. As urban environments become denser, noise pollution is becoming a major environmental issue as people deal with noise generated from flight paths, freeways, work environments, and neighborhoods. Which of the following is a TRUE statement?

   (A) Loud sound is not dangerous as long as you do not feel any pain in your ears.
   (B) Hearing loss after sound exposure is temporary.
   (C) Hearing loss is caused mostly by aging.
   (D) Loud sound damages only your hearing.
   (E) If you have a hearing loss already, you still have to protect your hearing.

46. What is meant by the term "enriched" uranium?

   (A) pure uranium, with no other elements present
   (B) all nuclei are U-238 in the sample
   (C) uranium that has been rinsed in heavy water to enrich it
   (D) uranium that has a higher proportion of U-235 nuclei than normal
   (E) uranium that has had plutonium added to it

47. Which of the following is NOT a benefit associated with integrated pest management?

   (A) More reliable and effective pest control
   (B) Reduction in the use of the most hazardous pesticides
   (C) Lessening the chance of pesticide resistance developing
   (D) Total elimination of pest species
   (E) All choices are benefits of integrated pest management

48. Which of the following statements about the role of carbon dioxide (CO_2) in the carbon cycle are TRUE?

   I.   Carbon dioxide is produced during photosynthesis
   II. Carbon dioxide concentration in the atmosphere decreases when trees are cut down and the trees decay
   III. The primary non-anthropomorphic source of carbon dioxide is outgassing from the Earth's interior

   (A) I only
   (B) II only
   (C) III only
   (D) II and III only
   (E) I, II, and III

Questions 49–50 refer to the following air pollutants

   (A) Nitrogen dioxide (NO_2)
   (B) Peroxyacl nitrates (PANs)
   (C) Volatile organic compounds (VOCs)
   (D) Ozone (O_3)
   (E) Particulates (PM_{10})

49. Powerful respiratory and eye irritants present in photochemical smog. They are secondary air pollutants, not directly emitted as exhaust from power plants or internal combustion engines, and formed from other pollutants by chemical reactions in the atmosphere.

50. Commonly found in paints and solvents. Examples include benzene, acetone, and formaldehyde.

51. Most commercial fish are caught in which ocean?

   (A) Atlantic
   (B) Indian
   (C) Arctic
   (D) Southern
   (E) Pacific

52. Why is the bioaccumulation of harmful chemicals especially destructive in species such as salmon?

   (A) Salmon migrate long distances, allowing them to spread harmful chemicals into many different ecosystems.
   (B) Salmon are tertiary consumers and are at the top of the food pyramid.
   (C) Salmon are a keystone species, meaning they are "key" in the health of the ecosystem(s) they inhabit.
   (D) Salmon are an increasingly endangered species.
   (E) All of the above are true.

53. Characteristics typical of an oligotrophic lake include which of the following?

    I.   High dissolved oxygen content
    II. High primary productivity
    III. Clear water

    (A) I only
    (B) II only
    (C) III only
    (D) I and III only
    (E) I, II, and III

54. Which federal agency listed below does NOT manage designated federal wilderness areas in the United States?

    (A) National Park Service
    (B) Forest Service
    (C) Bureau of Land Management
    (D) Fish and Wildlife Service
    (E) All share management

55. The greatest environmental impact to using compact fluorescent light-bulbs (CFL) is that they

    (A) are not energy efficient
    (B) contain mercury
    (C) take much longer to start than incandescent lightbulbs
    (D) are more expensive compared to incandescent lightbulbs
    (E) All of the above

56. A compact fluorescent lightbulb (CFL) used for lighting has an efficiency rating of 10%. If for every 10.00 joule of electrical energy consumed by the bulb, which of the following is produced?

    (A) 0.90 joules of light energy
    (B) 1.00 joule of light energy
    (C) 9.00 joules of light energy
    (D) 9.90 joules of heat energy
    (E) 9.90 joules of light energy

Questions 57–58

In 1940, ranchers introduced cattle into an area. The graph below shows the effect of cattle ranching on the populations of two organisms present in the area before the introduction of the cattle. Base your answers to questions 57 and 58 on the graph and on your knowledge of environmental science.

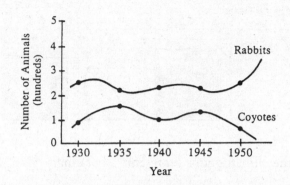

57. The most probable reason for the increase in the rabbit population after 1950 was

    (A) more food became available to the rabbits
    (B) the coyote population declined drastically
    (C) the cattle created a more favorable environment for the rabbits
    (D) the coyotes and cattle competed for the same food
    (E) the coyote population increased

58. If the interrelationship of rabbits and coyotes was once in balance, what is the most probable explanation for the decline of the coyotes?

    (A) Mutations
    (B) Starvation
    (C) Disease
    (D) Increase in reproductive rate
    (E) Removal by human beings

Questions 59–60

Questions 59 and 60 refer to the following graphs of temperature and rainfall for six major ecosystems. A year's temperature from January through December (the line) and rainfall pattern (the shaded area) of each ecosystem are shown.

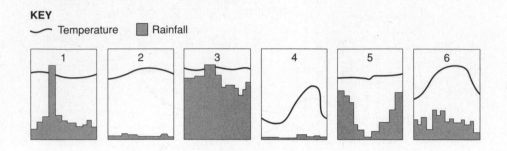

59. Which of the climatograms represents a savanna?

(A) 1
(B) 2
(C) 3
(D) 5
(E) 6

60. Which represents a tundra?

(A) 1
(B) 2
(C) 3
(D) 4
(E) 5

Question 61

Examine the timber cutting cycle below for Question 61.

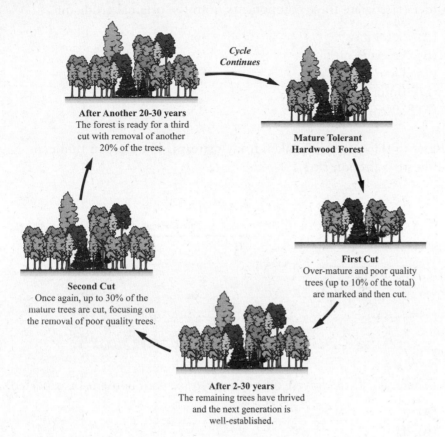

61. Which form of timber harvesting is represented above?

    (A) Clear-cutting
    (B) Shelterwood cutting
    (C) Selective cutting
    (D) Seed tree harvesting
    (E) None of the above

62. Rank the following biomes in order of most productive to least productive, as measured by biomass produced per acre.

    I. Desert
    II. Tropical rain forests
    III. Tundra
    IV. Grassland

    (A) I, II, III, IV
    (B) IV, III, II, I
    (C) III, I, IV, II
    (D) II, III, IV, I
    (E) II, IV, III, I

63. Kerosene, gasoline, motor oil, asphalt, tar, waxes, and diesel fuel all come from crude oil. Refineries take advantage of what physical property in order to separate these components from the original crude oil?

    (A) Solubility
    (B) Freezing point
    (C) Density
    (D) Boiling point
    (E) Viscosity

64. Refer to the data below taken from a stream. The Roman numerals indicate collection sites.

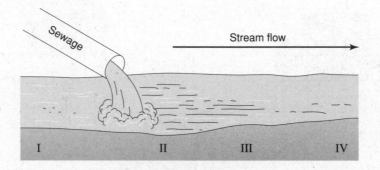

Where would the lowest DO (dissolved oxygen) content be expected?

    (A) I
    (B) II
    (C) III
    (D) IV
    (E) Dissolved oxygen would not be affected by sewage effluent. Therefore, DO would be equal at all points.

65. The theory that great disasters serve to maintain a population and its food supply balance was initially proposed by

    (A) Darwin
    (B) Wallace
    (C) Hardy and Weinberg
    (D) Malthus
    (E) Lyell

Questions 66–67

The following graph shows survival rates for five animal populations. When survival curves are calculated, the following assumptions are made:

    I. All individuals of a given population are the same age.

    II. No new individuals enter the population.

    III. No individuals leave the population.

These curves show the relationship of the number of individuals in a population to units of physiological life span. Base your answers to questions 66 and 67 on the graph and on your knowledge of environmental science.

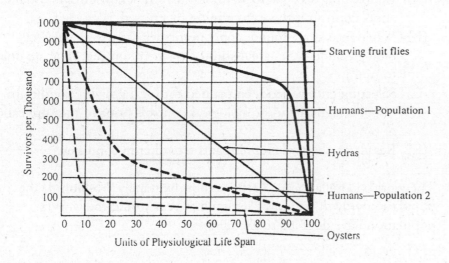

66. According to the data, it can be assumed that

    (A) fruit flies live longer than humans

    (B) oysters outlive fruit flies

    (C) the population of hydras is steadily declining

    (D) the life span of human populations is related to that of oysters

    (E) there is a high mortality rate among young oysters

67. The survival curves indicate that

    (A) starving fruit flies live out their full life span

    (B) human populations are more vulnerable than hydras

    (C) human population 2 has a greater rate of survival than human population 1

    (D) the hydra has a longer life span than the oyster

    (E) fruit flies die directly after pupation

68. The average requirement for drinking water per person per day is approximately

    (A) 1 pint
    (B) 1 quart
    (C) ½ gallon
    (D) 1 gallon
    (E) 3 gallons

69. Which of the following are NOT principles of Integrated Pest Management (IPM)?

    (A) Establishing acceptable pest levels, called action thresholds, and apply controls if those thresholds are crossed.
    (B) When insects are found, apply pesticides early so that insect biological life cycles are disrupted to the point that they are unable to multiply.
    (C) Selecting crop varieties best suited for local growing conditions.
    (D) Use mechanical and/or biological controls prior to the application of pesticides.
    (E) Regularly observe, monitor, and record crop conditions.

70. The world's population in 2010 was approximately 7 billion. If the population growth rate was 2%, in what year would the world's population have doubled to 14 billion?

    (A) 2020
    (B) 2045
    (C) 2090
    (D) 2100
    (E) 2110

71. In 1900 the amount of carbon dioxide gas released into the atmosphere by human activity was estimated to be 250 million metric tons per year. By the year 2000, this amount had increased to just over 350 million metric tons per year. What is the approximate percent increase in carbon dioxide concentration from 1900 to 2000?

    (A) 20%
    (B) 40%
    (C) 80%
    (D) 120%
    (E) 150%

72. Which of the following is NOT an acceptable mitigation technique used to reverse the process of desertification?

   (A) Fixating the soil through the use of shelter belts, woodlots, and windbreaks
   (B) Hyper-fertilizing the soil
   (C) Encourage large scale cultivation of the land to hold the soil in place
   (D) Reforestation
   (E) Provisioning of water

73. Which of the following statements is NOT consistent with "The Tragedy of the Commons" by Garret Hardin?

   (A) We will always add one too many cows to the village commons, destroying it.
   (B) The destruction of the commons will not be stopped by shame, moral admonitions, or cultural mores anywhere nearly so effectively as it will be by the will of the people expressed as a protective mandate, in other words, by government.
   (C) The "Tragedy of the Commons" is a modern phenomenon. Humans were not capable of doing too much damage until the population exceeded certain numbers and their technological tools became powerful beyond a certain point.
   (D) A free-market economy, based on capitalism, does not contribute to the "Tragedy of the Commons."
   (E) We will always opt for an immediate benefit at the expense of less-tangible values such as the availability of a resource to future generations.

74. The most common method of disposing of municipal solid wastes in the United States is

   (A) incineration
   (B) ocean dumping
   (C) sanitary landfills
   (D) recycling
   (E) exporting

75. Place the following events in sequential order

   I. Atmospheric CO_2 increases
   II. Warmer Ocean
   III. Warmer Atmosphere
   IV. Less CO_2 uptake by the oceans

   (A) I → II → III → IV
   (B) II → IV → I → III
   (C) IV → II → I → III
   (D) II → III → I → IV
   (E) III → II → I → IV

76. A new wide-range pesticide was being tested for efficacy on five species of insects. Which of the graphs below represents the insect species that would have the greatest potential over time for developing resistance to the pesticide?

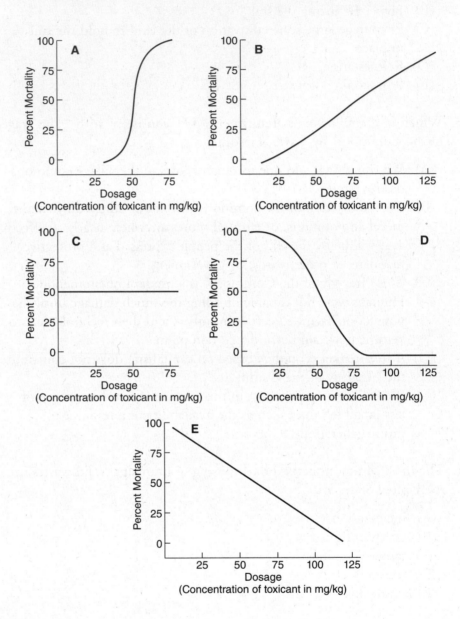

(A) A
(B) B
(C) C
(D) D
(E) E

77. The densest populations of most organisms that live in the ocean are found near the surface. The most probable explanation is that

    (A) the surface is less polluted
    (B) the bottom contains radioactive material
    (C) salt water has more minerals than freshwater
    (D) the light intensity that reaches the ocean decreases with increasing depth
    (E) the largest primary consumers are found near the surface

78. Of the following sources of pollution, which would NOT be classified as a point source of pollution?

    I.   Sulfur oxides released from an electrical generating plant
    II. Oil, grease, and toxic chemicals showing up in a stream
    III. Cyanide leaking from a heap leach pile at an abandoned gold mine site
    IV. Diesel exhaust soot coming from trucks on the interstate

    (A) I and III
    (B) I and IV
    (C) II and IV
    (D) I, III, and IV
    (E) I, II, III, and IV

79. Certain volcanoes are built almost entirely of fluid lava flows that pour out in all directions from a central summit vent or groups of vents. They build a broad, gently sloping, dome-shaped cone. These cones are built up slowly by the accretion of thousands of highly fluid basalt lava flows that spread widely over great distances and then cool as thin, gently, dipping sheets. These types of volcanoes are known as

    (A) Lava domes
    (B) Composite volcanoes
    (C) Cinder cones
    (D) Shield volcanoes
    (E) Stratovolcanoes

80. Of the causes of preventable deaths in the United States listed below, which one causes the MOST deaths per year?

    (A) AIDS
    (B) Illegal drug use
    (C) Alcohol use
    (D) Smoking
    (D) Motor vehicle accidents

Questions 81–83 refer to the locations marked by letters in the world map below.

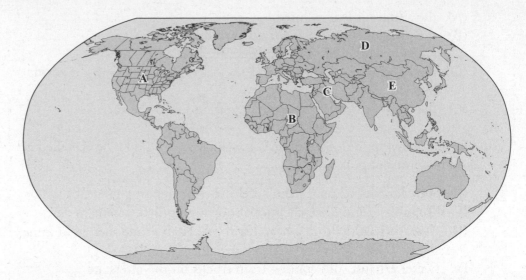

81. The location with the greatest oil reserves.

82. The area with the greatest coal reserves.

83. The area with the greatest natural gas reserves.

84. Which pair of compounds listed below reacts with H_2O to produce acid rain?

   (A) NH_3 and SO_2
   (B) CO and SO_2
   (C) CO_2 and H_2O_2
   (D) SO_3 and NO_3
   (E) SO_2 and NO_x

85. If current population trends continue, which regions of the world will experience the greatest population growth in the next 25 years?

   I.   Central and South America
   II. Africa
   III. Asia

   (A) I only
   (B) II only
   (C) III only
   (D) I and II
   (E) II and III

86. One of the greatest successes of the Endangered Species Act has been the

    (A) passenger pigeon
    (B) whooping crane
    (C) bald eagle
    (D) condor
    (E) auk

87. If global temperature warmed for any reason, atmospheric water vapor would increase due to evaporation. This would increase the greenhouse effect and thereby further raise the temperature. As a result, the ice caps would melt back, making the planet less reflective so that it would warm further. This process is an example of

    (A) positive coupling
    (B) negative coupling
    (C) synergistic feedback
    (D) positive-feedback loop
    (E) negative-feedback loop

88. The most abundant non-anthropogenic greenhouse gas is

    (A) water vapor
    (B) carbon dioxide
    (C) nitrogen
    (D) methane
    (E) tropospheric ozone

89. What is the most frequent cause of beach pollution?

    (A) Polluted runoff and storm water
    (B) Sewage spills from treatment plants
    (C) Oil spills
    (D) Ships dumping their holding tanks into coastal waters
    (E) People leaving their trash on the beach

90. Atmospheric carbon dioxide levels

    (A) are about 10% higher than they were at the time of the Industrial Revolution
    (B) are about 25% higher than they were at the time of the Industrial Revolution
    (C) are about the same as they were at the time of the Industrial Revolution
    (D) are slightly lower than they were at the time of the Industrial Revolution
    (E) are significantly lower than they were at the time of the Industrial Revolution

Question 91

For question 91 use the following information: A scientific study took ice core samples to measure the concentration of carbon dioxide that was present in the atmosphere for the last 1,000 years. Their findings are shown below:

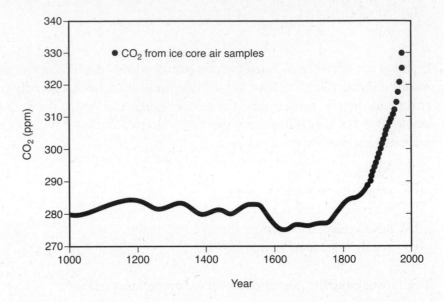

91. Which of the following mathematical setups could be used to determine the approximate percentage increase in carbon dioxide concentration over the last 1,000 years (use the year 1000 as the base year for comparison)?

(A) $\dfrac{330\text{ ppm} - 280\text{ ppm}}{280\text{ ppm}} \times 100\%$

(B) $\dfrac{330\text{ ppm} + 280\text{ ppm}}{2} \times 100\%$

(C) $\dfrac{330\text{ ppm} - 280\text{ ppm}}{50\text{ ppm}} \times 100\%$

(D) $\dfrac{330\text{ ppm} + 280\text{ ppm}}{330\text{ ppm}} \times 100\%$

(E) $\dfrac{330\text{ ppm} + 280\text{ ppm}}{50\text{ ppm}} \times 100\%$

92. The major sink for phosphorus is

(A) marine sediments
(B) atmospheric gases
(C) seawater
(D) plants
(E) animals

93. A large agricultural company that manufactures pesticides had developed a new insecticide to kill a certain species of aphid that was damaging citrus crops. After repeated testing using rats, the results are presented below. Which of the following statements listed below is the most accurate statement regarding this insecticide?

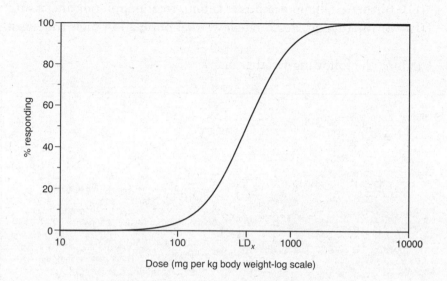

Dose (mg per kg body weight-log scale)

(A) Any aphid receiving 400 mg of this insecticide will die.
(B) For every 100 rats exposed to 400 mg of this insecticide, 50 will die.
(C) For every one kilogram of aphids receiving 400 mg of this insecticide, 50% of them will die.
(D) Out of 100 rats each receiving a dose of 400 mg of this insecticide per kilogram of body weight, 50 will die.
(E) Out of 100 people exposed to 400 mg of this insecticide, 50 will become acutely ill and die.

94. Which of the following would NOT be a likely location for seismic activity?

(A) Along mid-oceanic ridges
(B) Faults associated with volcanic activity
(C) Boundaries between oceanic and continental plates
(D) Interior of continental plates
(E) Boundaries between continental plates

95. Which of the following statements is correct for the third stage of the demographic transition model?

    (A) Birthrate increasing, death rate falling, total population increasing
    (B) Birthrate low, death rate low, total population high and constant
    (C) Birthrate falling, death rate high, total population increasing
    (D) Birthrate falling, death rate falling, total population increasing
    (E) Birthrate high, death rate low, total population high and constant

96. Examine the following weather map.

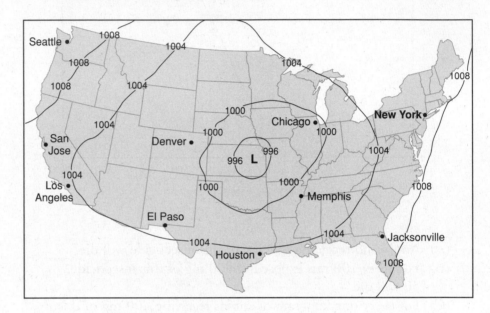

    Which of the following would be TRUE?

    (A) It is likely to be fair weather in the Midwest.
    (B) It is likely to be raining in the Midwest.
    (C) It is likely to be raining in the Northeast.
    (D) Rain would be expected in the western United States.
    (E) Not enough information is provided.

97. The concentration of H^+ ions in a solution with a pH value of 3 is how many times as great as the concentration of H^+ ions in a solution with a pH of 6?

    (A) 2
    (B) 3
    (C) 100
    (D) 1,000
    (E) 10,000

98. Causes of sick building syndrome include all of the following EXCEPT

    (A) radon and asbestos
    (B) chemical contaminants from indoor sources
    (C) chemical contaminants from outdoor sources
    (D) biological contaminants
    (E) inadequate ventilation

99. Which of the following statements is FALSE?

    (A) Evidence exists of a dose-response relationship between nonmelanoma skin cancer and cumulative exposure to UVB radiation.
    (B) Individuals, usually those living in areas with limited sunlight and long, dark winters, may suffer severe photo-allergies to the UVB in sunlight.
    (C) Increased absorption of UVB triggers a thickening of the superficial skin layers and an increase in skin pigmentation.
    (D) There is a relationship between skin cancer prevalence and increases in ultraviolet radiation due to the depletion of tropospheric ozone.
    (E) Acute exposure to UVB causes sunburn, and chronic exposure results in loss of elasticity and increased skin aging.

100. Of the following choices listed below, which constitutes the greatest percent of global freshwater use?

    (A) Domestic use other than drinking (cooking, flushing toilets, showering, etc.)
    (B) Use in energy production
    (C) Agriculture
    (D) Industry
    (E) Drinking

SECTION II (FREE-RESPONSE QUESTIONS)

Time: 90 minutes

No calculators allowed

4 questions

> **Directions:** Answer all four questions, which are weighted equally. The suggested time is about 22 minutes for answering each question. Write all your answers on the pages following the questions in the pink booklet. Where calculations are required, clearly show how you arrived at your answer. Where explanation or discussion is required, support your answers with relevant information and/or specific examples.

1. The environmental impact of washing a load of clothes in an electric washing machine is different than washing the same clothes by hand. Use the information below to answer the questions that follow. Show your calculations.

   Assume the following:

   1. All of the clothes can be washed in one load in the washing machine.

   2. The water entering the water heater is 60°F.

   3. The water leaving the water heater is 130°F.

   4. The electric washing machine uses 20 gallons of water. It uses 110 volts of electricity at an average of 1,500 watts for 30 minutes.

   5. Washing the clothes by hand requires 35 gallons of hot water.

   Other information:

   1 gallon of water = 8 pounds
   1 Btu = amount of energy required to raise the temperature of 1 pound of water by 1°F
   1 kilowatt-hour = 3,400 Btus

   (a) Calculate the total amount of energy (in Btus) to wash the clothes using the washing machine.
   (b) Calculate the total amount of energy (in Btus) to wash the clothes by hand.
   (c) Discuss the economic and environmental costs and benefits of using a washing machine in terms of

       (i)   its manufacture and disposal
       (ii)   selecting one and purchasing it (specifically, how can consumers compare various appliances in terms of their energy use)
       (iii) steps consumers can take to reduce its environmental impact

2. An AP Environmental Science class did an investigation on competition. Part I of the investigation focused on intraspecific competition to assess the effect of growth among radish plants at different population densities. Part II of the investigation focused on the relative competitiveness of two species of plants (radish and wheat) when they were planted together. The results are presented below:

Part I: Intraspecific Competition Among Radish Plants

Seeds per Pot	Total Biomass per Pot (g)
1	5.0
10	70.0
20	75.0

Part II: Interspecific Competition Between Radish and Wheat Plants

Seeds per Pot	Total Biomass per Pot (g)
1 radish	5.0
1 wheat	3.0
10 radish	50.0
10 wheat	25.0
20 radish	75.0
20 wheat	40.0

(a) Discuss the results of Part I of the investigation.

 (i) At what population density was the biomass per plant highest?
 (ii) Identify and describe two resources that may have been limited.
 (iii) Discuss the results obtained from the class for Part I in terms of two biological laws or principles.

(b) Discuss the results of Part II of the investigation.

 (i) Which plant was most affected by the competition between the two species?
 (ii) Describe TWO possible reasons why the plant chosen in (i) above may have been more successful.
 (iii) Discuss the results obtained from the class for Part II in terms of two biological laws or principles.

3. Examine the age-structure diagrams of Sweden and Kenya below, and answer the following questions.

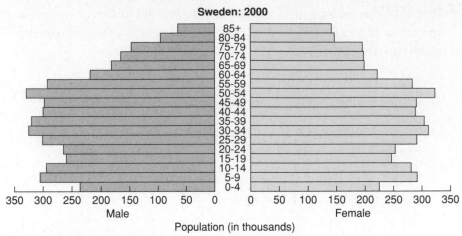

Sweden: 2000

Population (in thousands)

Source: U.S. Census Bureau, International Data Base.

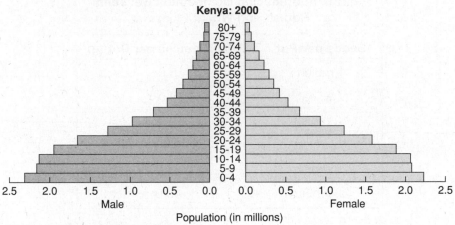

Kenya: 2000

Population (in millions)

Source: U.S. Census Bureau, International Data Base.

(a) Compare and contrast the two age structure diagrams in terms of two population dynamics—birthrate and death rate.

(b) What factors affect birthrates and death rates?

(c) Discuss methods that have been employed in another country to curb population growth.

4. Pesticides have become the most widely used form of pest management and are a controversial topic in environmental science.

(a) Explain what a pesticide is.

　(i) Describe TWO major categories of pesticides. For example: Insecticide—kills insects and other arthropods. (Do NOT use insecticide as one of your categories.)

　(ii) Identify and discuss TWO positive effects of pesticide use.

　(iii) Identify and discuss TWO negative effects of pesticide use.

(b) Discuss TWO alternatives to the use of pesticides.

(c) Name and describe ONE U.S. federal law *OR* ONE international treaty that focuses on the use of pesticides.

ANSWER KEY

Practice Test 1

Section I (Multiple-Choice Questions)

1. C	26. D	51. E	76. B
2. D	27. C	52. C	77. D
3. C	28. B	53. D	78. C
4. E	29. B	54. E	79. D
5. B	30. D	55. B	80. D
6. D	31. C	56. B	81. C
7. E	32. B	57. B	82. A
8. B	33. C	58. E	83. D
9. D	34. D	59. A	84. E
10. E	35. A	60. D	85. E
11. B	36. E	61. C	86. C
12. D	37. D	62. E	87. D
13. A	38. E	63. D	88. A
14. D	39. D	64. C	89. A
15. C	40. A	65. D	90. B
16. C	41. D	66. E	91. A
17. C	42. A	67. A	92. A
18. D	43. C	68. C	93. D
19. B	44. A	69. B	94. D
20. A	45. E	70. B	95. D
21. B	46. D	71. B	96. B
22. C	47. D	72. C	97. D
23. A	48. C	73. D	98. A
24. C	49. B	74. C	99. D
25. C	50. C	75. B	100. C

PREDICT YOUR SCORE ON THE APES EXAM

Place a check mark (✔) next to the multiple-choice questions you got correct on page 469. Then fill in the blanks below to predict your overall score on the APES exam. Essay questions are not used in this prediction as they require subjective grading. You can also use this page to determine your areas of weakness. For example, if you got 5 out of 5 questions correct on the topic The Earth, but only 2 out of 5 questions correct on the topic The Atmosphere, spend some more time reviewing The Atmosphere.

Unit I EARTH SYSTEMS AND RESOURCES (10–15%)

Chapter 1 The Earth
(#2, 14, 23, 79, 94)   _____ correct/5 = _____%

Chapter 2 The Atmosphere
(#7, 36, 38, 44, 96)   _____ correct/5 = _____%

Chapter 3 Global Water Resources and Use
(#8, 39, 53, 68, 100)   _____ correct/5 = _____%

Unit II THE LIVING WORLD (10–15%)

Chapter 4 Ecosystems
(#1, 15, 19, 22, 26, 27, 30, 31, 32, 33,
59, 60)   _____ correct/12 = _____%

Chapter 5 Natural Biogeochemical Cycles
(#5, 48, 92)   _____ correct/3 = _____%

Unit III POPULATIONS (10–15%)

Chapter 6 Populations
(#29, 35, 57, 58, 65, 66, 67, 70, 85, 95)   _____ correct/10 = _____%

Unit IV LAND AND WATER USE (10–15%)

Chapter 7 Land and Water Use
(#6, 11, 17, 25, 28, 40, 47, 51, 54, 61,
69, 72, 73, 76)   _____ correct/14 = _____%

Unit V ENERGY RESOURCES AND CONSUMPTION
(10–15%)

Chapter 8 Energy
(#24, 42, 43, 46, 55, 56, 63, 81, 82, 83) _____ correct/10 = _____%

Unit VI POLLUTION
(25–30%)

Chapter 9 Pollution
(#9, 13, 16, 18, 20, 21, 37, 41, 45, 49, 50, 64, 74, 78, 84, 89, 97, 98) _____ correct/18 = _____%

Chapter 10 Impacts on the Human Environment
(#4, 12, 34, 52, 80, 93, 99) _____ correct/7 = _____%

Unit VII GLOBAL CHANGE
(10–15%)

Chapter 11 Stratospheric Ozone and Global Warming
(#3, 10, 62, 71, 75, 77, 86, 87, 88, 90, 91) _____ correct/11 = _____%

Total Number Correct _____ / 100 = _____%

PREDICTED AP SCORE*

Less than 50 correct: 1 or 2 (not passing)

50–60 correct: 3 on the APES exam

61–75 correct: 4 on the APES exam

76+ correct: 5 on the APES exam

*Please note this is a rough estimate and is not intended to be an indicator of an actual AP score.

MULTIPLE-CHOICE EXPLANATIONS

1. **(C)** An increase in the population of squirrels would increase the competition for food and space.

2. **(D)** Approximately 3.5 billion years ago, the earliest life appeared on Earth, possibly derived from self-reproducing RNA molecules. DNA molecules eventually evolved and the first genomes soon developed inside enclosed membranes, which provided a stable physical and chemical environment conducive to their replication.

3. **(C)** For (C) to be true, it would have read, "The Kyoto Protocol would have required the United States to *reduce* its greenhouse gas emissions—primarily carbon dioxide (CO_2), methane (CH_4), and nitrous oxide (N_2O)—to 7% below 1990 levels by the year 2012."

4. **(E)** External costs are the costs that are borne by people other than the producer of a product.

5. **(B)** Nitrogen is the fourth most abundant element in living things, being a major constituent of proteins and nucleic acids. Ammonia is produced by the breakdown of these organic sources of nitrogen when organisms die. Nitrification is the process by which this ammonia is converted to nitrites (NO_2^-) by bacteria of the genus *Nitrosomonas* and then to nitrates (NO_3^-) by bacteria of the genus *Nitrobacter*.

6. **(D)** Mitigation involves taking steps to lessen risk by lowering the probability of a risk event's occurrence or reducing its effect should it occur. Organizing a beach cleanup is remediation—reacting after the beach has been polluted.

7. **(E)** The easterly trade winds are driven by a surface pressure pattern of higher pressure in the eastern Pacific and lower pressure in the west. When this pressure gradient weakens, so do the trade winds. The weakened trade winds allow warmer water from the western Pacific to surge eastward leading to a buildup of warm surface water and a sinking of the thermocline in the eastern Pacific. The deeper thermocline limits the amount of nutrient-rich deep water tapped by upwelling processes. These nutrients are vital for sustaining the large fish populations normally found in the region and any reduction in the supply of nutrients means a reduction in the fish population. Convective clouds and heavy rains are fueled by increased buoyancy of the lower atmosphere resulting from heating by the warmer waters below. As the warmer water shifts eastward, so do the clouds and thunderstorms associated with it, resulting in dry conditioning in Indonesia and Australia, while more flood-like conditions exist in Peru and Ecuador. The air-sea interaction that occurs during an El Niño event feed off of each other. As the pressure falls in the east and rises in the west, the surface pressure gradient is reduced and the trade winds weaken. This allows more warm surface water to flow eastward, which brings with it more rain, which leads to a further decrease of pressure in the east because the latent condensation warms the air.

8. **(B)** Refer to page 135. Of the total freshwater on Earth, over 68% is locked up in ice and glaciers. However, as ice and glaciers melt, this number will change significantly. Another 30% of freshwater is in the ground. Rivers are the source of most of the fresh surface water people use, but they only constitute about 0.0002% of the total freshwater on Earth.

9. **(D)** About 38% of municipal solid waste (before recycling) by weight is paper and paper products. Papermaking has an effect on the environment in how and where raw materials are acquired and processed and its impact on waste disposal. About 90% of paper is made of wood and accounts for about 35% of felled trees. Trees grown specifically for paper production account for 16% of all trees commercially grown and 9% of old growth forests. The manufacture and use of recycled paper products results in 35% less water pollution and 74% less air pollution than producing paper from trees. The average per capita paper use worldwide was 110 pounds (50 kg). Recycling 1 ton (0.91 m.t) of paper saves 17 mature trees, 7,000 gallons (26 m^3) of water, 3 cubic yards (2.3 m^3) of landfill space, 2 barrels of oil (84 US gal or 320 l), and 4,100 kilowatt-hours (15 GJ) of electricity—enough energy to power the average American home for six months. Packaging is the single largest category of paper use at 41% of all paper used. Around 115 billion sheets of paper are used annually for personal computers while the average web user prints 28 pages daily.

10. **(E)** Sea level is rising worldwide and is caused by both natural and human factors. It is predicted that within 100 years, there will be a net loss of 17–43% of coastal wetlands due to rising sea levels.

11. **(B)** Ground fires generally kill large and small trees because of the long and high temperature heat pulse generated. They release considerable amounts of nutrients from the burned fuels, destroy many small organisms and fungi that live in the humus and organic layers, consume seed stored in the litter, and kill roots in all but deep soil layers. They increase the chance of surface flow and erosion on slopes, and leave a baked and hardened seedbed that may prevent rapid revegetation. Increased surface runoff across the exposed surface may carry away ash and dissolved nutrients, making conditions even less favorable for plant growth.

12. **(D)** Primary economic activities (III) are at the beginning of the production cycle where humans are in closest contact with resources and the environment. Primary economic activities are located at the site of the natural resources being exploited. In many developing nations, approximately 3/4 of the labor force engages in subsistence farming or herding. By contrast, in highly developed countries only a small fraction of the labor force is directly employed in agriculture (less than 3% in the U.S.).

Secondary economic activities (I) use raw materials to produce or manufacture something new and more valuable. Examples of secondary activities include manufacturing, processing, producing power, and construction. Secondary economic activities are located either at the resource site or in close proximity to the market for the manufactured/processed good. Location depends upon whether the raw material or finished product costs more to ship. The major share of global secondary manufacturing activity is found within a relatively small number of major industrial concentrations. Rather than manufacturing/producing within its own country, richer nations often "set up shop" in developing nations because costs are significantly cheaper.

Tertiary economic activities (IV) include all activities that amount to doing services for others. Doctors, teachers, and secretaries provide personal and professional services. Restaurant staff, store clerks, and hotel personnel provide retail and wholesale services.

Quaternary economic activities (II) are not connected to resources, access to a market, or the environment. Rather, they include professions that process, administer, and disseminate information. Computer engineers, professors, and lawyers serve as examples of white-collar professionals who specialize in the collection and manipulation of information. With vast advancements in technology, quaternary economic activities could potentially exist anywhere. However, typically they exist in nations that have access to research centers, universities, efficient transportation and communication networks, and a pool of highly trained, skilled, flexible workers. Quaternary economic activities loom large in highly advanced, developed societies.

13. **(A)** Plastic makes up 90% of all trash floating in the oceans, enough that, in some areas, plastic outweighs plankton by a ratio of 6 to 1. A circular pattern of currents, called the North Pacific Subtropical Gyre, has corralled an enormous vortex of floating garbage, often referred to as the Great Pacific Garbage Patch. About 80% of the debris in the Great Pacific Garbage Patch, which ranges from bottles and fishing gear to toothbrushes and packaging scraps, came from land.

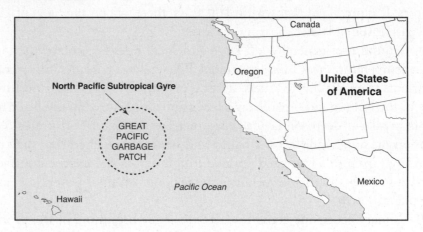

14. **(D)** Refer to Figure 1.3 (page 78).

15. **(C)** Coral reefs are among the most ancient of ecosystem types, dating back to the Mesozoic era some 225 million years ago. The release of carbon dioxide from coral reefs is very small (probably less than 100 million tons of carbon per year) relative to emissions due to fossil fuel combustion (about 5.7 billion tons of carbon per year). Coral reefs store very little organic carbon and are not very effective sinks for carbon dioxide from the atmosphere. Forests are more effective sinks for atmospheric carbon. Although tropical rain forests contain more species than coral reefs, reefs contain more phyla than rain forests. Covering less than 0.2% of the ocean floor, coral reefs contain perhaps 1/4 of all marine species. Despite their limited area, coral reefs may be home to up to 25% of the fish catch of developing countries or 10% of the total amount of fish caught globally for human consumption as food. Coral reefs in 93 of the 109 countries containing them have been damaged or destroyed by human activities. In addition, human impacts may have directly or indirectly caused the death of 5–10% of the world's living reefs. If the pace of destruction is maintained, another 60% could be lost in the next 20 to 40 years. The most important short-term threats to coral reefs are sedimentation (from poor land

use such as clear-cutting on steep slopes and other activities such as dredging, eutrophication (overfertilization and sewage pollution), and destructive fishing methods (dynamiting and overharvesting).

16. **(C)** Green taxes are taxes levied by the government on industries for each unit of pollution and are a source of governmental income. Government subsidies for reducing pollution are limited when industries surpass a break-even point or optimum level of pollution—the point at which cleanup costs exceed the harmful costs of pollution. Regulation is a command-and-control governmental approach that incurs costs to enact and enforce laws, set standards, regulate and monitor potentially harmful activities, and prosecute violators. Furthermore, regulation often focuses on cleanup instead of prevention, discourages innovation by mandating prescribed pollution control strategies, and is often unrealistic with the realities of a competitive global business environment. Charging a user fee provides income to the government by charging industries to utilize a natural resource. Trading pollution or resource use rights occurs between companies. Permits are allocated by the government for certain levels of pollution, and companies are free to trade unused pollution allocations. Once permits are sold or auctioned by the government, further financial impacts occur between the companies, not between the companies and the government. Permits often allow the largest and most financially secure companies to pollute the most, may concentrate pollutants at the most-polluting sites, and may not create economic incentives to reduce pollution since pollution levels are simply moved, not reduced.

17. **(C)** The Green Revolution refers to a series of research, development, and technology transfer initiatives, occurring between the 1940s and the late 1970s, that increased agriculture production around the world. The crops developed during the Green Revolution were high yield varieties, which were bred specifically to respond to fertilizers and produce an increased amount of grain per acre planted. In the 1940s, research began in Mexico that developed new disease-resistant, high-yield varieties of wheat. By combining new wheat varieties with new mechanized agricultural technologies, Mexico was able to produce more wheat than was needed by its own citizens, leading to its becoming an exporter of wheat by the 1960s. Prior to the use of these varieties, the country was importing almost half of its wheat supply. Due to the success of the Green Revolution in Mexico, its technologies spread worldwide in the 1950s and 1960s. The United States, for instance, imported about half of its wheat in the 1940s but after using Green Revolution technologies, it became self-sufficient in the 1950s and became an exporter by the 1960s.

India was on the brink of mass famine in the early 1960s because of its rapidly growing population. Agricultural research resulted in a new variety of rice, IR8, which produced more grain per plant when grown with irrigation and fertilizers. Today, India is one of the world's leading rice producers and growing of IR8 rice has spread throughout Asia.

18. **(D)** According to a 1992 study by the National Oceanographic and Atmospheric Administration, 50% of the spilled oil underwent biodegradation and photolysis (chemical decomposition by the action of radiant electromagnetic energy, especially light). Cleanup crews recovered about 14% of the oil, and approximately 13% sunk to the seafloor. About 2% (some 216,000 gallons) remained on the beaches.

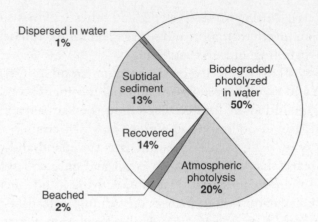

19. **(B)** Asexual reproduction produces organisms that are genetically identical to the parent. Speciation is the result of genetic variability. There are several types of asexual reproduction. Fission is the simplest form of asexual reproduction and involves the division of a single organism into two complete organisms, each genetically identical to the other and to the parent. A similar form of asexual reproduction is regeneration, in which an entire organism may be generated from a part of its parent. Budding occurs when a group of self-supportive cells sprouts from and then detaches from the parent organism.

20. **(A)** The Resource Conservation and Recovery Act (RCRA) of 1976 addresses comprehensive management of nonhazardous and hazardous solid waste, sets minimal standards for all waste disposal facilities and for hazardous wastes, and regulates treatment, storage, and transport.

21. **(B)** In 1980, Congress passed the Superfund statute providing broad authorities both to respond to chemical emergencies and to clean up toxic waste sites for long-term protection.

22. **(C)** Environmental resistance or limiting factors consists of all factors in an environment that limit a population's ability to increase in numbers. Predation, competition for resources, disease, limited food and space, etc., all work to hold a population in check. Environmental resistance can be either density-dependent or density-independent. Factors that are density-dependent are stronger when a population has a higher density (more crowded). Examples would include predation, parasitism, disease, and competition for space or food. Density-independent factors (usually abiotic) will kill organisms—whether they are crowded or not. Examples would include floods, storms, earthquakes, fire, etc. In nature, biotic potential and environmental resistance work together to level out population numbers to an amount that can be supported by the environment. When environmental resistance "pushes down" on J curve growth, the curve levels into what is known as an S curve (or sigmoid curve) and reflects a balanced community.

23. **(A)** Refer to "Plate Tectonics" in Chapter 1.

24. **(C)** The Environmental Protection Agency (EPA) estimates that the emissions to the atmosphere from coal-burning power plants contain 84 of the 187 hazardous air pollutants identified by the EPA as posing a threat to human health and the environment. Coal-fired power plants in the United States release 386,000 tons of hazardous air pollutants into the atmosphere each year and account for 40% of all hazardous air pollutant point sources, more than any other point source category. Coal-burning power plants are the largest

point source category for release of hydrochloric acid, mercury, and arsenic and are a major source of several criteria air pollutants such as sulfur dioxide, oxides of nitrogen, and particulate matter.

25. **(C)** Also known as Superfund, it is a law designed to clean up sites contaminated with hazardous substances. The law authorized the Environmental Protection Agency (EPA) to identify parties responsible for contamination of sites and compel the parties to clean up the sites. Where responsible parties cannot be found, the EPA is authorized to clean up sites itself, using a special trust fund. CERCLA authorizes two kinds of response actions: (1) Removal actions. These are typically short-term response actions, where actions may be taken to address releases or threatened releases requiring prompt response and (2) Remedial actions. These are usually long-term response actions. Remedial actions seek to permanently and significantly reduce the risks associated with releases or threats of releases of hazardous substances that are serious. Choice (E) addresses coal mining.

26. **(D)** Green plants are the producers that can store enough energy to provide the basis upon which a community is built. The least amount of energy is found at the level of the hawks.

27. **(C)** Snakes feed on rodents. At each trophic level, energy is lost. Therefore the snakes must consume more than their biomass to survive.

28. **(B)** Alley cropping is used to enhance or diversify farm products, reduce surface water runoff and erosion, improve utilization of nutrients, reduce wind erosion, modify the microclimate for improved crop production, and improve wildlife habitat. Hardwood trees, like walnut, oak, ash, and pecan, are favored species in alley cropping systems and can potentially provide high-value lumber. Nut crops can be another intermediate product. Common examples of alley cropping plantings include wheat, corn, soybeans, or hay planted in between rows of black walnut or pecan trees.

29. **(B)** The population had grown by 1,300 (4,000 births – 3,000 deaths + 500 immigrants – 200 emigrants = 1,300).

$$\text{Net annual growth rate} = \frac{1,300}{100,000} \times 100\% = 1.3\%$$

30. **(D)** Less energy is available at each trophic level because energy is lost by organisms through respiration and incomplete digestion of food sources. Therefore, fewer herbivores can be supported by the vegetative material.

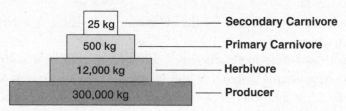

31. **(C)** Secondary consumers are carnivores that feed upon the flesh of other animals (for example, a snake eating a rat).

32. **(B)** Both lizards and arachnids eat grasshoppers. Competition is the struggle between different species for the same resources.

33. **(C)** Saprophytic agents gain nourishment from dead organic matter. They are the organisms of decay. Bacteria and fungi are saprophytes.

34. **(D)** When resources are scarce, the price goes up (law of supply and demand). The increase in prices causes people to conserve resources and use less because they are saving money.

35. **(A)** Diagram A, also known as an expansive pyramid, shows a broad base, indicating a high proportion of children, a rapid rate of population growth, and a low proportion of older people. A steady upward narrowing shows that more people die at each higher age band. This type of pyramid indicates a population in which there is a high birth rate, a high death rate, and a short life expectancy. This is the typical pattern for less economically developed countries, due to little access to and incentive to use birth control, negative environmental factors (e.g., lack of clean water), and poor access to health care.

36. **(E)** The thermohaline circulation plays an important role in supplying heat to the polar regions, and thus in regulating the amount of sea ice in these regions, although poleward heat transport outside the tropics is considerably larger in the atmosphere than in the ocean. Insofar as the thermohaline circulation governs the rate at which deep waters are exposed to the surface, it may also play an important role in determining the concentration of carbon dioxide in the atmosphere.

37. **(D)** Car A travels 500 miles at 15 miles to the gallon and therefore uses about 500/15 = 33.3 gallons of gasoline. Car B travels the same 500 miles but uses about 500/30 = 16.7 gallons of gasoline. Car A uses 33.3 − 16.7 = 16.6 more gallons of gasoline than car B. If each gallon of gasoline that is burned contributes 5 pounds of carbon (in the form of CO_2) to the atmosphere, then car A will contribute 5 × 16.6 = 83 *more* pounds of carbon than car B.

38. **(E)** Earth is actually closest to the sun during the Northern Hemisphere winter.

39. **(D)** As the ocean warms, it expands and sea level rises, accounting for about one-third of the approximately 8 inches (20 cm) sea level rise seen in the past 100 years. Water released by melting glaciers and ice sheets accounts for the other two-thirds of sea level rise.

40. **(A)** Besides the effects already listed, additional effects include:
    - Flattens the ocean bottom, filling in holes, leveling humps, and knocking down protruding organisms, all of which reduce biodiversity
    - Initiates erosion, causing long-term sedimentation problems
    - Changes the microhabitat, affecting primary productivity
    - Damages or kills both commercially harvested species and noncommercial species
    - Unlike clear-cutting in forests, the effects are not visible, thereby not getting much attention

41. **(D)** Pressure-treated wood (sometimes called CCA), which usually has a greenish tint to it, contains arsenic, a highly toxic element, as well as chromium and copper. Dog and cat feces may carry diseases that can infect humans. Morning glory, sheep sorrel, ivy, several kinds of grasses, and some other plants can resprout from their roots and/or stems in the compost pile. Bones (and the fat and marrow) are very attractive to pests such as rats. In addition, fatty food wastes can be very slow to break down because the fat can exclude the air that composting microbes need to do their work. Manure, which is defined as barnyard or stable dung often containing discarded animal bedding, typically contains large amounts of nitrogen (the fresher the manure, the more nitrogen

it contains) and is considered a green ingredient. Fresh manures can heat a compost pile quickly and will accelerate the decomposition of woody materials, autumn leaves, and other browns.

42. **(A)** Sustainability is a concept long recognized and utilized by many cultures. It results from recognition of the need for a harmonious existence between the environment, society, and economy. Sustainability focuses on improving the quality of life for all people without increasing the use of natural resources beyond the capacity of the environment to supply them indefinitely. In order for hydrogen to be a true, long-term, renewable alternative to fossil fuels (it runs more efficiently than gasoline but it costs twice to three times as much to produce), it will need to be produced with a clean, low-cost electricity source such as hydroelectric power. Most oil-reliant countries simply do not have access to vast amounts of clean electricity, but Iceland and Norway do.

43. **(C)** In a fossil fuel power plant, the chemical energy stored in chemical bonds in the fossil fuels (such as coal, fuel oil, or natural gas) and oxygen in the air is converted successively into thermal energy (heat), mechanical energy (spinning turbines), and finally electrical energy for continuous use and distribution across a wide geographic area. Each of these processes is less than 100% efficient, with energy lost (usually as heat) at each stage.—The Second Law of Thermodynamics states that the quality of a particular amount of energy—i.e., the amount of work or action that it can do—diminishes each time this energy is used and is true for all forms of energy (mechanical, chemical, etc.)

44. **(A)** If you missed this question, review Hadley, Ferrel, and polar cells in Chapter 2. In land masses influenced by polar cells, temperature rather than precipitation is the critical factor in plant distribution and soil development. Land masses in equatorial regions primarily influenced by Hadley cells are characterized by high humidity and lots of precipitation (e.g., tropical rain forests). Subtropical land masses primarily influenced by Hadley cells are characterized by low relative humidity and deserts.

45. **(E)** The threshold for pain is at about 120 to 140 dB, but sound begins to damage hearing when it is above 85 dB. Some of the hearing loss after exposure to excessive noise will be permanent. Indication of damage is ringing and noise in the ears (called tinnitus) after sound exposure. Research shows that cumulative exposure to loud sounds, not age, is the major cause of hearing loss.

46. **(D)** Enriched uranium is uranium whose uranium-235 content has been increased through the process of isotope separation. Natural uranium consists mostly of the U-238 isotope, with about 0.72% by weight as U-235, the only isotope existing in nature in any appreciable amount that is fissionable by thermal neutrons. Enriched uranium is a critical component for both nuclear power generation (electricity) and military nuclear weapons. The International Atomic Energy Agency attempts to monitor and control enriched uranium supplies and processes in its efforts to ensure nuclear power generation safety and to curb nuclear weapons proliferation.

47. **(D)** Integrated pest management (IPM) is a broad based ecological approach to agricultural pest control that integrates pesticides/herbicides into a management system incorporating a range of practices for economic control of a pest. In IPM, the purpose is to prevent infestation, to observe patterns of infestation when they occur, and to intervene (without poisons) when necessary. An IPM system is designed around six basic components: (1) acceptable pest levels;

(2) preventive cultural practices; (3) monitoring; (4) mechanical controls; (5) biological controls; and (6) responsible pesticide use.

48. **(C)** Carbon dioxide is produced during cellular respiration, not photosynthesis. When trees (natural sinks for carbon) are cut down, the carbon in the tree is sequestered (remains) until decay processes begin, at which time, some of the carbon is released back into the environment as CO_2. The primary source of CO_2 is outgassing from the Earth's interior at mid-ocean ridges, hot-spot volcanoes, and volcanic arcs. Much of the CO_2 released at subduction zones is derived from the metamorphism of carbonate rocks subducting with the ocean crust. Much of the overall outgassing of CO_2, especially at mid-ocean ridges and hot-spot volcanoes, was stored in the mantle when the Earth formed. Some of the outgassed carbon remains as CO_2 in the atmosphere, some is dissolved in the oceans, some carbon is held as biomass in living or dead and decaying organisms, and some is bound in carbonate rocks. Carbon is removed into long-term storage by burial of sedimentary strata, especially coal and shale that store organic carbon from undecayed biomass and carbonate rocks such as limestone also known as calcium carbonate—$CaCO_3$.

49. **(B)** Peroxyacyl nitrates, or PANs, are powerful respiratory and eye irritants present in photochemical smog. PANs are both toxic and irritating, as they dissolve more readily in water than ozone. They cause eye irritation at concentrations of only a few parts per billion. At higher concentrations they cause extensive damage to vegetation. PANs are mutagenic and can be a factor causing skin cancer. PANs are classified as secondary pollutants as they are not directly emitted as exhaust from power plants or internal combustion engines, but they are formed from other pollutants by chemical reactions in the atmosphere. Since they dissociate quite slowly in the atmosphere, PANs are able to transport NO_x and other unstable compounds far away from their urban and industrial origins and form tropospheric ozone.

50. **(C)** Volatile organic compounds (VOCs) are emitted as gases from certain solids or liquids with high vapor pressures (evaporate easily). Concentrations of many VOCs are often up to ten times higher indoors than outdoors. Examples of VOCs include paints and lacquers, paint strippers, cleaning supplies, pesticides, building materials and furnishings, office equipment such as copiers and printers, glues, and markers.

51. **(E)** Oceans occupy 71% of the Earth's surface. They are divided into five major oceans, which in decreasing order of size are: the Pacific Ocean, Atlantic Ocean, Indian Ocean, Southern Ocean, and Arctic Ocean. Over 70% of the world catch from the sea comes from the Pacific Ocean.

Ocean	Amount of Fish Caught (Million tons)	% Fish Caught
Pacific	83	71
Atlantic	24	20
Indian	10	9
Southern (Antarctic)	0.147	0.1
Arctic	0	0
Overall	116	100%

52. **(C)** Salmon are inseparable from their freshwater, saltwater, and estuarine ecosystems. They are extremely sensitive to changes in water quality, changes in river flow, water turbidity, and temperature. Juvenile salmon feed on freshwater invertebrates that are also indicators of water quality. Generally, the more pristine, diverse, and productive the freshwater ecosystem is, the healthier the salmon stocks. Salmon bring large amounts of marine nutrients upstream to the headwaters of otherwise low productivity rivers. Salmon carcasses are the primary food for aquatic invertebrates and fish, as well as terrestrial fauna ranging from marine mammals to birds—eagles, ducks, and songbirds—to terrestrial mammals, especially bears and humans. Salmon and other anadromous fish bring biomass and nutrients (nitrogen, phosphorus, carbon, and micronutrients) from the sea into freshwater and terrestrial ecosystems.

53. **(D)** Oligotrophic lakes contain very low concentrations of nutrients required for plant growth and thus the overall productivity of these lakes is low. Only a small quantity of organic matter is present; e.g., phytoplankton, zooplankton, algae, aquatic weeds, bacteria, and fish. With so little production of organic matter, there is very little accumulation of organic sediment on the bottom and the water is clear. They are seldom in good agricultural areas since water runoff is low in nutrients.

54. **(E)** (1) National Park Service (Interior Department) provides for the use and enjoyment of the parks by people and to preserve the land in its original state. It manages 13% of federal lands and 40% of the acreage within the National Wilderness Preservation System. (2) The Forest Service (Department of Agriculture) manages public lands in national forests and grasslands. It manages 30% of federal lands and 33% of the acreage within the National Wilderness Preservation System. (3) Bureau of Land Management (Interior Department) initially managed range lands for grazing, oil and gas development, and mining. Their role expanded to include multiple use resources such as wildlife, watersheds, recreation, wilderness, and other conservation values with the passage of the Federal Land Policy and Management Act in 1976. It manages 42% of federal lands and 7% of the acreage within the National Wilderness Preservation System. (4) The Fish and Wildlife Service (Interior Department) administers a national network of lands and waters for the conservation, management, and, where appropriate, restoration of the fish, wildlife, and plant resources and their habitats within the United States. It manages 15% of federal lands and 18% of the acreage within the National Wilderness Preservation System.

55. **(B)** CFLs, like all fluorescent lightbulbs, contain mercury as vapor inside the glass tubing. Most CFLs contain 3–5 mg per bulb, with the bulbs labeled "eco-friendly" containing as little as 1 mg. Because mercury is poisonous, even these small amounts are a concern for landfills and waste incinerators where the mercury from lightbulbs may be released and contribute to air and water pollution. In 2008 the U.S. Environmental Protection Agency (EPA) published a data sheet stating that the net system emission of mercury for CFL lighting was lower than for incandescent lighting of comparable lumen output. This was based on the average rate of mercury emission for electricity production from a coal-fed power plant and the average estimated escape of mercury from a CFL put into a landfill. In the United States, the EPA estimated that if all 270 million compact fluorescent lightbulbs sold in 2007 were sent to landfill

sites, that this would represent around 0.13 metric tons, or 0.1% of all U.S. emissions of mercury, which was 104 metric tons.

56. **(B)** Lighting efficiency ranges of 9–11% for CFLs and 1.9–2.6%, for incandescent lightbulbs. Because of their higher efficacy, CFLs use between one-quarter to one-third of the power of an equivalent incandescent lightbulb. About 50%–70% of the world's total lighting market sales are incandescent. Replacing incandescent lighting with CFLs would save 2.5% of the world's yearly electricity consumption. Since CFLs use much less energy than incandescent lightbulbs, a phase-out of incandescent lightbulbs would result in less CO_2 being emitted into the atmosphere. The efficiency rating is calculated by comparing the amount of energy input to the amount of energy output (work accomplished). Efficiency is usually given as a percentage and can be computed with the following formula:

$$\frac{\text{Useful output}}{\text{Energy input}} = \text{efficiency rating (\textit{in decimal})}$$

Useful output = 10.0 joules $\times$ 0.10
Useful output = 1 joule of light and therefore 9 joules of heat energy (wasted)

57. **(B)** The most probable explanation for the increase in the rabbit population after 1950 was that the coyote population declined drastically. The relationship indicated by the graph is one of predator-prey. The coyotes prey upon rabbits, keeping the population in check.

58. **(E)** The most probable explanation for the decline of the coyotes is removal by humans. Since the decline in coyotes occurred with cattle introduction and since cattle and coyotes do not compete for the same food, the individuals that introduced the cattle must have interfered with the coyote population. Because coyotes are predators, they were perceived as dangerous to the cattle and were hunted and poisoned by humans. Attempts to control coyotes by poisoning may deplete the numbers of their natural prey and lead to increasing attacks by coyotes on farm animals.

59. **(A)** Savannas are warm year-round with a prolonged dry season and scattered trees. The environment is intermediate between grassland and forest. An extended dry season followed by a rainy season occurs in Australia; Central, Eastern, and South Africa; India; Madagascar; central South America; Southeast Asia; and Thailand. Savannas consist of grasslands with stands of deciduous shrubs and trees that do not grow more than 30 meters high. Trees and shrubs generally shed leaves during the dry season, which reduces the need for water. Food is limited during the dry season so that many animals migrate during this season. Soils are rich in nutrients. Savannas contain large herds of grazing animals and browsing animals that provide resources for predators.

60. **(D)** The tundra is located at 60° north latitude and farther north. The weather is influenced by the polar cell. Alpine tundra is located in mountainous areas, above the tree line, with well-drained soil and where dominant animals are small rodents and insects. Arctic tundra is frozen treeless plains, low rainfall, low average temperatures (summers average <10°C), and many bogs and ponds. Frozen ground prevents drainage. The growing season lasts 50–60 days. Tundra is found in Alaska, Canada, Europe, Greenland, and Russia. Dominant vegetation includes flowering dwarf shrubs, grasses, lichens, mosses, and sedges. The soil has few nutrients due to low vegetation and little

decomposition. There are between 60 to 100 frost-free days per year. Arctic tundra is higher latitude than alpine tundra.

61. **(C)** Selective cutting is used for the majority of shade-tolerant hardwood forests. About every 20 to 30 years, individual mature and declining (diseased or unhealthy) trees are cut. The growth rate and quality of the remaining trees improves, and young trees of the shade-tolerant species become established in the mostly-shaded understory. Selection cutting imitates minor natural disturbances like wind and disease, and perpetuates an all-aged tolerant hardwood forest.

62. **(E)** Swamps and marshes are the most productive biomes, producing the most biomass per year, while tropical rain forests have the most standing biomass. Extreme deserts are the least-productive biomes. Lakes and streams are equivalent to extreme deserts for having the least standing biomass.

63. **(D)** The boiling point of a crude oil component or fraction, which is the temperature at which it evaporates, is dependent on the length of the carbon chain in the molecule. Those fractions with shorter chains (such as gasoline) evaporate more easily than those with longer chains (such as waxes).

64. **(C)** At the point of discharge, there is a sudden increase in the amount of toxins, suspended solids, and dissolved organic compounds. The water becomes unsuitable for any human use. Aerobic bacteria increase rapidly, reducing the organic pollutants but using up dissolved oxygen. As the oxygen level falls, anaerobic bacteria multiply, resulting in an unpleasant smell or stench. Eventually, as their food supply is used up, the decomposers reduce and algae thrive, reoxygenating the water. Fish and other animal life reappear, and the stream returns to its previous condition.

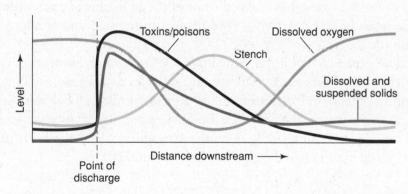

65. **(D)** Thomas Malthus recognized that once the carrying capacity of an area was exceeded, organisms would die of starvation. Darwin's theory of evolution based on natural selection may have been influenced by Malthus's concept.

66. **(E)** Note that only 10% of the population of young oysters reaches the 20th unit of life span.

67. **(A)** Note that almost all of the fruit flies reach the 90th unit of life span before the entire population dies.

68. **(C)** On a cool, inactive day, the average man loses about 12 eight-ounce cups of water but consumes only about 9 cups of water (about half of that from the water in fruits, vegetables, and other solid foods).

69. **(B)** Integrated pest management (IPM) is a broad based ecological approach to agricultural pest control that integrates pesticides/herbicides into a manage-

ment system incorporating a range of practices for economic control of a pest. In IPM, one attempts to prevent infestation, to observe patterns of infestation when they occur, and to intervene when necessary. Synthetic pesticides are generally only used as required and only at specific times in a pest's life cycle.

70. **(B)** To calculate doubling time, divide the growth rate into the number 70. Thus a growth rate of 2% (0.02) will cause a population to double in number in 35 years (70 ÷ 2 = 35). 2010 + 35 years = 2045.

71. **(B)** 350 − 250 = 100   100 / 250 = 0.40   0.40 × 100% = 40%

72. **(C)** Desertification is the degradation of land in any dryland. Caused by a variety of factors, such as climate change and human activities, desertification is a significant global environmental problem. Drylands occupy approximately 40% of Earth's land area and are home to more than 2 billion people, and it has been estimated that some 10–20% of drylands are already degraded. The most common cause of desertification is the overcultivation of desert lands. Overcultivation causes the nutrients in the soil to be depleted faster than they are restored. Furthermore, improper irrigation practices result in salinated soils and the depletion of aquifers.

73. **(D)** A pure capitalistic economy, in which financial gain is the primary societal motivator, leads to always taking just one more. Be careful of the double negative in the question.

74. **(C)** Industrial wastes of unknown content are often commingled with domestic wastes in sanitary landfills. Groundwater infiltration and contamination of water supplies with toxic chemicals have recently led to more active control of landfills and industrial waste disposal. Careful management of sanitary landfills, such as providing for leachate and runoff treatment as well as daily coverage with topsoil, has stopped most of the problems of open dumping. In many areas, however, space for landfills is running out, and alternatives must be found.

75. **(B)** The oceans are an important sink for CO_2 through absorption of the gas into the water surface. As atmospheric CO_2 levels increase, it increases the warming potential of the atmosphere, a process called global warming. If air temperatures warm, they increase ocean temperatures. However, the ability of the ocean to remove CO_2 from the atmosphere decreases with increasing ocean temperatures based on principles of solubility.

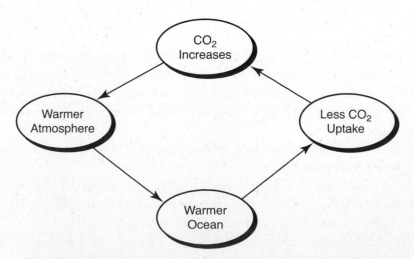

76. **(B)** The slope of the dose-response curves gives a good indication of how a target population will respond to the pesticide. When all members of the population react in a similar way, their dose-response curve is quite steep (Choice A) and the population is said to be fairly homogeneous. On the other hand, if some members of the population are much more sensitive to the pesticide than others, the population is said to be more heterogeneous and the dose-response curve is flatter (Choice B). Populations that are more heterogeneous or diverse tend to survive better through natural selection in a changing environment.

77. **(D)** The depth at which plankton can exist depends on the penetration of sunlight. Sunlight provides the energy for photosynthesis. Since the phytoplankton is abundant near the surface, the region of sunlight penetration, the consumers are also found in this region.

78. **(C)** A point source of pollution is a single, identifiable, localized source of air, water, thermal, or noise or light pollution and is generally considered to be fixed (immobile).

79. **(D)** Shield volcanoes are built almost entirely of fluid lava flows. Flow after flow pours out in all directions from a central summit vent, or group of vents, building a broad, gently sloping cone with a flat, domed shape. They are built up slowly by the accumulation of thousands of flows of highly fluid basaltic lava that spread widely over great distances and then cool as thin, gently dipping sheets. Examples of shield volcanoes include those found in the Hawaiian Island chain, Galapagos Islands, Iceland, and East Africa.

80. **(D)** More preventable deaths in the United States are caused each year by tobacco use than by all deaths from AIDS, illegal drug use, alcohol use, motor vehicle injuries, suicides, and murders combined; however, obesity and its effects are dramatically increasing. Cigarette smoking causes about 1 of every 5 deaths in the United States each year. On average, adults who smoke cigarettes die 14 years earlier than nonsmokers. Based on current cigarette smoking patterns, an estimated 25 million Americans who are alive today will die prematurely from smoking-related illnesses, including 5 million people younger than 18 years of age.

81. **(C)** Countries in the Middle East own 56% of the world's proven oil reserves. A 2008 U.S. Geological Survey estimates that areas north of the Arctic Circle have 13% of the expected undiscovered oil in the world. See Figure 8.10, page 316 to see a list of the world's oil reserves.

82. **(A)** The United States still has the world's largest coal reserves with a slightly more than one-quarter of the total (28%). Russia is second with 18%, and China is third with 13%. Existing supplies are estimated to last for almost 200 years at current rates of consumption. Coal provides 44% of the world's electricity and 27% of its overall energy. The figures for the U.S. are very similar at 48% of electricity and 21% overall. China generates 80% of its electricity from coal and is actually importing coal, despite its vast reserves. Estimates are that 86% of the incremental demand for coal over the next twenty years will come from China and India. See Figure 8.9, page 315 to see a list of the world's coal reserves.

83. **(D)** As of 2009, proven gas reserves are dominated by three countries: Russia (24%), Iran (16%), and Turkmenistan (15%). Due to constant announcements of shale gas recoverable reserves, as well as drilling in Central Asia,

South America, and Africa, deepwater drilling, estimates are undergoing frequent updates. See Figure 8.11, page 316 to see a list of the world's natural gas reserves.

84. **(E)** Chemical formulas involved in producing acid rain include:

(i) $H_2O_{(l)} + CO_{2(g)} \rightleftharpoons H_2CO_{3(aq)}$

(ii) SO_2: $S_{(s)} + O_{2(g)} \rightarrow SO_{2(g)}$

The $SO_{2(g)}$ combines with water to produce sulfurous acid:

$$H_2O_{(l)} + SO_{2(g)} \rightleftharpoons H_2SO_{3(g)}.$$

Sulfur dioxide is however not readily oxidized to sulfur trioxide in dry, clean air. Water droplets and dust particles, however, catalyse the reaction between O_2 and SO_2 in the air producing sulfur trioxide, SO_3. This dissolves in water and produces sulfuric acid, which is a much stronger acid:

$$SO_{2(g)} + \tfrac{1}{2}O_{2(g)} \rightarrow SO_{3(g)} \qquad H_2O_{(l)} + SO_{3(g)} \rightarrow H_2SO_{4(aq)}$$

This can cause considerable damage to buildings, vegetation, and fish populations by destroying fish eggs.

(iii) Sources of NO_x are more widespread. At temperatures over 2370°F (1300°C), nitrogen combines with oxygen to form nitrogen monoxide:

$$N_{2(g)} + O_{2(g)} \rightarrow 2NO_{(g)}$$

These high temperatures can be achieved by: (i) the internal combustion engine; and (ii) lightning in the atmosphere. The nitrogen monoxide slowly combines with oxygen to form soluble nitrogen dioxide gas:

$$2NO_{(g)} + O_{2(g)} \rightarrow 2NO_{2(g)}$$

Nitrogen dioxide readily dissolves in water producing a mixture of nitric and nitrous acids:

$$2NO_{2(g)} + H_2O_{(l)} \rightarrow HNO_{3(aq)} + HNO_{2(g)}$$

85. **(E)** Between 1999 and the start of 2012 when the world's population increased from 6 billion to 7 billion people, the Asian population, which represents 60% of the current world population, grew by an estimated 579 million people making up about 57% of that increase. The African population meanwhile contributed 28% to the population increase.

86. **(C)** From a population of tens of thousands in the 19th century, the bald eagle had declined to only 417 pairs in the lower 48 states by 1963. By 1994, after 20 years of protection through the Endangered Species Act, the population had grown to more than 4,000 breeding pairs.

87. **(D)** Couplings are one-way linkages in which one component of a system affects another component. In positive coupling between components A and B, a change in A causes a change of the same sign in B. In negative coupling, a change in A causes a change of the opposing sign in B. Two or more couplings acting in sequence form a closed loop. In a positive-feedback loop, a positive or negative change in one of the components of the system is amplified by the sequence of couplings. In a negative-feedback loop, the change is reduced in amplitude.

88. **(A)** Non-anthropogenic means not caused by humans. When the choices are ranked by their contribution to the greenhouse effect, they are: (1) water vapor, which contributes 36–72%; (2) carbon dioxide, which contributes 9–26%; (3) methane, which contributes 4–9%, and (4) tropospheric ozone, which contributes 3–7%. Pound for pound, ozone is about 3,000 times stron-

ger as a greenhouse gas than CO_2. Even though there is much less ozone at middle altitudes than CO_2, its impact is considerable, with ozone trapping up to one-third as much heat as CO_2. Nitrogen is not a greenhouse gas.

89. **(A)** In 2000, beach pollution prompted at least 11,270 closings and swimming advisories in the United States, twice the number that occurred in 1999. The increase was due largely to improved and increased monitoring and reporting. The most frequent pollution sources are polluted runoff and storm water, which led to more than 4,102 closings in 2000.

90. **(B)** Atmospheric carbon dioxide levels are rising rapidly. Currently, they are 25% above where they stood before the Industrial Revolution. Earth's atmosphere now contains some 200 gigatons (GT) more carbon than it did two centuries ago.

91. **(A)** % change = ((new value − original value) / original value) × 100

92. **(A)** Marine sediments contain an estimated 4×10^9 Tg (1 Tg = 10^{12} grams with an estimated turnover time of 2×10^8 years). Turnover time is the time it takes for a complete cycle to occur within a system or the ratio of the mass of a reservoir to the rate of its removal from that reservoir.

93. **(D)** The median lethal dose, LD_{50} of a toxin, radiation, or pathogen is the dose required to kill half the members of a tested population after a specified test duration. LD_{50} figures are frequently used as a general indicator of a substance's acute toxicity. An LD_{50} level determined for rats cannot be interpreted as the LD_{50} level for insects or humans.

94. **(D)** In (A) activity is low, and it occurs at very shallow depths. The point is that the lithosphere is very thin and weak at these boundaries, so the strain cannot build up enough to cause large earthquakes. The San Andreas fault is a good example of (B) in which two mature plates are scraping by one another. The friction between the plates can be so great that very large strains can build up before they are periodically relieved by large earthquakes. In (C) one plate is thrust or subducted under the other plate so that a deep ocean trench is produced. In (E) shallow earthquakes are associated with high mountain ranges where intense compression is taking place. In (D) the interiors of the plates themselves are largely free of large earthquakes, that is, they are aseismic.

95. **(D)** In stage three of the demographic stage model, birthrates fall due to access to contraception resulting in smaller family sizes, increases in wages, urbanization, a reduction in subsistence agriculture, an increase in the status and education of women, a reduction in the value of children's work, an increase in parental investment in the education of children and other social changes, and population growth rates begin to decrease. Although birthrates are declining, the population is still increasing. The death rate continues to fall from stage two due to greater access to health care, smaller family sizes, and lower childhood mortality rates. Examples of countries in stage three include Mexico, the Philippines, Indonesia, and India.

96. **(B)** Differences in air pressure help cause winds and affect air masses. They are also factors in the formation of storms such as thunderstorms, tornadoes, and hurricanes. Differences in air pressure are shown on a weather map with lines called isobars. The weather map illustrates isobars marking areas of high and low pressure. High-pressure areas generally have dry, good weather, and areas of low pressure have precipitation. On the weather map, the only area of low pressure is centered in the Midwest.

97. **(D)** The pH scale is logarithmic, and as a result, each whole pH value below 7 is ten times more acidic than the next higher value.

98. **(A)** SBS (sick-building syndrome) and BRI (building-related illness) are associated with acute or immediate health problems. Radon and asbestos cause long-term diseases that occur years after exposure and are therefore not considered to be among the causes of sick buildings. Indicators of SBS include headache; eye, nose, or throat irritation; dry cough; dry or itchy skin; dizziness and nausea; difficulty in concentrating; fatigue; and sensitivity to odors.

99. **(D)** Tropospheric ozone would be ozone found in the air that we breathe, generally produced by burning fossil fuels. No relationship has been found that skin cancer is increased by decreases in smog.

100. **(C)** Agriculture accounts for 92% of all freshwater use globally. Specifically, water-intensive cereal grains like wheat, rice, and corn account for 27%; meat production 22%; and dairy 7%. China, India, and the United States are responsible for nearly 38% of global water consumption. The United States, which has a much smaller population than China or India, led the world in per capita consumption of freshwater: 2842 m^3 each year, compared with the global average of 1385 m^3 per year.

FREE-RESPONSE EXPLANATIONS

Question 1

(a) 5 points maximum

Restatement: Total amount of energy (in Btus) to wash the clothes using the washing machine.

Mass of water:

$$20 \text{ gallons H}_2\text{O} \times \frac{8 \text{ pounds}}{1 \text{ gallon H}_2\text{O}} = 160 \text{ pounds H}_2\text{O} \quad \text{(1 point)}$$

Energy to heat water:

$$160 \text{ pounds H}_2\text{O} \times \frac{1 \text{ Btu/°F}}{1 \text{ pound H}_2\text{O}} \times (130°\text{F} - 60°\text{F}) = 11,200 \text{ Btu} \quad \text{(1 point)}$$

Energy to run washing machine (kWh):

$$1500 \text{ Watts} \times \frac{1 \text{ kW}}{1000 \text{ Watts}} \times 30 \text{ min.} \times \frac{1 \text{ hr.}}{60 \text{ min.}} = 0.75 \text{ kWh} \quad \text{(1 point)}$$

Energy to run washing machine (Btus):

$$0.75 \text{ kWh} \times \frac{3,400 \text{ Btu}}{1 \text{ kWh}} = 2,550 \text{ Btu} \quad \text{(1 point)}$$

Total energy for washing machine:

$$11,200 \text{ Btu} + 2,550 \text{ Btu} = \textbf{13,750 Btu} \quad \text{(1 point)}$$

(b) 2 points maximum

Restatement: Total amount of energy (in Btus) to wash clothes by hand.

$$35 \text{ gallons H}_2\text{O} \times \frac{8 \text{ pounds}}{1 \text{ gallon H}_2\text{O}} = 280 \text{ pounds H}_2\text{O} \quad \text{(1 point)}$$

Energy to heat water:

$$280 \text{ pounds H}_2\text{O} \times \frac{1 \text{ Btu/°F}}{1 \text{ pound H}_2\text{O}} \times (130°\text{F} - 60°\text{F}) = \textbf{19,600 Btu} \quad \text{(1 point)}$$

NOTE:

1. If you do NOT show calculations, no points are awarded.
2. No penalty assessed if you do not show units. However, you risk setting up the problem incorrectly if you do not show units so that they cancel properly.
3. If your setup is correct but you make an arithmetic error, no penalty.

(c) 3 points maximum

Restatement: Environmental impact of washing clothes.

NOTE: There are no points for just listing ideas. Each idea MUST be explained. The following are ideas that you could use to write your paragraph(s). Take these ideas and create an outline of the order in which you wish to answer them. You do NOT need to use all ideas. Before you begin, decide on the format of how you wish to answer your question (pros versus cons, chart format with explanations within the chart, compare and contrast, etc.).

(i) Restatement: Discuss the economic and environmental costs and benefits of using a washing machine in terms of its manufacture and disposal. 1 point

- Manufacturing washing machines provides jobs in mining, smelting, designing, engineering, manufacturing, transportation, administration, advertising, repair, and sales.
- Use of natural resources in the manufacture and distribution of washing machines. Examples: use of renewable or less polluting energy sources in manufacturing; use of less polluting methods of transportation.
- Advantages of repairing washing machines rather than discarding them. Discarding them creates landfill issues. Recycling of metal parts and packaging material.

(ii) Restatement: Discuss the economic and environmental costs and benefits of selecting one and purchasing it (specifically, how can consumers compare various appliances in terms of their energy use). 1 point

- Energy efficiency of different models. The U.S. Department of Energy and the U.S. Protection Agency have joined with appliance manufacturers to institute the "Energy Star" program. Appliances displaying an Energy Star label have been certified to significantly exceed the federal efficiency standards.
- Choosing a machine that uses the fewest gallons of water per pound of clothes and one that has high, medium, and low water level controls and an automatic cold rinse cycle.
- Consider high efficiency (HE) washing machines. Advantages of HE washing machines include:
  - A traditional washing machine will use 40 to 47 gallons (151 to 178 liters) of water per cycle, while high efficiency washing machines use 11 to 32 gallons (42 to 121 liters) of water per cycle.

- Energy savings with high efficiency washers are estimated at 50% to 60% per each load. Much of the energy savings afforded by high efficiency washing machines are a result of the significant reduction in hot water they require.
- The horizontal drums utilized in a high efficiency washing machine spin at a much faster rate than other machines. This allows more water to be extracted from the clothes during the spin cycle. Consequently, clothes contain less moisture at the end of the wash and require less time and energy to dry.
- A high efficiency washing machine will use two to four clean rinses per cycle reducing levels of detergent residue left on clothing as well as improving cleaning efficiency by not having clothes sitting in dirty water.
- The tumbling motion in a high efficiency washing machine causes clothes to circulate freely resulting in a more effective cleaning process as well as less wear and tear on garments. The tumble motion is gentler with less stress and pulling on fabric than the agitation motion.
- Tumbling and increased circulation of items in the drum requires that a high efficiency detergent be used in these machines. High efficiency detergents have fewer suds, requiring less water to then rinse away.

(iii) Discuss the economic and environmental costs and benefits of using a washing machine and other appliances in terms of steps consumers can take to reduce its environmental impact. 1 point

- Use appliances during off-peak hours
- Use biodegradable detergents that use minimal amounts of phosphates
- Pre-soak heavily soiled clothes
- Try to use only with full loads
- Use as low a water temperature for the washing cycle as will provide satisfactory results
- Always use a cold rinse cycle
- Follow maintenance schedule and instructions found in owner's manual

Question 2

(a) (i) 1 point maximum

Restatement: Biomass as a function of population density

To determine the optimum population density that produced the greatest biomass would require *dividing the total biomass for each planting by the total amount of seeds planted* per pot. This would provide the average biomass and produce the following results:

Seeds per Pot	Average Biomass (g)
1	5.0
10	7.0
20	3.8

From the data table, it appears that *10 radish seeds per pot produced the largest average biomass*. When graphed, the results look like:

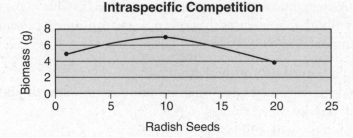

Intraspecific Competition

(a) (ii) 2 points maximum

Restatement: Limited resources

The maximum biomass was achieved at 10 seeds per pot. After that, as population density increased, biomass decreased. Factors that may have limited biomass might have included:

1. *Competition for soil nutrients.* As root density increased, less nutrients would have been available for growth.
2. *Competition for light.* As the density increased, less light would have been available for smaller seedlings.
3. *Competition for water.* Given a fixed amount of water provided, as plant density increased, less water would have been available for additional seedlings.

(a) (iii) 2 points maximum

Restatement: Results obtained from the class for Part I in terms of biological laws or principles

The effect of intraspecific competition in plant populations is usually examined by planting the species over a range of densities. The most common result is that a point is reached where the mean weight per plant decreases as density increases, so that the maximum yield approaches some constant value. This result is called the *law of constant yield*. The maximum plant productivity of a particular environment is called the *carrying capacity*. In this case, the carrying capacity was reached when the biomass reached 7.0 grams for the size of the pot. Environmental variables (nutrients, space, water, and light) that restrict the realized niche are called limiting factors. *Liebig's law of the minimum* states that an organism is most limited by the essential factor that is in least supply. In terms of *survivorship*, the radish seeds represent Type III characteristics that include large amounts of seeds being produced with few surviving, small size invaders of disturbed environments, and rapid growth.

(b) (i) 1 point maximum

Restatement: Plant species most affected by competition

To determine which plant species was most negatively affected by interspecific competition would require *dividing the total biomass per pot by the total number of seeds per pot to determine average biomass*. The results are provided below:

Seeds per Pot	Average Biomass (g)
1 radish	5.0
1 wheat	3.0
10 radish	5.0
10 wheat	2.5
20 radish	3.8
20 wheat	2.0

From the table, it appears that the *wheat seedlings were negatively impacted by the presence of the radish seeds*. When graphed, the results are:

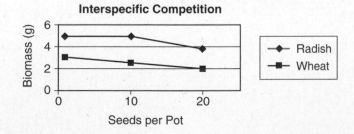

(b) (ii) 2 points maximum

Restatement: Possible reasons why the plant chosen in (b) (i) may have been more successful

Two different species of plants were grown in the *same* medium. *Nutrient requirements may have been different* between the two species. For example, the amount and ratios of minerals contained in the potting soil may have met the metabolic requirements of the radish plants more than the wheat seedlings. *Water and light availability* (either too much or too little) may also have affected outcome. *Differences in growth patterns* (radish may sprout earlier than wheat) may have existed. In this case, the radish seedlings may have *sprouted earlier* and established a population that was taking up a fixed amount of nutrients at a faster rate or had established a canopy that decreased the light available for the wheat seedlings. Simultaneous to the interspecific competition was intraspecific competition. Not only were radish seedlings in competition with wheat seedlings, they were also in competition with other radish seedlings for limited nutrients.

(c) (iii) 2 points maximum

Restatement: Results obtained from the class for Part II in terms of biological laws or principles

In Part II, plant density was held constant in a substitutive or replacement experimental design in which the total density of both species was kept constant while the relative densities were varied (1, 10, and 20). This approach allows investigation of the effects of interspecific competition from the effects of increasing overall plant density on growth or production of a species. However, it requires first quantifying the background effect of intraspecific competition on growth of each species individually. Such an experiment has three possible outcomes. First, one species prospers at the expense of the other (competitive exclusion). Second, one species outperforms the other but only when in higher proportion (coexistence). Third, the two species have no measurable effect on each other (no competition). The latter would be the null hypothesis (H_o) by which the class would judge whether or not one of the other two scenarios happened. Another scenario, which was not tested for, could have been instability—that is, two species are in a constant state of dynamic tension where at one time one species dominates and at other times the other species dominates. Different species of plants are able to coexist within the same biome through resource partitioning, which functions through evolution of different metabolic pathways, having different tolerances to shade, taking up water and nutrients at different depths, and so on.

Question 3

(a) 4 points maximum

List any two characteristics from the column labeled "Sweden" and any two characteristics from the column labeled "Kenya."

Restatement: Given two age-structure diagrams, Sweden and Kenya for 2000, compare and contrast the diagrams in terms of population dynamics.

Age structure diagrams are basically divided into three major age categories:

- Prereproductive (0–15 years old)
- Reproductive (16–45 years old)
- Postreproductive (46 years old–death)

Population Characteristics

	Affected By	Sweden	Kenya
Birth Rate	• Importance of children as a part of the labor force • Urbanization • Cost of raising and educating children • Educational and employment opportunities for women • Infant mortality rate • Average age at marriage • Pensions • Abortions • Birth control • Religious beliefs	• Population has nearly equal proportions of prereproductive and reproductive individuals. • Little growth over a long period of time will produce a population with about equal numbers of people in all age groups. • Children not required or necessary to support parents. • Availability and acceptance of birth control.	• Population has pyramid–shaped age structures, with large numbers of prereproductive individuals. • Population momentum results from large numbers of prereproductive children becoming reproductive within short period of time. • High population rate due to high birth mortality rates. • Children viewed as status symbol. • Resistance to birth control.
Death Rate	• Increased food supply • Better nutrition • Improved medical and public health technology • Improvements in sanitation and personal hygiene • Safer water supplies	• Elderly survive longer due to advances in medical technology and availability. • Social welfare programs ensure that elderly are taken care of.	• Elderly do not survive due to lack of available medical technology. • Disease (for example, malaria or AIDS) and lack of nutritious food decreases life span.

(b) 4 points maximum

Restatement: Factors that affect both birth and death rates

List any four characteristics from the column labeled "Affected By."

See preceding chart.

(c) 2 points maximum

Restatement: Methods that have been employed by another country to curb population growth

List any two of the six methods mentioned.

China: Between 1958 and 1962, an estimated 30 million people died from famine in China. Since then, China has made good progress in trying to feed its people and bring its population growth under control. Much of this reduced population growth was brought about by a drop in the birth rate from 32 to 18 per 1,000 between 1972 and 1985. China instituted one of the most rigorous population control programs in the world at an estimated cost of about $1 per person. Some features of the program included:

1. Strong encouragement for couples to postpone marriage.
2. Providing married couples with free access to sterilization, contraceptives, and abortion.
3. Giving couples who sign pledges to have no more than one child economic rewards such as salary bonuses, extra food, larger pensions, better housing, free medical care and school tuition for their child, and preferential treatment in employment when the child grows up.
4. Requiring those who break the pledge to return all benefits.
5. Exerting pressure on women pregnant with a third child to have abortions.
6. Requiring one of the parents in a two-child family to be sterilized.

Question 4

(a) 1 point maximum (1 point for correct explanation)

Restatement: Definition of a pesticide

A pesticide is any substance or mixture of substances intended for preventing, destroying, repelling, or mitigating any pest. Pests can be insects, rodents or other animals, unwanted plants (weeds), fungi, or microorganisms such as bacteria or viruses.

(i) 2 points maximum (1 point for each correct category of pesticide and description)

Describe TWO major categories of pesticides. For example: Insecticide—kills insects and other arthropods. (Do NOT use insecticide as one of your categories.)

Algicides	Control algae in lakes, canals, swimming pools, water tanks, and other sites.
Antifouling agents	Kill or repel organisms that attach to underwater surfaces, such as boat bottoms.
Antimicrobials	Kill microorganisms (such as bacteria and viruses).
Attractants	Attract pests (for example, to lure an insect or rodent to a trap). Food is not considered a pesticide when used as an attractant.
Biocides	Kill microorganisms.
Defoliants	Cause leaves or other foliage to drop from a plant, usually to facilitate harvest.
Desiccants	Promote drying of living tissues, such as unwanted plant tops.
Disinfectants and sanitizers	Kill or inactivate disease-producing microorganisms on inanimate objects.
Fungicides	Kill fungi (including blights, mildews, molds, and rusts).
Fumigants	Produce gas or vapor intended to destroy pests in buildings or soil.
Herbicides	Kill weeds and other plants that grow where they are not wanted.
Insect growth regulators	Disrupt the molting, maturing from pupal stage to adult, or other life processes of insects.
Insecticides	Kill insects and other arthropods.

Miticides (acaricides)	Kill mites that feed on plants and animals.
Microbial pesticides	Microorganisms that kill, inhibit, or outcompete pests, including insects or other microorganisms.
Molluscicides	Kill snails and slugs.
Nematicides	Kill nematodes (microscopic, wormlike organisms that feed on plant roots).
Ovicides	Kill eggs of insects and mites.
Pheromones	Biochemicals used to disrupt the mating behavior of insects.
Plant growth regulators	Substances (excluding fertilizers or other plant nutrients) that alter the expected growth, flowering, or reproduction rate of plants.
Repellents	Repel pests, including insects (such as mosquitoes) and birds.
Rodenticides	Control rats, mice, and other rodents.

(ii) 2 points maximum (1 point for each correct positive effect and explanation)

Restatement: Two positive effects of pesticide use. Identify and discuss TWO positive effects of pesticide use.

(1) Plants are directly and indirectly humankind's main source of food. They are attacked by tens of thousands of diseases caused by viruses, bacteria, fungi, and other organisms. There are over 30,000 kinds of weeds competing with crops worldwide; thousands of nematode species reduce crop vigor; and some 10,000 species of insects devour crops. It is estimated that one-third of the world's food crop is destroyed by these pests annually. *Pesticides increase the world food supply.*

(2) There are an estimated 300–500 million cases of malaria per year. The majority of these occur in Africa. The vast majority of the estimated 1 million annual deaths from the disease occur among children and mainly among poor African children. Malaria is above all a disease of the poor, impacting at least three times more greatly on the poor than any other disease. Africa's GDP would be up to $100 billion greater if malaria had been eliminated years ago. Mosquitoes have been estimated to be responsible for half of all human deaths due to transmission of disease. *Pesticides improve human health by destroying disease-carrying organisms* (West Nile virus, encephalitis, African sleeping sickness, malaria, yellow fever, plague, etc.).

(iii) 2 points maximum (1 point for each correct negative effect and explanation)

Restatement: Two negative effects of pesticide use. Identify and discuss TWO negative effects of pesticide use.

(1) If a pesticide is continually applied to a population of the pest species, most susceptible individuals will be killed, leaving only resistant individuals. These resistant individuals breed and multiply. Eventually a high proportion of the individuals from that pest species are now resistant to the pesticide. We have simply *caused pest populations with a higher tolerance for poisons to survive and breed*. More toxic pesticides in turn are developed and utilized.

(2) *Pesticides can accumulate in living organisms.* An example of accumulation is the uptake of a water-insoluble pesticide, such as chlordane, by a creature living in water. Since this pesticide is stored in the organism, the pesticide accumulates and increases over time. If this organism is eaten by an organism higher in the food chain that can also store this pesticide, levels can reach higher values in the higher organism than is present in the water in which it lives. Levels in fish, for example, can be tens to hundreds of thousands of times greater than ambient water levels of the same pesticide. This type of accumulation is called bioaccumulation. In this regard, it should be remembered that humans are at the top of the food chain and so may be the most vulnerable to bioaccumulation.

(b) 2 points maximum (1 point for each correct alternative and explanation)

Restatement: Two alternatives to the use of pesticides. Discuss TWO alternatives to the use of pesticides.

GROWING PEST-RESISTANT CROPS

When landscaping a yard or planning a garden, choose plant varieties that are native to the region and climate. Hearty, native plants resist disease and infestation, and they often use less water.

CROP ROTATION

Plant rotation and interplanting prevent the buildup and spread of pests in one area or among specific plant types.

BENEFICIAL INSECTS AND ANIMALS

Protect and encourage the presence of insect-feeding birds, bats, spiders, praying mantises, ladybugs, predatory mites, parasitic flies, and wasps. Beneficial insect species can often be purchased in volume.

PHEROMONES

Pheromones are chemical signals produced by animals to communicate with others of the same species. In insects, they consist of highly specific perfumes, generally derivatives of natural fatty acids closely related to the aromas of fruit. They are nontoxic. Pheromones may be used to attract insects to traps or to deter insects from laying eggs. However, the most widespread and effective application of pheromones is for mating disruption.

(c) 1 point maximum. (1 point for naming a correct U.S. federal law or international treaty along with a correct description.) Note: Credit will be awarded for choosing *any* applicable U.S. federal law or international treaty that focuses on the use of pesticides (e.g., Clean Water Act; Resource Conservation and Recovery Act; Comprehensive Environmental Response, Compensation, and Liability Act (Superfund); etc.).

Restatement: Name and description of one U.S. federal law **OR** one international treaty that focuses on the use of pesticides.

The U. S. federal government first regulated pesticides when Congress passed the Insecticide Act of 1910. This law was intended to protect farmers from adulterated or misbranded products. Congress broadened the federal government's control of pesticides by passing the original Federal Insecticide, Fungicide and Rodenticide Act (FIFRA) of 1947. FIFRA required the Department of Agriculture to register all pesticides prior to their introduction in interstate commerce. A 1964 amendment authorized the secretary of agriculture to refuse registration to pesticides that were unsafe or ineffective and to remove them from the market. In 1970, Congress transferred the administration of FIFRA to the newly created Environmental Protection Agency (EPA). This was the initiation of a shift in the focus of federal policy from the control of pesticides for reasonably safe use in agricultural production to control of pesticides for reduction of unreasonable risks to humans and the environment. This new policy focus was expanded by the passage of the Federal Environmental Pesticide Control Act of 1972 (FEPCA) that amended FIFRA by specifying methods and standards of control in greater detail. In general, there has been a shift toward greater emphasis on minimizing risks associated with toxicity and environmental degradation and away from pesticide efficacy issues. Under FIFRA, no one may sell, distribute, or use a pesticide unless it is registered by the EPA. Registration includes approval by the EPA of the pesticide's label, which must give detailed instructions for its safe use. The EPA must classify each pesticide as either general use, restricted use, or both. General-use pesticides may be applied by anyone. However, restricted-use pesticides may be applied only by certified applicators or persons working under the direct supervision of a certified applicator. Because there are only limited data for new chemicals, most pesticides are initially classified as restricted use. Applicators are certified by a state if the state operates a certification program approved by the EPA.

Practice Exam 2

With Answers and Analysis

SECTION I (MULTIPLE-CHOICE QUESTIONS)

Time: 90 minutes

100 questions

60% of total grade

No calculators allowed

This section consists of 100 multiple-choice questions. Mark your answers carefully on the answer sheet.

General Instructions

Do not open this booklet until you are told to do so by the proctor.

   Be sure to write your answers for Section I on the separate answer sheet. Use the test booklet for your scratch work or notes. Remember, though, that no credit will be given for work, notes, or answers written only in the test booklet. Once you have selected an answer, thoroughly blacken the corresponding circle on the answer sheet. To change an answer, erase your previous mark completely, and then record your new answer. Mark only one answer for each question.

Example   **Sample Answer**

The Pacific is   Ⓐ Ⓑ ● Ⓓ Ⓔ

   (A) a river
   (B) a lake
   (C) an ocean
   (D) a sea
   (E) a gulf

There is no penalty for wrong answers on the multiple-choice section, so you should answer all multiple-choice questions. Even if you have no idea of the correct answer, you should try to eliminate any obvious incorrect choices, and then guess.

   Because it is not expected that all test takers will complete this section, do not spend too much time on difficult questions. First answer the questions you can answer readily. Then, if you have time, return to the difficult questions later. Do not get stuck on one question. Work quickly but accurately. Use your time effectively.

Answer Sheet

PRACTICE EXAM 2

1 Ⓐ Ⓑ Ⓒ Ⓓ Ⓔ	26 Ⓐ Ⓑ Ⓒ Ⓓ Ⓔ	51 Ⓐ Ⓑ Ⓒ Ⓓ Ⓔ	76 Ⓐ Ⓑ Ⓒ Ⓓ Ⓔ
2 Ⓐ Ⓑ Ⓒ Ⓓ Ⓔ	27 Ⓐ Ⓑ Ⓒ Ⓓ Ⓔ	52 Ⓐ Ⓑ Ⓒ Ⓓ Ⓔ	77 Ⓐ Ⓑ Ⓒ Ⓓ Ⓔ
3 Ⓐ Ⓑ Ⓒ Ⓓ Ⓔ	28 Ⓐ Ⓑ Ⓒ Ⓓ Ⓔ	53 Ⓐ Ⓑ Ⓒ Ⓓ Ⓔ	78 Ⓐ Ⓑ Ⓒ Ⓓ Ⓔ
4 Ⓐ Ⓑ Ⓒ Ⓓ Ⓔ	29 Ⓐ Ⓑ Ⓒ Ⓓ Ⓔ	54 Ⓐ Ⓑ Ⓒ Ⓓ Ⓔ	79 Ⓐ Ⓑ Ⓒ Ⓓ Ⓔ
5 Ⓐ Ⓑ Ⓒ Ⓓ Ⓔ	30 Ⓐ Ⓑ Ⓒ Ⓓ Ⓔ	55 Ⓐ Ⓑ Ⓒ Ⓓ Ⓔ	80 Ⓐ Ⓑ Ⓒ Ⓓ Ⓔ
6 Ⓐ Ⓑ Ⓒ Ⓓ Ⓔ	31 Ⓐ Ⓑ Ⓒ Ⓓ Ⓔ	56 Ⓐ Ⓑ Ⓒ Ⓓ Ⓔ	81 Ⓐ Ⓑ Ⓒ Ⓓ Ⓔ
7 Ⓐ Ⓑ Ⓒ Ⓓ Ⓔ	32 Ⓐ Ⓑ Ⓒ Ⓓ Ⓔ	57 Ⓐ Ⓑ Ⓒ Ⓓ Ⓔ	82 Ⓐ Ⓑ Ⓒ Ⓓ Ⓔ
8 Ⓐ Ⓑ Ⓒ Ⓓ Ⓔ	33 Ⓐ Ⓑ Ⓒ Ⓓ Ⓔ	58 Ⓐ Ⓑ Ⓒ Ⓓ Ⓔ	83 Ⓐ Ⓑ Ⓒ Ⓓ Ⓔ
9 Ⓐ Ⓑ Ⓒ Ⓓ Ⓔ	34 Ⓐ Ⓑ Ⓒ Ⓓ Ⓔ	59 Ⓐ Ⓑ Ⓒ Ⓓ Ⓔ	84 Ⓐ Ⓑ Ⓒ Ⓓ Ⓔ
10 Ⓐ Ⓑ Ⓒ Ⓓ Ⓔ	35 Ⓐ Ⓑ Ⓒ Ⓓ Ⓔ	60 Ⓐ Ⓑ Ⓒ Ⓓ Ⓔ	85 Ⓐ Ⓑ Ⓒ Ⓓ Ⓔ
11 Ⓐ Ⓑ Ⓒ Ⓓ Ⓔ	36 Ⓐ Ⓑ Ⓒ Ⓓ Ⓔ	61 Ⓐ Ⓑ Ⓒ Ⓓ Ⓔ	86 Ⓐ Ⓑ Ⓒ Ⓓ Ⓔ
12 Ⓐ Ⓑ Ⓒ Ⓓ Ⓔ	37 Ⓐ Ⓑ Ⓒ Ⓓ Ⓔ	62 Ⓐ Ⓑ Ⓒ Ⓓ Ⓔ	87 Ⓐ Ⓑ Ⓒ Ⓓ Ⓔ
13 Ⓐ Ⓑ Ⓒ Ⓓ Ⓔ	38 Ⓐ Ⓑ Ⓒ Ⓓ Ⓔ	63 Ⓐ Ⓑ Ⓒ Ⓓ Ⓔ	88 Ⓐ Ⓑ Ⓒ Ⓓ Ⓔ
14 Ⓐ Ⓑ Ⓒ Ⓓ Ⓔ	39 Ⓐ Ⓑ Ⓒ Ⓓ Ⓔ	64 Ⓐ Ⓑ Ⓒ Ⓓ Ⓔ	89 Ⓐ Ⓑ Ⓒ Ⓓ Ⓔ
15 Ⓐ Ⓑ Ⓒ Ⓓ Ⓔ	40 Ⓐ Ⓑ Ⓒ Ⓓ Ⓔ	65 Ⓐ Ⓑ Ⓒ Ⓓ Ⓔ	90 Ⓐ Ⓑ Ⓒ Ⓓ Ⓔ
16 Ⓐ Ⓑ Ⓒ Ⓓ Ⓔ	41 Ⓐ Ⓑ Ⓒ Ⓓ Ⓔ	66 Ⓐ Ⓑ Ⓒ Ⓓ Ⓔ	91 Ⓐ Ⓑ Ⓒ Ⓓ Ⓔ
17 Ⓐ Ⓑ Ⓒ Ⓓ Ⓔ	42 Ⓐ Ⓑ Ⓒ Ⓓ Ⓔ	67 Ⓐ Ⓑ Ⓒ Ⓓ Ⓔ	92 Ⓐ Ⓑ Ⓒ Ⓓ Ⓔ
18 Ⓐ Ⓑ Ⓒ Ⓓ Ⓔ	43 Ⓐ Ⓑ Ⓒ Ⓓ Ⓔ	68 Ⓐ Ⓑ Ⓒ Ⓓ Ⓔ	93 Ⓐ Ⓑ Ⓒ Ⓓ Ⓔ
19 Ⓐ Ⓑ Ⓒ Ⓓ Ⓔ	44 Ⓐ Ⓑ Ⓒ Ⓓ Ⓔ	69 Ⓐ Ⓑ Ⓒ Ⓓ Ⓔ	94 Ⓐ Ⓑ Ⓒ Ⓓ Ⓔ
20 Ⓐ Ⓑ Ⓒ Ⓓ Ⓔ	45 Ⓐ Ⓑ Ⓒ Ⓓ Ⓔ	70 Ⓐ Ⓑ Ⓒ Ⓓ Ⓔ	95 Ⓐ Ⓑ Ⓒ Ⓓ Ⓔ
21 Ⓐ Ⓑ Ⓒ Ⓓ Ⓔ	46 Ⓐ Ⓑ Ⓒ Ⓓ Ⓔ	71 Ⓐ Ⓑ Ⓒ Ⓓ Ⓔ	96 Ⓐ Ⓑ Ⓒ Ⓓ Ⓔ
22 Ⓐ Ⓑ Ⓒ Ⓓ Ⓔ	47 Ⓐ Ⓑ Ⓒ Ⓓ Ⓔ	72 Ⓐ Ⓑ Ⓒ Ⓓ Ⓔ	97 Ⓐ Ⓑ Ⓒ Ⓓ Ⓔ
23 Ⓐ Ⓑ Ⓒ Ⓓ Ⓔ	48 Ⓐ Ⓑ Ⓒ Ⓓ Ⓔ	73 Ⓐ Ⓑ Ⓒ Ⓓ Ⓔ	98 Ⓐ Ⓑ Ⓒ Ⓓ Ⓔ
24 Ⓐ Ⓑ Ⓒ Ⓓ Ⓔ	49 Ⓐ Ⓑ Ⓒ Ⓓ Ⓔ	74 Ⓐ Ⓑ Ⓒ Ⓓ Ⓔ	99 Ⓐ Ⓑ Ⓒ Ⓓ Ⓔ
25 Ⓐ Ⓑ Ⓒ Ⓓ Ⓔ	50 Ⓐ Ⓑ Ⓒ Ⓓ Ⓔ	75 Ⓐ Ⓑ Ⓒ Ⓓ Ⓔ	100 Ⓐ Ⓑ Ⓒ Ⓓ Ⓔ

Directions: For each question or statement, select the one lettered choice that is the best answer and fill in the corresponding circle on the answer sheet.

1. The diagram below represents a phylogenetic tree of the evolution of even-toed ungulates.

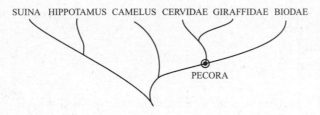

The most likely explanation for the branching pattern seen in the circled region is that

(A) environmental changes caused extinction
(B) inbreeding led to speciation
(C) no speciation occurred
(D) speciation was influenced by environmental change
(E) only the best-adapted organisms survived from generation to generation

2. Two lightbulbs are for sale—a 15-watt compact fluorescent light (CFL) and a 60-watt incandescent. You know that a 15-watt compact fluorescent lightbulb will produce the same amount of light as a 60-watt incandescent lightbulb. If electricity costs $0.10 per kWh and you run each lightbulb for 8,000 hours, how much money will you save in the cost of electricity by buying the compact fluorescent lightbulb?

(A) $4.00
(B) $8.00
(C) $12.00
(D) $36.00
(E) $48.00

3. Winds are primarily caused by

(A) differences in air pressure
(B) the Coriolis effect
(C) ocean currents
(D) seasons
(E) the Earth's rotation on its axis

4. Which of the following processes does NOT occur within current catalytic converters?

    I. Conversion of nitrogen oxides to nitrogen and oxygen
    II. Removal of carbon dioxide
    III. Conversion of unburned hydrocarbons to carbon dioxide and water

    (A) I only
    (B) II only
    (C) III only
    (D) I and II
    (E) I and III

5. You and a partner run an automobile salvage yard where you offer used automobile parts to customers, offering them substantial savings over buying new parts. Which of the following auto parts in your salvage yard would pose the greatest risk to stratospheric ozone depletion?

    (A) Tires, should they catch on fire.
    (B) Oil, should it leak from the crank case and enter the groundwater.
    (C) Air conditioners, if they should leak Freon™.
    (D) Leaking gasoline, should it catch on fire.
    (E) Heavy metals from batteries such as lead should it enter the food web.

6. Suppose that the ecological efficiency at each trophic level of a particular ecosystem is 20%. If the green plants of the ecosystem capture 100 units of energy, about _____ units of energy will be available to support herbivores, and about _____ units of energy will be available to support primary carnivores.

    (A) 120, 140
    (B) 120, 240
    (C) 20, 2
    (D) 20, 4
    (E) 20, 1

7. The water table is at the

    (A) top of the zone of saturation
    (B) bottom of the zone of saturation
    (C) top of the zone of aeration
    (D) bottom of the zone of aeration
    (E) middle of the zone of infiltration

8. A soil test report recommends 8 pounds of 8-0-24 fertilizer per 1,000 square feet. How much phosphorus does it recommend if the area planted is equal to 10,000 square feet?

   (A) 0 pounds
   (B) 8 pounds
   (C) 24 pounds
   (D) 80 pounds
   (E) 240 pounds

9. Today, most of the world's energy comes from

   (A) natural gas, coal, and oil
   (B) oil, wood, and hydroelectric
   (C) hydroelectric, solar, and biomass
   (D) coal, oil, and nuclear
   (E) natural gas, hydroelectric, and oil

10. An AP Environmental Science class conducted an experiment to illustrate the principles of Thomas Malthus. On day 1, three male and three female fruit flies were placed in a flat-bottom flask that contained a cornmeal/banana medium. No other flies were added or removed during the course of this experiment. The students counted the number of flies in the flask each week. The graph shows the results that the class obtained after 55 days.

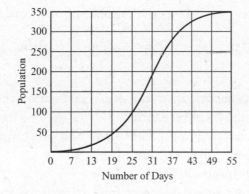

The rate of reproduction is equal to the rate of death on day

   (A) 1
   (B) 25
   (C) 43
   (D) 55
   (E) 1 and 55

11. The concept of net primary productivity

   (A) is the rate at which producers manufacture chemical energy through photosynthesis
   (B) is the rate at which producers use chemical energy through respiration
   (C) is the rate of photosynthesis plus the rate of respiration
   (D) can be thought of as the basic food source for decomposers in an ecosystem
   (E) is usually reported as the energy output of an area of producers over a given time period

12. Which of the following is NOT a unit of energy?

   (A) Joule
   (B) Calorie
   (C) Watt
   (D) Kilowatt-hour
   (E) Btu

13. Which of the following is NOT a phenomenon associated with La Niña?

   (A) Unusually cold ocean temperatures in the eastern equatorial Pacific
   (B) Wetter-than-normal conditions across the Pacific Northwest
   (C) Dryer and warmer-than-normal conditions in the southern states
   (D) Warmer-than-normal winter temperatures in the southeastern United States
   (E) Substantial decrease in the number of hurricanes

14. Which of the following is NOT a possible cause/effect of increasing ocean temperatures?

   (A) A significant increase in the ocean circulation that transports warm water to the North Atlantic
   (B) Large reductions in the Greenland and West Antarctic ice sheets
   (C) Accelerated global warming
   (D) Decreases in upwelling
   (E) Releases of terrestrial carbon from permafrost regions and methane from hydrates in coastal sediments

15. Which of the following would NOT be a chemical property of a persistent organic pollutant such as DDT?

   I. Highly soluble in water
   II. High molecular masses
   III. Usually contain a halogen like chlorine, bromine, or fluorine

   (A) I only
   (B) II only
   (C) III only
   (D) I and II
   (E) II and III

16. The geologic time scale was constructed based upon

    I. fossil records
    II. theory of superposition
    III. global correlation of strata or layers of rocks

    (A) I only
    (B) II
    (C) III
    (D) I and II
    (E) I, II, and III

17. Which biome, found primarily in the eastern United States, central Europe, and eastern Asia, is home to some of the world's largest cities and has probably endured the impact of humans more than any other biome?

    (A) Desert
    (B) Coniferous forest
    (C) Temperate deciduous forest
    (D) Grassland
    (E) Chaparral

18. How does the great ocean conveyor belt moderate the Earth's climate?

    I. It brings more precipitation to the lower latitudes
    II. It brings heat from tropical regions to northern latitudes
    III. It brings cold air from the polar regions to the tropics

    (A) I only
    (B) II only
    (C) III only
    (D) I and II
    (E) II and III

19. An agricultural consequence of higher CO_2 levels in the atmosphere for most crops is:

    I. Decreased nitrogen levels in the crop and their seeds
    II. Increased nitrogen levels in the crop and their seeds
    III. Larger crop yields

    (A) I only
    (B) II only
    (C) III only
    (D) I and III
    (E) II and III

20. Cars, trucks, and buses account for _____ of U.S. greenhouse gas emissions.

    (A) less than 5%
    (B) between 10% and 20%
    (C) between 20% and 33%
    (D) between 35% and 75%
    (E) more than 75%

21. The amount of the environment necessary to produce the goods and services necessary to support a particular lifestyle is known as

    (A) a resource partition
    (B) an ecological footprint
    (C) the per capita index requirement
    (D) the lifestyle index
    (E) commons

22. Which of the organisms listed below are uniquely sensitive to air pollution, making them valuable as early warning indicators of reduced air quality?

    (A) Algae
    (B) Fungus
    (C) Lichen
    (D) Phytoplankton
    (E) Humans

23. Which of the following threats to biodiversity would fall under the category of the introduction of an exotic or invasive species?

    I. Reintroducing black-footed ferrets into Wyoming
    II. Sea lampreys swimming up the Erie Canal to enter the Great Lakes
    III. Bringing wolves into Yellowstone National Park

    (A) I only
    (B) II only
    (C) III only
    (D) I and III
    (E) II and III

24. Which act listed below set criteria for municipal solid waste landfills and prohibited the open dumping of solid wastes?

    (A) CERCLA—Comprehensive Environmental Response, Compensation, and Liability Act (Superfund)
    (B) TOSCA—Toxic Substances Control Act
    (C) RCRA—Resource Conservation and Recovery Act
    (D) Clean Water Act
    (E) None of the above

25. The greatest threat to global biodiversity is

    (A) introduction of invasive species
    (B) poaching
    (C) pollution
    (D) emerging viruses
    (E) deforestation

26. Certain volcanoes have a bowl-shaped crater at the summit and grow to only about a thousand feet. They are usually made of piles of lava, not ash. During the eruption, blobs of lava are blown into the air and break into small fragments that fall around the opening to the volcano. Parícutin in Mexico and the middle of Crater Lake, Oregon, are examples. These are called

    (A) cinder cones
    (B) shield volcanoes
    (C) composite volcanoes
    (D) mud volcanoes
    (E) spatter cones

27. Which U.S. executive agency, commission, service, or department listed below is involved in regulations and protection of endangered species?

    I. Department of the Interior
    II. Fish and Wildlife Service
    III. Environmental Protection Agency

    (A) I only
    (B) II only
    (C) III only
    (D) I and II
    (E) I, II, and III

28. Which of the following is NOT a method used to disinfect municipal water supplies?

    (A) Using chlorine or chlorine compounds
    (B) Using ultraviolet light
    (C) Using ozone
    (D) Using fluorine or fluoride compounds
    (E) Using hydrogen peroxide

29. Excavating and hauling soil off-site to an approved soil disposal/treatment facility would be an example of

    (A) sustainability
    (B) remediation
    (C) conservation
    (D) preservation
    (E) mitigation

Question 30 refers to the locations marked by letters on the world map below.

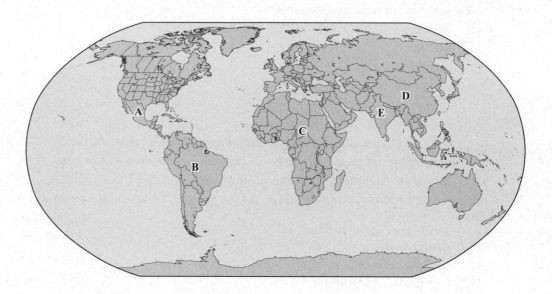

30. The location where fertility rates are highest.

    (A) A
    (B) B
    (C) C
    (D) D
    (E) E

31. Wood in developing countries is primarily used for

    (A) construction
    (B) making furniture
    (C) wood pulp to make paper products
    (D making cardboard
    (E) cooking and heating

Questions 32–36

Select from the following locations.

(A) Bhopal, India
(B) Chernobyl, Ukraine
(C) Love Canal, New York
(D) Minamata Bay, Japan
(E) Three Mile Island, Pennsylvania

32. Site of a hazardous chemical dumping ground over which homes and a school were built.

33. Site of mercury poisoning.

34. The most serious commercial nuclear accident in U.S. history.

35. Leakage of poisonous gases from a pesticide manufacturing plant.

36. Nuclear power plant accident that released 30 to 40 times the radiation of the atomic bombs dropped on Hiroshima and Nagasaki.

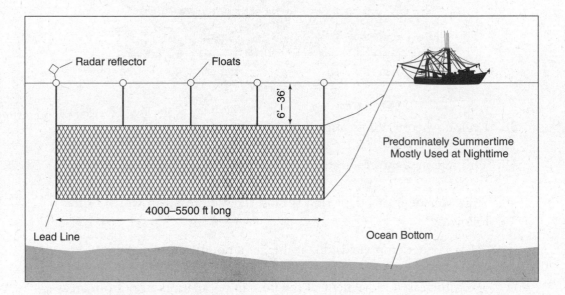

37. The type of fishing seen in the diagram above would be best described as

(A) bottom trawling
(B) longline fishing
(C) drift gill net
(D) angling
(E) cast netting

Questions 38–39 refer to the layers of the Earth's atmosphere as shown below.

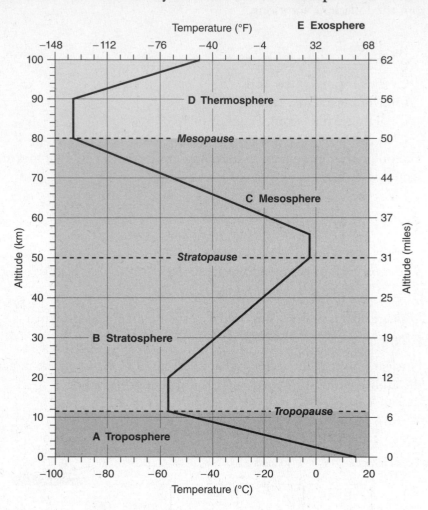

38. Region where UVC is filtered out.

39. This region is mostly heated by transfer of energy from the surface.

40. Stratospheric ozone depletion is most likely to result in which of the following?

    (A) Higher crop yields due to higher amounts of sunlight
    (B) Disruption of photosynthesis in plants
    (C) Increased movement of the human population away from areas influenced by the hole in the ozone
    (D) Less clouds in the sky
    (E) Warmer temperatures

41. Viability is the capacity of living organisms to live, develop, or germinate under favorable conditions. Which of the following factors decreases a population's viability?

    I. Specialized niche
    II. Few competitors
    III. Ability to migrate

    (A) I only
    (B) II only
    (C) III only
    (D) I and III
    (E) II and III

42. Most of the world's electronic wastes end up in what country?

    (A) United States
    (B) India
    (C) Mexico
    (D) China
    (E) South Korea

43. Which one of the following is a secondary sewage treatment process?

    (A) Lagooning
    (B) Filter (oxidizing) beds
    (C) Screening
    (D) Nutrient removal
    (E) Constructed wetlands

44. The largest user of freshwater worldwide is

    (A) mining
    (B) irrigation
    (C) industry
    (D) home use
    (E) production of electrical power

45. Choose the statement that is FALSE.

    (A) Domestic fruits and vegetables are less likely to have pesticide residues than imported ones.
    (B) Cancer may not be the primary risk from chronic, long-term exposure to pesticides.
    (C) When the EPA looks at a pesticide to decide whether to register it for use in the United States, its primary concern is to ensure that there is no significant human health or environmental risk presented by the chemical.
    (D) The federal government does not prohibit the use of pesticides known to cause cancer.
    (E) Washing and peeling fruits and vegetables does not remove all or most pesticide residues.

46. Of the choices listed, cap-and-trade policies have historically been used in controlling

   (A) overfishing
   (B) dumping trash into the open sea
   (C) carbon dioxide emissions
   (D) the storage of nuclear wastes
   (E) the release of dioxin and other pollutants into groundwater systems

Questions 47 and 48 refer to the following laboratory investigation.

An APES class was doing field research using a mark-recapture technique to estimate the population size of a particular species of butterfly. The size of the field was exactly 2 hectares (ha) or 20,000 square meters. Initially, the class caught and marked 150 of these butterflies in the field and then released them. Later, the class recaptured 100 of this species of butterfly, of which 20 were marked.

47. What is the estimate of the population size of this species of butterfly?

   (A) 13
   (B) 30
   (C) 750
   (D) 1,500
   (E) 3,000

48. What was the population density of the butterflies?

   (A) 188 butterflies per ha
   (B) 375 butterflies per ha
   (C) 750 butterflies per ha
   (D) 1,500 butterflies per ha
   (E) 3,000 butterflies per ha

49. An APES class went on a field trip into a coniferous forest. They discovered a very large section of land that had been completely logged. There were just stumps where large coniferous trees had once stood. There was also very little animal life in the area. Which method of logging had most likely been used in this section of land?

   (A) Strip cutting
   (B) Clear cutting
   (C) Seed tree cutting
   (D) Shelterwood cutting
   (E) Selective cutting

50. You have been placed in charge of rebuilding a salmon population in a river basin that contains a hydroelectric dam. Which of the following remediation techniques would NOT be effective?

    (A) Reduce silt runoff
    (B) Build fish ladders
    (C) Decrease water flow from the dam
    (D) Release juvenile salmon from hatcheries
    (E) Use trucks and barges to transport salmon around the dam

51. Which one of the following proposals would NOT increase the sustainability of ocean fisheries management?

    (A) Establish fishing quotas based on past harvests.
    (B) Set quotas for fisheries well below their estimated maximum sustainable yields.
    (C) Sharply reduce fishing subsidies.
    (D) Shift the burden of proof to the fishing industry to show that their operations are sustainable.
    (E) Strengthen integrated coastal management programs.

52. The most abundant element in the Earth's crust is

    (A) silicon
    (B) aluminum
    (C) oxygen
    (D) hydrogen
    (E) carbon

53. Smog levels in large, urban cities are generally highest during the

    (A) summer
    (B) spring
    (C) fall
    (D) winter
    (E) season or time of year does not matter

54. In general, parasites tend to

    (A) become more virulent as they live within the host
    (B) destroy the host completely
    (C) become deactivated as they live within the host
    (D) be only mildly pathogenic
    (E) require large amounts of oxygen

55. The population of a country in 1994 was 200 million. Its rate of growth was 1.2%. Assuming that the rate of growth remains unchanged and all other factors remain constant, in what year will the population of the country reach 400 million?

    (A) 2004
    (B) 2024
    (C) 2040
    (D) 2054
    (E) 2104

56. Which of the following contributes LEAST to acid deposition?

    (A) Sulfur dioxide
    (B) Ammonia
    (C) Nitrogen oxides
    (D) Lead
    (E) Ozone

57. The circulation of air in Hadley cells results in

    (A) low pressure and rainfall at the equator
    (B) high pressure and rainfall at the equator
    (C) low pressure and dry conditions at about 30° north and south of the equator
    (D) high pressure and wet conditions at about 30° north and south of the equator
    (E) both (A) and (C)

58. All of the following are characteristics of *K*-strategists EXCEPT

    (A) mature slowly
    (B) low juvenile mortality rate
    (C) niche generalists
    (D) Type I or II survivorship curve
    (E) intraspecific competition due to density-dependent limiting factors

59. If a city of population 100,000 experiences 1,000 births, 600 deaths, 500 immigrants, and 100 emigrants in the course of a year, what is its net annual percentage growth rate?

    (A) 0.4%
    (B) 0.8%
    (C) 1.6%
    (D) 3.2%
    (E) 8%

Question 60 refers to the diagram below.

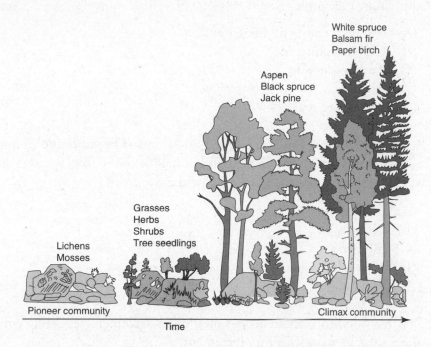

60. The greatest species diversity would be found in

    (A) the pioneer community represented by the lichens and mosses.
    (B) the community composed primarily of grasses, herbs, shrubs, and tree seedlings.
    (C) the community composed primarily of aspen, black spruce, and jack pine.
    (D) the community composed primarily of white spruce, balsam fir, and paper birch.
    (E) all communities equally.

61. A new road connecting a planned community to an existing urban area was needed due to high traffic volume on the existing roadway. The least expensive option was to build the road through a forested area owned by the county. An environmental impact study was conducted and the developers agreed to the following conditions in exchange for the contract: (1) use selective cutting and clearing practices; (2) replace or restore forested areas that will be affected during the construction of the new highway; (3) preserve as much existing vegetation as possible during the construction of the new highway; and (4) the developer would purchase a similar amount of privately held land and place it in conservation easements. This trade-off approach to addressing an environmental issue is known as

    (A) sustainability
    (B) remediation
    (C) preservation
    (D) restoration
    (E) mitigation

62. Which mobile source pollutant cannot be currently controlled by emission control technology?

    (A) Ozone-forming hydrocarbons
    (B) Carbon monoxide
    (C) Carbon dioxide
    (D) Air toxics
    (E) Particulate matter

63. A new cancer drug was being tested on rabbits. The study found that it took 400 milligrams of the drug to kill half of the rabbits whose average weight was 4 kilograms. The LD_{50} for the drug was

    (A) 0.01%
    (B) 0.1 g/kg
    (C) 50 mg/kg
    (D) 10 mg/kg
    (E) 100 g/kg

64. The 1997 Kyoto Conference resulted in an agreement regarding emissions of greenhouse gasses. In this agreement,

    (A) all countries would be subject to the same amount of reductions in their emissions using metric tons as the international standard.
    (B) all countries would be subject to the same percentage reduction in their emissions.
    (C) only the underdeveloped countries would be subject to a reduction in their emissions.
    (D) the United States would be exempt from any mandated reductions.
    (E) the developed countries would set goals for the percentage reduction of their emissions.

65. The largest contaminant (by weight and volume) of inland surface waters such as lakes, rivers, and streams in the United States is

    (A) sediment
    (B) organic wastes
    (C) pathogenic organisms
    (D) heavy metals
    (E) oil

66. Currently, the United States recycles about what percentage of the municipal solid waste it generates?

    (A) 3%
    (B) 15%
    (C) 35%
    (D) 50%
    (E) 75%

Questions 67, 68, and 69 refer to the following diagram of a heavy water moderated nuclear reactor.

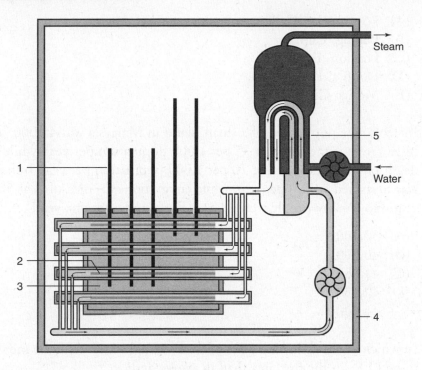

67. "1" in the diagram above represents the

    (A) control rods
    (B) moderator
    (C) fuel rods
    (D) concrete shields
    (E) condensers

68. "3" in the diagram above represents the

    (A) control rods
    (B) moderator
    (C) fuel rods
    (D) concrete shield
    (E) heat exchanger

69. Which one of the statements below is TRUE regarding the moderator and the control rods in a moderated nuclear reactor?

    (A) The moderator and control rods absorb neutrons.
    (B) The moderator and control rods slow the neutrons down.
    (C) The moderator slows the neutrons down, while the control rods absorb the neutrons.
    (D) The control rods slow the neutrons down, while the moderator absorbs the neutrons.
    (E) The moderator slows the neutrons down, while the control rods speed up the neutrons.

70. Which of the cities listed below currently has the HIGHEST level of air pollution?

   (A) Los Angeles, California
   (B) Beijing, China
   (C) London, England
   (D) Donora, Pennsylvania
   (E) Mexico City, Mexico

71. In 1995, the population of a small island in Malaysia was 40,000. The birth rate was measured at 35 per 1,000 population per year, while the death rate was measured at 10 per 1,000 population per year. Immigration was measured at 100 per year, while emigration was measured at 50 per year. How many people would be on the island after one year?

   (A) 39,100
   (B) 40,000
   (C) 41,050
   (D) 42,150
   (E) 44,500

72. In which of the following pairs of items would biodiversity be more likely to be LESS in the first area than in the second?

   (A) savanna, taiga
   (B) large lake, small lake
   (C) temperate area, the Arctic
   (D) temperate area, tropical area
   (E) intertidal zone, benthic zone

73. Which choice listed below makes up 60% of the world's human food energy intake?

   (A) Meat, fish, and milk or milk products
   (B) Rice, corn (maize), and wheat
   (C) Millet, sorghum, roots, and tubers
   (D) Rice and fish
   (E) Beans, rice, corn

74. Among adults, which of the following represents the most preventable cause of death?

   (A) Cardiovascular disease
   (B) AIDS
   (C) Alcoholism
   (D) Use of tobacco
   (E) Traffic accidents

75. The process in which ammonia is oxidized to nitrite (NO_2^-) and nitrate (NO_3^-), the forms most usable by plants, through the action of *Nitrosomonas* and *Nitrobacter* bacteria is known as

    (A) nitrogen fixation
    (B) nitrification
    (C) assimilation
    (D) ammonification
    (E) denitrification

76. Which of the following sections of the Earth contains most of the Earth's mass, is mostly solid, and is primarily composed of iron, magnesium, aluminum, and silicon-oxygen compounds?

    (A) Continental crust
    (B) Oceanic crust
    (C) Core
    (D) Moho
    (E) Mantle

77. Examine the following graph, which shows the predator-prey relationship between hares and lynxes.

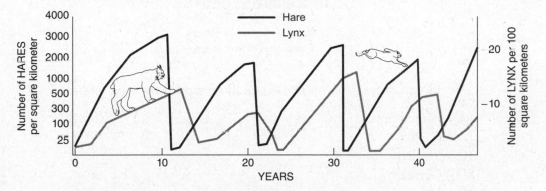

    The population cycles of the snowshoe hare and its predator, the lynx, reveal that

    (A) the prey population is controlled by predators alone
    (B) hares and lynxes are so mutually dependent that each species cannot survive without the other
    (C) the relationship is primarily independent; in other words, the number of hares does not have any effect on the number of lynxes and vice versa.
    (D) both hare and lynx populations are regulated mainly by abiotic factors
    (E) the hare population is *r*-selected and the lynx population is *K*-selected

78. "We can burn coal to produce electricity to operate a refrigerator" is an example of the _____ and "If we burn coal to produce electricity to operate a refrigerator, we lose a great deal of energy in the form of heat" is an example of the _____

    (A) first law of thermodynamics, first law of thermodynamics
    (B) second law of thermodynamics, first law of thermodynamics
    (C) first law of thermodynamics, second law of thermodynamics
    (D) first law of thermodynamics, third law of thermodynamics
    (E) third law of thermodynamics, first law of thermodynamics

79. Refer to the following age-structure diagram:

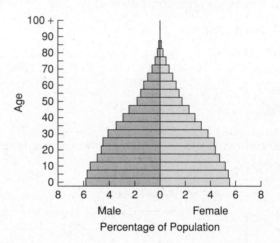

    The age-structure diagram above would be typical for what country?

    (A) United States
    (B) Canada
    (C) Mexico
    (D) Germany
    (E) Japan

80. Why is territoriality an adaptive behavior for songbirds maintaining populations at or near their carrying capacity?

    (A) Because songbirds expend a tremendous amount of energy defending territories, they spend less time feeding their young and fledgling mortality increases.
    (B) Only the fittest males defend territories, and they attract the fittest females; therefore, the best genes are conveyed to the next generation.
    (C) Songbird males defend territories commensurate with the size from which they can derive adequate resources for themselves, their mates, and their chicks.
    (D) Many individuals are killed in the ritualistic conflicts that go along with territorial defense.
    (E) Songbirds make improvements to the territories they inhabit so that they can all enjoy larger clutches and successfully fledged chicks.

81. Butterflies and moths both feed on flowers. Butterflies feed during the day, and moths feed at night. This is an example of

    (A) *r*-strategy
    (B) *K*-strategy
    (C) resource partitioning
    (D) commensalism
    (E) mutualism

82. It costs a copper smelter $200 to reduce its emissions by 1 ton and $250 for an additional ton. It costs an electric utility $100 to reduce its emissions by 1 ton and $150 for an additional ton. What is the cheapest way of reducing total emissions by 2 tons?

    (A) Legislate that each firm must reduce emissions by two tons.
    (B) Charge both firms $251 for every ton they emit.
    (C) Allow each firm to buy a permit to pollute that costs $151.
    (D) File an injunction to halt production until the firms reduce emissions by 2 tons.
    (E) None of the above is correct.

83. Which of the following organisms listed below would be MOST at risk for accumulating fat soluble pesticides?

    (A) Aquatic plants
    (B) Cattle
    (C) Hawks
    (D) Protozoa
    (E) Bees

84. Which of the following examples does/do NOT demonstrate a negative feedback loop?

    (A) Increased technology in agricultural science produces more food resources, resulting in a higher birth rate.
    (B) Frequent large fires can increase atmospheric greenhouse gas concentrations, causing more warming and drought, and leading to more fires, which reduce the trees available to store carbon.
    (C) A thermostat working to keep a home at 68°F during the winter.
    (D) A nuclear chain reaction.
    (E) A, B, and D.

85. Demographic transition leads to stabilizing population growth and is generally characterized as having four separate stages. Place the following stages in the proper order as they would most likely occur.

    I. Birthrates equal mortality rates, and zero population growth is achieved. Birthrates and death rates are both relatively low, and the standard of living is much higher than during the earlier periods. In some countries, birthrates may actually fall below mortality rates and result in net losses in population.

    II. Living conditions are severe, medical care is poor or nonexistent, and the food supply is limited due to poor agricultural techniques, poor preservation, and pestilence. Birthrates are high to replace individuals lost through high mortality rates. The net result is little population growth.

    III. Urbanization decreases the economic incentives for large families. The cost of supporting an urban family grows, and parents are more actively discouraged from having large families. Educational and work opportunities for women decrease birth rates. Obtaining food is not a major focus of the day. Leisure time is available, and retirement safety nets are in place, reducing the need for extra children to support parents. In response to these economic pressures, the birthrate starts to drop, ultimately coming close to the death rate.

    IV. Occurs after the start of industrialization. Standards of hygiene and more modern medical techniques begin to drive the death rate down, leading to a significant upward trend in population size. Mortality rates drop as a result of advances in medical care, improved sanitation, cleaner water supplies, vaccination, and higher levels of education. The net result is a rapid increase in population.

    (A) I, II, III, IV
    (B) IV, III, II, I
    (C) I, IV, II, III
    (D) II, IV, III, I
    (E) III, I, II, IV

86. The concentration of which gas can be reduced by preventing forest depletion?

    (A) Carbon dioxide
    (B) Nitrous oxide
    (C) Oxygen
    (D) Methane
    (E) CFCs

87. Heap leaching could be a process used in the extraction of

    (A) copper
    (B) nickel
    (C) gold
    (D) uranium
    (E) All of the above

88. After a recent storm, an APES class took a field trip to a storm drain out-
    let entering the Pacific Ocean at Ballona Creek, in southern California.
    They carefully collected water samples and brought the samples back to
    the lab. Alpha group took 100.00 mL of the collected water and filtered
    it through a 1.2 μm Millipore glass fiber filter. The filter was carefully
    transferred to a premassed stainless steel crucible and placed into a
    105°C oven for 24 hours. The following is a hypothetical set of data
    from alpha group:

    Weight of crucible + filter + residue after 24 hrs at 105°C = 100.000 g
    Weight of crucible + filter = 80.000 g

    What is the total suspended solid amount measured in mg · L^{-1}?

    (A) $(80.000/100.00) \times 100\%$

    (B) $100.00 - 80.00$

    (C) $\dfrac{(100.00 - 80.00) \times 1{,}000 \times 1{,}000}{100.00}$

    (D) $\dfrac{(100.00 - 80.00) \times 1{,}000}{100}$

    (E) $\dfrac{(100.00 + 80.00) \times 100}{1{,}000}$

89. If a plant stores 1,000 units of energy from the sun, and a field mouse eats
    the plant and is then eaten by a snake, who is then eaten by a bird, who is
    then eaten by a carnivore, how much of the original 1,000 units of energy
    will be available to the carnivore?

    (A) 90
    (B) 9
    (C) 1
    (D) 0.9
    (E) 0.1

90. Which of the following is an advantage of using green manure?

    I. Leguminous green manures such as clover contain nitrogen-fixing symbiotic bacteria in root nodules that fix atmospheric nitrogen in a form that plants can use.

    II. It increases the percentage of organic matter (biomass) in the soil, thereby improving water retention, aeration, and other soil characteristics.

    III. The root systems of some varieties of green manure grow deep in the soil and bring up nutrient resources unavailable to shallower-rooted crops.

(A) I
(B) II
(C) III
(D) I and III
(E) I, II, and III

91. Which is probably the principal source of nitrates and phosphates?

(A) The water cycle
(B) Nitrogen fixation of lighting
(C) Bacterial decay
(D) Changes in environmental temperature
(E) Changes in light intensity

92. Ocean water temperatures fluctuate less than air temperatures over land. The primary reason for this phenomenon is

(A) land absorbs more heat than water
(B) winds distribute heat that builds up over the land
(C) there is more water on Earth than land
(D) the high heat capacity of water
(E) water both cools down and heats up faster than land

93. Which of the following pollutants does NOT occur naturally in nature?

(A) Nitrous oxide
(B) Radon
(C) Sulfur dioxide
(D) Formaldehyde
(E) All pollutants listed above occur naturally in nature

94. Which of the following sequences correctly lists the different wave arrivals from first to last during an earthquake?

(A) P waves then S waves and then finally surface waves
(B) Surface waves then P waves and then finally S waves
(C) P waves then surface waves and then finally S waves
(D) S waves then P waves and then finally surface waves
(E) Surface waves then S waves and then finally P waves

95. Harmful effects of increased UV radiation on Earth's surface include all of the following EXCEPT

    (A) reduction in crop production

    (B) reduction in the growth of phytoplankton and its cumulative effect on food webs

    (C) cooling of the stratosphere

    (D) warming of the stratosphere

    (E) increases in sunburns and damage to the skin

96. Which of the following choices shows the correct order of events that might occur in the process of primary succession?

    (A) Grasses → small bushes → small herbaceous plants → lichen → conifers → long-lived hardwoods → short-lived hardwoods → bare rock

    (B) Bare rock → mosses → grasses → small herbaceous plants → small bushes → conifers → short-lived hardwoods → long-lived hardwoods

    (C) Long-lived hardwoods → short-lived hardwoods → conifers → small bushes → small herbaceous plants → grasses → bare rock

    (D) Bare rock → grasses → mosses → small herbaceous plants → small bushes → conifers → short-lived hardwoods → long-lived hardwoods

    (E) Conifers → bare rock → long-lived hardwoods → small herbaceous plants → lichen → grasses → small bushes → short-lived hardwoods

97. Which of the following indoor pollutants would NOT be contributed by carpeting?

    (A) Formaldehyde

    (B) Styrene

    (C) Mold

    (D) Methylene chloride

    (E) Mites

98. Which of the following groups would be most likely to exhibit uniform dispersion?

    (A) Red squirrels, which actively defend territories

    (B) Cattails, which grow primarily at edges of lakes and streams

    (C) Dwarf mistletoes, which parasitize particular species of forest tree

    (D) Moths in a city at night

    (E) Lake trout, which seek out deep water

99. All of the following are correct statements about the regulation of populations EXCEPT

    (A) a logistic equation reflects the effect of density-dependent factors, which can ultimately stabilize a population around the carrying capacity
    (B) density-independent factors have a greater effect as a population's density increases
    (C) high densities in a population may cause physiological changes that inhibit reproduction
    (D) because of the overlapping nature of population-regulating factors, it is often difficult to determine their cause-and-effect relationships precisely
    (E) the occurrence of population cycles in some populations may be the result of crowding or lag times in the response to density-dependent factors

100. Which of the following energy sources has the lowest quality?

    (A) High-velocity water flow
    (B) Fuelwood
    (C) Food
    (D) Dispersed geothermal energy
    (E) Saudi Arabian oil deposits

SECTION II (FREE-RESPONSE QUESTIONS)

Time: 90 minutes

No calculators allowed

4 questions

> **Directions:** Answer all four questions, which are weighted equally. The suggested time is about 22 minutes for answering each question. Write all your answers on the pages following the questions in the pink booklet. Where calculations are required, clearly show how you arrived at your answer. Where explanation or discussion is required, support your answers with relevant information and/or specific examples.

1. A family is building a new home in Buffalo, New York. Buffalo experiences severe winters. Assume the following.

   - The house has 4,000 square feet.
   - 100,000 Btus of heat per square foot are required to heat the house for the winter.
   - Natural gas sells for $5.00 per thousand cubic feet.
   - 1 cubic foot of natural gas supplies 1,000 Btus of heat energy.
   - 1 kilowatt-hour of electricity supplies 10,000 Btus of heat energy.
   - Electricity costs $50 per 500 kWh.

   (a) Calculate the following, showing all the steps of your calculations, including units.

      (i) The number of cubic feet of natural gas required to heat the house for the winter.
      (ii) The cost of heating the house using natural gas.
      (iii) The cost of heating the house using electricity.

   (b) The homeowners are discussing with their architect the possibility of using either active or passive solar design to reduce their heating costs.

      (i) Define "active solar systems" and describe ONE method of how they are used in heating a home.
      (ii) Define "passive solar systems" and describe ONE method of how they are used in heating a home.
      (iii) Identify and describe ONE environmental benefit *or* ONE disadvantage of using solar power systems to heat buildings.

   (c) The homeowners wish to incorporate "green design" into their new home.

      (i) Define "green design."
      (ii) Describe ONE example of "green design" architecture.

2.

> **Legal Notice**
>
> **Initial Public
> Offering of
> Common Stock**
>
> The Fremont Seafood Company, headquartered in New Orleans, Louisiana, is offering 5,000,000 shares of common stock at $3.00 per share to investors for the purpose of raising capital to construct commercial aquafarms in the Mississippi Delta area to commercially raise shrimp, clams, and salmon. This is a fantastic offer to the right individuals who see aquafarming as a solution to world hunger and the ever dwindling supplies of catching these species in the open ocean. No guarantee of return on investment is implied. For further information on this investment opportunity, contact L.D. Breckenridge, Ltd. of Shreveport, LA.

(a) Define "aquafarming."

   (i) Describe TWO advantages of aquafarming over traditional wild harvesting.
   (ii) Describe TWO negative environmental impacts of aquafarming.

(b) Describe TWO methods that can be employed in aquaculture to lessen the environmental impact.

(c) How is raising and harvesting kelp and bivalve mollusks a more sustainable method of aquafarming than raising and harvesting higher trophic order fish such as salmon?

(d) The newspaper announcement claims that aquafarming is the answer to world hunger. Give TWO examples of how this statement may be contrary to environmental sustainability.

3.   The diagrams below show the growth in the hole in the strato-
      spheric ozone layer over the Antarctic over the last 30 years.

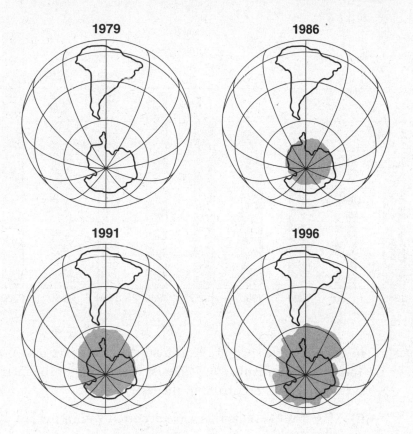

1979   **1986**

1991   **1996**

(a) Describe the role and importance of stratospheric ozone to life on
    Earth.

(b) Suggest ONE possible reason for the appearance of the hole in the
    ozone layer, which was first observed in the 1980s.

(c) Ultraviolet (UV) radiation occurs in three forms, known as UVA,
    UVB, and UVC.

    (i)   Describe at least ONE characteristic of EACH form of ultraviolet
          radiation.
    (ii) Describe the effects that increased UV radiation would have on:
          (a) Human health
          (b) Terrestrial ecosystems
          (c) Marine ecosystems

(d) Identify and describe ONE alternative to ozone-destroying chemical
    compounds.

(e) Identify and briefly describe ONE international treaty that addresses
    the release of chemical compounds that destroy stratospheric ozone.

4. The chart below shows the average American life span for the past 160 years.

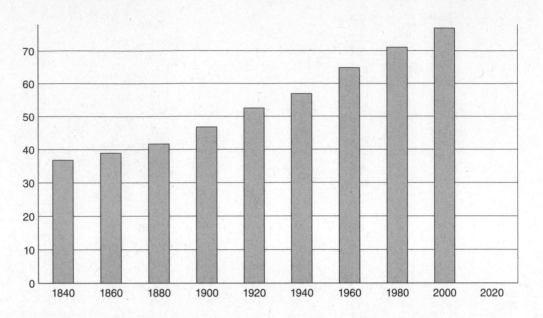

(a) In 1840, the life span of the average American was just 37 years. Today, it is approximately 77 years, with much of the increase in average life span occurring in the last 50 years.

   (i) List TWO events that have occurred within the last 100 years that could account for the increase in human life span.

  (ii) Choose ONE of the events you selected in (i) above and describe in detail its role in increasing average human life span.

(b) Currently, about 3.3 billion people—half of the world's population—are at risk of malaria. Every year, this leads to about 250 million malaria cases and nearly 1 million deaths, with people living in the poorest countries being the most vulnerable. Malaria is especially serious in Africa, where 20% of childhood deaths are caused by the disease (i.e., every 30 seconds a child dies from the disease).

   (i) Explain how malaria is transmitted through the human population.

  (ii) Identify TWO environmental factors that contribute to the emergence of malaria and explain how those factors influence the increase in the incidence of the disease.

 (iii) Traditionally, the spread of malaria was controlled by the use of pesticides such as DDT. However, DDT and other pesticides often have negative environmental impacts. A promising alternative to the use of pesticides is biological control. Describe biological control and provide ONE biological control method for controlling the spread of the disease.

ANSWER KEY

Practice Test 2

Section I (Multiple-Choice Questions)

1. D	26. A	51. A	76. E
2. D	27. E	52. C	77. B
3. A	28. D	53. A	78. C
4. B	29. B	54. D	79. C
5. C	30. C	55. D	80. C
6. D	31. E	56. D	81. C
7. A	32. C	57. A	82. C
8. A	33. D	58. C	83. C
9. A	34. E	59. B	84. E
10. E	35. A	60. D	85. D
11. E	36. B	61. E	86. A
12. C	37. C	62. C	87. E
13. E	38. B	63. B	88. C
14. A	39. A	64. E	89. E
15. A	40. B	65. A	90. E
16. E	41. A	66. C	91. C
17. C	42. D	67. A	92. D
18. B	43. B	68. B	93. E
19. D	44. B	69. C	94. A
20. C	45. A	70. B	95. D
21. B	46. C	71. C	96. B
22. C	47. C	72. D	97. D
23. B	48. B	73. B	98. A
24. C	49. B	74. D	99. B
25. E	50. C	75. B	100. D

PREDICT YOUR SCORE ON THE APES EXAM

Place a check mark (✔) next to the multiple-choice questions you got correct on page 533. Then fill in the blanks below to predict your overall score on the APES exam. Essay questions are not used in this prediction as they require subjective grading. You can also use this page to determine your areas of weakness. For example, if you got 5 out of 5 questions correct on the topic The Earth, but only 2 out of 5 questions correct on the topic The Atmosphere, spend some more time reviewing The Atmosphere.

Unit I EARTH SYSTEMS AND RESOURCES (10–15%)

Chapter 1 The Earth
(#16, 26, 52, 76, 94)   _____ correct/5 = _____%

Chapter 2 The Atmosphere
(#3, 13, 38, 39, 57)   _____ correct/5 = _____%

Chapter 3 Global Water Resources and Use
(#7, 18, 44, 92)   _____ correct/4 = _____%

Unit II THE LIVING WORLD (10–15%)

Chapter 4 Ecosystems
(#1, 6, 11, 17, 54, 58, 60, 80, 81, 89, 96, 98)   _____ correct/12 = _____%

Chapter 5 Natural Biogeochemical Cycles
(#8, 75, 91)   _____ correct/3 = _____%

Unit III POPULATIONS (10–15%)

Chapter 6 Populations
(#10, 30, 41, 47, 48, 55, 59, 77, 79, 85, 99)   _____ correct/11 = _____%

Unit IV LAND AND WATER USE (10–15%)

Chapter 7 Land and Water Use
(#15, 19, 31, 35, 37, 45, 49, 51, 61, 73, 84, 86, 87, 90)   _____ correct/14 = _____%

Practice Exam 2

Unit V ENERGY RESOURCES AND CONSUMPTION (10–15%)

Chapter 8 Energy
(#2, 9, 12, 34, 36, 50, 67, 68, 69,
78, 100)

_____ correct/11 = _____%

Unit VI POLLUTION (25–30%)

Chapter 9 Pollution
(#4, 20, 22, 24, 28, 32, 33, 43, 53, 56,
62, 65, 66, 70, 83, 88, 93, 97)

_____ correct/18 = _____%

Chapter 10 Impacts on the Human
Environment
(#21, 29, 42, 46, 63, 74, 82)

_____ correct/7 = _____%

Unit VII GLOBAL CHANGE (10–15%)

Chapter 11 Stratospheric Ozone and
Global Warming
(#5, 14, 23, 25, 27, 40, 64, 71, 72, 95)

_____ correct/10 = _____%

Total Number Correct _____ / 100 = _____%

PREDICTED AP SCORE*	
Less than 50 correct:	1 or 2 (not passing)
50–60 correct:	3 on the APES exam
61–75 correct:	4 on the APES exam
76+ correct:	5 on the APES exam

*Please note this is a rough estimate and is not intended to be an indicator of an actual AP score.

MULTIPLE-CHOICE EXPLANATIONS

1. **(D)** Speciation occurs when environmental factors change the composition of the gene pool.

2. **(D)** Incandescent: $\dfrac{60 \text{ Watt}}{1} \times \dfrac{\$0.10}{1,000 \text{ Watt} \cdot \text{hr}} \times \dfrac{8,000 \text{ hr}}{1} = \$48.00$

   Compact Fluorescent: $\dfrac{15 \text{ Watt}}{1} \times \dfrac{\$0.10}{1,000 \text{ Watt} \cdot \text{hr}} \times \dfrac{8,000 \text{ hr}}{1} = \$12.00$

   Therefore, the saving is $\$48.00 - \$12.00 = \$36.00$

3. **(A)** Wind is primarily caused by the uneven heating of Earth's surface; i.e., the equator receives more solar radiation than the poles. This uneven heating results in differences in air pressure and sets convection currents (winds) in motion.

4. **(B)** The three reactions that occur in catalytic converters are:
   (a) Reduction of nitrogen oxides to nitrogen and oxygen:
   $$NO_x \rightarrow O_2 + N_2$$
   (b) Oxidation of carbon monoxide to carbon dioxide:
   $$CO + O_2 \rightarrow CO_2$$
   (c) Oxidation of unburned hydrocarbons to carbon dioxide and water:
   $$C_xH_{2x} + O_2 \rightarrow CO_2 + H_2O$$

   Although catalytic converters are effective at removing hydrocarbons and other harmful emissions, they do not reduce the emission of carbon dioxide produced when fossil fuels are burnt. Additionally, the U.S. Environmental Protection Agency (EPA) has stated catalytic converters are a significant and growing cause of global warming, because of their release of nitrous oxide (N_2O), a greenhouse gas over 300 times more potent than carbon dioxide. The EPA states that motor vehicles contribute approximately 50% of nitrous oxide emissions, and nitrous oxide makes up 7% of greenhouse gases.

5. **(C)** Air conditioners that contain chlorine-fluorine based refrigerants are known as CFCs (chloro-fluorocarbons). One chlorine atom released from a CFC can ultimately destroy over 100,000 stratospheric ozone molecules.

6. **(D)** 100 units of energy are found in the green plants. $100 \times 0.20 = 20$ units available for herbivores. Of the 20 units now available, $20 \times 0.20 = 4$ units are left for the primary carnivores.

7. **(A)** The water table is the upper surface of an area filled with groundwater, separating the zone of aeration (the subsurface region of soil and rocks in which the pores are filled with air and usually some water) from the zone of saturation (the subsurface region in which the pores are filled only with water). Water tables rise and fall with seasonal moisture, water absorption by vegetation, and the withdrawal of groundwater from wells, among other factors. See Figure 3.7 on page 140.

8. **(A)** The three numbers on the side of the fertilizer bag refer to the amount of nitrogen, phosphorus, and potassium that the fertilizer contains. The rest of the bag contains minor nutrients and filler material. Since the number representing phosphorus is 0, no phosphorus is required.

9. **(A)** Notice that all three forms of energy are nonrenewable sources derived from fossil fuels. Oil is the primary energy source for the world today with the United States at approximately 40% dependency on this single source. Coal and natural gas are each around 22% dependency in the United States and the world.

10. **(E)** Wherever on this graph the slope of the line is zero (horizontal), there is no population growth; i.e., the rate of reproduction is equal to the rate of death.

11. **(E)** Primary productivity is the amount of biomass produced through photosynthesis per unit area per time by plants. Gross primary productivity is the total energy fixed by plants through photosynthesis. A portion of the energy of gross primary productivity is used by plants for respiration. Respiration provides a plant with the energy needed for various activities. Subtracting respiration from gross primary productivity gives net primary productivity, which represents the rate of production of biomass that is available for consumption by heterotrophic organisms.

12. **(C)** A watt is a unit of power.

13. **(E)** The effects of La Niña are opposite to those of El Niño. A La Niña effect may be defined as a drop in average sea-surface temperatures to more than 0.7°F (0.4°C) below normal, lasting at least six months, across parts of the eastern tropical Pacific. When La Niña forms, the hurricane season is affected as the cooling water creates dramatic changes in the upper-level air currents that play a major role in storm development. During La Niña, high-level westerly winds either weaken or shift to come from the east, allowing *more* storms to develop.

14. **(A)** In order to balance the excess heating near the equator and cooling at the poles of Earth, both atmosphere and ocean transport heat from low to high latitudes. Warmer surface water is cooled at high latitudes, releasing heat to the atmosphere, which is then radiated away to space. This heat engine operates to reduce equator-to-pole temperature differences and is a prime moderating mechanism for climate on Earth.

15. **(A)** An important chemical property of persistent organic pollutants is their lipid (oil or fat) solubility, which results in their ability to pass through biological phospholipid membranes and bioaccumulate in the fatty tissues of living organisms.

16. **(E)** The geologic time scale is a system describing the timing and relationships between events that have occurred throughout Earth's history. The identification of strata or rock layers by the fossils they contained enabled geologists to correlate strata between continents. If two strata (however far apart or different in composition) contained the same fossils, it was likely they had been laid down at the same time. The principle of superposition states that a sedimentary rock layer in an undisturbed sequence is younger than the one beneath it and older than the one above it. This principle allows sedimentary layers to be viewed as a form of vertical time line, a partial or complete record of the time elapsed from deposition of the lowest layer to deposition of the highest bed.

17. **(C)** Temperate deciduous forests have warm summers and cold winters. Deciduous trees escape these winters by losing their leaves. Typical mammals are bears, badgers, squirrels, woodchucks, insectivores, rodents, wolves, wildcats, and deer. These forests are rich in birds. The climate that is suitable for temperate deciduous forests is most suited for humans—hence forest destruction.

18. **(B)** The great ocean conveyor belt transfers warm water from the Pacific Ocean to the Atlantic as a shallow current and returns cold water from the Atlantic to the Pacific as a deep current that flows further south. As it passes Europe, the surface water evaporates and the ocean water cools, releasing heat to the atmosphere. This release of heat is largely responsible for the relatively warm temperatures enjoyed by western Europe. As the water becomes colder, it increases in salinity and becomes dense, sinking thousands of feet below the surface. The deep water slowly travels south through the oceanic abyss, eventually mixing upward to the surface in different parts of the world up to 1,000 years later. However, global climate changes could alter, or even halt, the current as we know it today. As the Earth heats up, there could be an increase in precipitation and a melting of freshwater ice in the Arctic Ocean (when saltwater freezes it leaves the salt behind), which would flow into the Atlantic Ocean. This additional freshwater could dilute the Atlantic Gulf Stream to the point where it would not continue to sink into the depths of the ocean.

19. **(D)** Crops generally have higher yields when more CO_2 is available. However, while crops may be more productive, the resulting crop will generally be of lower nutritional quality. Nutritional quality declines because while the plants produce more seeds under higher CO_2 levels, the seeds contain less nitrogen. Under the rising CO_2 scenario, livestock and humans would have to increase their intake of plants to compensate for the loss. In a recent study, specific plants grown at higher CO_2 levels had 19% more flowers; 16% more seeds; 4% greater individual seed weight; 25% greater total seed, and a 14% lower concentration of nitrogen in the seeds than those grown at current levels of atmospheric CO_2. On the flip side, legumes increased the amount of nitrogen they take in under higher than normal CO_2 concentration. Changes in the number of flowers, fruits, and seeds and their nutritional quality could have far-reaching consequences as changes in the amount of nutrients in seeds could affect reproductive success and seedling survival. Such changes could also have long-term effects on ecosystem functioning.

20. **(C)** Automobiles and other forms of transportation are responsible for approximately one-third of human-made nitrogen oxides and volatile organic compound emissions, one-fifth of particulate emissions, two-thirds of carbon monoxide emissions, and less than 5% of sulfur dioxide emissions.

21. **(B)** The ecological footprint has emerged as the world's premier measure of humans' demand on nature. It measures how much land and water area a human population requires to produce the resource it consumes and to absorb its carbon dioxide emissions. It currently takes the Earth one year and six months to regenerate what we use in a year.

22. **(C)** Lichens are organisms formed by the mutualistic combination of an alga and a fungus. Lacking roots, stems, and leaves, lichens can grow almost anywhere, but rely on nutrients they accumulate from the air. Lichens appear to function like a natural filter, accumulating airborne pollutants as they are deposited on the lichen surface. Scientists have used them as biomonitors for many years, including an effort to estimate the amount of nuclear fallout from the Chernobyl meltdown in the late 1980s. Using lichens to reflect air pollution levels eliminates the need for installation and maintenance of expensive

air sampling equipment that collects airborne particulates using filters, which are later removed and analyzed in the lab.

23. **(B)** It is thought that improvements to the Erie Canal in 1919 allowed sea lampreys to spread from Lake Ontario to Lake Erie, and eventually spread to Lake Michigan, Lake Huron, and Lake Superior, where it decimated indigenous fish populations in the 1930s and 1940s. Sea lampreys have created a problem with their aggressive parasitism on key predator species and game fish, such as lake trout. The lake trout is considered an apex predator, which means that the entire system relies on its presence to be diverse and healthy. With the removal of an apex predator from a system, the entire system is affected. The sea lamprey is an aggressive predator by nature, which gives it a competitive advantage in a lake system where it has no predators and its prey lacks defenses against it. Lamprey introduction along with poor unsustainable fishing practices caused the lake trout populations to decline drastically. This resulted in the relationship between predators and prey in the Great Lakes' ecosystem becoming unbalanced. Control efforts and research include electric current, chemical lampricides, physical barriers, control through genetic research into the sea lamprey's immune system, and pheromones. Reintroducing animals into a habitat that they once were plentiful in is not an example of introducing an invasive species.

24. **(C)** The Resource Conservation and Recovery Act (RCRA) gives the Environmental Protection Agency (EPA) the authority to control hazardous waste from the "cradle-to-grave." This includes the generation, transportation, treatment, storage, and disposal of hazardous waste. RCRA also sets forth a framework for the management of nonhazardous solid wastes. The 1986 amendments to RCRA enabled the EPA to address environmental problems that could result from underground tanks storing petroleum and other hazardous substances. The Federal Hazardous and Solid Waste Amendments are the 1984 amendments to RCRA that focused on waste minimization and phasing out land disposal of hazardous waste.

25. **(E)** Scientists have identified nearly 2 million individual species, and more than 9 million more species may yet remain undiscovered (the majority of which will be insects). The greatest of all threats to Earth's biodiversity, however, is deforestation. While deforestation threatens ecosystems across the globe, it's particularly destructive to tropical rain forests. In terms of Earth's biodiversity, rain forests are hugely important; though they cover only 7% of the Earth, they house more than half the world's species. Through logging, mining, and farming, humans destroy approximately 2% of the Earth's rain forests every year, often damaging the soil so badly in the process that the forest has a difficult time recovering. In recent history, deforestation has led to approximately 36% of all extinctions, and as the habitat loss accelerates, that number is bound to increase. However, recently in Brazil, satellite imagery revealed that the rate of deforestation fell by 49% compared to the previous year, due in part to stricter environmental regulations and increased enforcement. Recent studies have also shown that as a country's economic conditions improve, its deforestation rate slows considerably as the indigenous populations rely less on the rain forest's resources for survival.

26. **(A)** Cinder cones are one of the most common types of volcanoes. A steep, conical hill of volcanic fragments called cinders accumulates around a vent. The rock fragments, often called cinders, are glassy and contain numerous gas bubbles "frozen" into place as magma explodes into the air and then cools quickly. Cinder cones range in size from tens to hundreds of meters tall and usually occur in groups.

27. **(E)** The U.S. Fish and Wildlife Service is a bureau within the Department of the Interior. As the principal federal partner responsible for administering the Endangered Species Act (ESA), the Fish and Wildlife Service takes the lead in recovering and conserving imperiled species by fostering partnerships, employing scientific research, and developing a workforce of conservation leaders. The Environmental Protection Agency's Endangered Species Protection Program (ESPP) helps promote the recovery of listed species. The ESPP is a program designed to determine whether pesticide use in a certain geographic area may affect any listed species.

28. **(D)** In many areas, fluoride (F^-) is added to water after the disinfection process with the goal of preventing tooth decay, not to disinfect the water. The U.S. Centers for Disease Control and Prevention states that the addition of fluoride to drinking water supplies is "one of 10 great public health achievements of the 20th century." Some areas have excessive levels of natural fluoride in the source water and excessive levels of fluoride can be toxic or cause undesirable cosmetic effects such as staining of teeth. Public opposition to fluoridation has existed since its initiation in the 1940s and stems from perceived safety concerns.

29. **(B)** Remediation technologies are those that render harmful or hazardous substances harmless after they enter the environment.

30. **(C)** Refer to the map below.

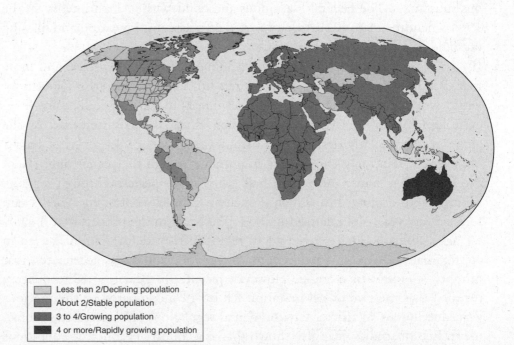

- Less than 2/Declining population
- About 2/Stable population
- 3 to 4/Growing population
- 4 or more/Rapidly growing population

31. **(E)** For the poor in developing countries (urban and rural) wood is usually the principal source of energy for cooking food and for keeping warm. In these countries an estimated 86% of all the wood consumed annually is used as fuel. As populations have grown, this dependence on wood has led inexorably to destruction of the forests. Wood fuels account for two-thirds of all energy sources used in Africa and one-third in Asia. This compares with the 0.3% of total energy use in developed countries. In developing countries, the principal noncommercial fuels other than wood are animal dung and crop residues.

32. **(C)** During the 1940s and 1950s, the Hooker Chemical Company dumped approximately 21,000 tons of organic solvents, acids, and pesticides as well as their by-products, many of them carcinogenic (cancer causing) or teratogenic (creating birth defects), into an abandoned canal in New York State (near Niagara Falls). A school and homes were built over the site. Chemicals began to leak from the ground and cause illness. Since the disaster, various levels of government have spent around $250 million (Superfund) and 20 years cleaning up the site with much of the waste still buried there. New York State has since rebuilt homes in the area at reduced prices to attract new residents.

33. **(D)** From 1932 to 1968, Chisso Corporation (a petrochemical and plastics manufacturer) dumped an estimated 27 tons of mercury compounds into Minamata Bay, Japan. Thousands of people, whose normal diet included fish from the bay, unexpectedly developed symptoms of methyl mercury poisoning. The illness became known as Minamata Disease. Victims were diagnosed as having a degeneration of their nervous systems, numbness occurring in their limbs and lips, their speech becoming slurred, and their vision being constricted; 12,615 people have been officially recognized as patients affected by mercury.

34. **(E)** On March 28, 1979, a minor malfunction occurred in the system that fed water to the steam generators at the Three Mile Island Nuclear Generating Station near Harrisburg, Pennsylvania. This event eventually led to the most serious commercial nuclear accident in U.S. history and caused fundamental changes in the way nuclear power plants were operated and regulated. Despite the severity of the damage, no injuries due to radiation occurred. Eleven days after the events of Three Mile Island, the movie *The China Syndrome* was released—a film about a nuclear accident.

35. **(A)** On the night of December 2–3, 1984, 40 tons of cyanide and other lethal gases began spewing from Union Carbide Corporation's pesticide factory in Bhopal, India. Over half a million people were exposed to the deadly gases. The gases burned the tissues of the eyes and lungs, crossed into the bloodstream, and damaged almost every system in the body. With an estimated 10–15 people continuing to die each month, the number of deaths to date is put at close to 20,000. Today more than 120,000 people are still in need of urgent medical attention.

36. **(B)** On April 26, 1986, a reactor exploded in the town of Chernobyl in the Ukraine and released 30 to 40 times the radioactivity of the atomic bombs dropped on Hiroshima and Nagasaki. Hundreds of thousands of Ukrainians, Russians, and Belorussians had to abandon entire cities and settlements within a 20-mile zone of extreme contamination. Some 3 million people are still living in contaminated areas.

37. **(C)** Drift netting is a fishing technique in which nets, called drift nets, are allowed to drift free in a sea or lake. Usually a drift net is a gill net with floats attached to a rope along the top of the net, and weights attached to another rope along the foot of the net. With gill drift nets, fish try to swim through certain-sized mesh openings but are unable to squeeze through swimming forward. Once in this position, they are prevented from backing out due to the tendency of their gills to become entangled in the net. Drift nets can range in length from 100 feet (30 m) to 2.5 miles (4 km). Nets of up to 30 miles (50 km) have been set in recent times.

38. **(B)** The ozone layer is a layer in Earth's atmosphere that contains relatively high concentrations of ozone (O_3). This layer absorbs 93–99% of the sun's high-frequency ultraviolet light, which is potentially damaging to life on earth. More than 91% of the ozone in Earth's atmosphere is present here. It is mainly located in the lower portion of the stratosphere approximately 42,650–65,620 feet (13–20 km) above Earth, though the thickness varies seasonally and geographically.

39. **(A)** The troposphere begins at the surface of the Earth and extends to between 23,000 feet (7 km) at the poles to 56,000 feet (17 km) at the equator, with some variation due to weather. The troposphere is mostly heated by transfer of energy from the surface, so on average the lowest part of the troposphere is warmest, and temperature decreases with altitude, promoting vertical mixing. The troposphere contains approximately 80% of the mass of the atmosphere.

40. **(B)** Stratospheric ozone depletion describes two distinct but related phenomena observed since the late 1970s: a steady decline of about 4% per decade in the total volume of ozone in the Earth's stratosphere (the ozone layer), and a much larger springtime decrease in stratospheric ozone over Earth's polar regions ("ozone hole"). Effects of decreases in stratospheric ozone resulting in increased UV radiation reaching the Earth include: (a) a 6% to 12% reduction in the growth rate of phytoplankton. An increase in UV radiation also induces DNA damage and DNA synthesis delay in phytoplankton; (b) microorganisms (zooplankton) that feed on phytoplankton are affected if the DNA composition of phytoplankton changes; (c) due to the decreased number of phytoplankton, which is the basis of the food chain in the ocean, krill face a shortage of food, which in turn is the main source of food for various birds and whales in the Antarctic region; (d) increases in UV radiation affects the growth of leaves in plants observed by a reduction in leaf elongation and a reduction in leaf area; and (e) increased UV radiation causes a decrease in photosynthetic activity (photoinhibition).

41. **(A)** Organisms adapted to generalized niches (generalists) are those organisms capable of surviving within a wide range of environmental circumstances with multiple food sources. An example of a generalist would be mice, which can adapt to practically any environment and consume a variety of different seeds, grains, and nuts. Should one food source become scarce, organisms that live in generalized niches have the ability to switch to another. The downside of living in a generalized niche is competition from other species competing for the same resources. Organisms that live in specialized niches on the other hand are not adapted to a wide range of environmental circumstances and food sources. An example of an animal living in a specialized niche would be

the koala, which lives in eucalyptus trees and exclusively consumes eucalyptus leaves. Should anything affect the eucalyptus trees, the viability of the koala population is threatened. The advantage of living in a specialized niche is that there is little interspecific competition for the same resource. Generalists tend to evolve somewhat more quickly than specialists, being put under greater evolutionary pressures.

42. **(D)** China is the dominant recipient of the world's electronic waste (e-waste), with an estimated one billion tons of electronic waste being shipped there each year, mostly from the United States, Canada, and Japan. About 90% of e-waste from the United States is exported to China and Nigeria. E-waste recycling operations are extremely toxic and are usually conducted without protective clothing and with almost 90% of workers suffering from neurological, digestive abnormalities, or skin diseases and many developing breathing problems. Besides skin contact with toxic materials including heavy metals, workers burn plastics coatings to separate them from metals, resulting in inhalation of poisonous gases. Workers also use highly corrosive and dangerous cyanide (heap leaching) and acid baths along riverbanks, which eventually seeps into the groundwater to extract gold from microchips. Studies of China e-waste facilities have shown lead and copper were 371 and 115 times higher, respectively, than established human exposure safety limits.

43. **(B)** Settled sewage liquor is spread onto the surface of a trickling filter (oxidizing) bed, a deep bed made up of coal, limestone chips, or specially fabricated plastic media. Such media must have high surface areas to support the biofilms that form. The sewage liquor is distributed through perforated rotating arms radiating from a central pivot. The distributed sewage liquor trickles through the bed and is collected in drains at the base. These drains also provide a source of air, which percolates up through the bed, keeping it aerobic. Biological films of bacteria, protozoa, and fungi form on the media's surfaces and eat or otherwise reduce the organic content. This biofilm is grazed by insect larvae and worms, which help maintain an optimal thickness. Screening is a primary sewage treatment process. Lagooning, nutrient removal, and constructed wetlands are tertiary sewage treatment processes.

44. **(B)** Almost 60% of all the world's freshwater withdrawals go toward irrigation purposes. Rice production uses the most water; soybeans and oats use the least. Producing electrical power is also a major use of water in the United States. In 1995, 189,700 million gallons of water each day were used to produce electricity—to cool the power-producing equipment.

45. **(A)** In (A), a 1999 study by *Consumer Reports* found that two-thirds of domestic produce had more toxic pesticide residues than those products that were imported. In (B), risks to the human immune, reproductive, and endocrine systems, as well as neurotoxicity, may be equally or even more significant than cancer. (C) is true as written. The legal standard for registration set down by the Federal Insecticide, Fungicide, and Rodenticide Act (FIFRA) is a risk-benefit standard. The EPA must register pesticides if they do not pose "unreasonable risk to man or the environment, taking into account the economic, social and environmental costs and benefits of the use of any pesticide." In (D), 12 of the 26 most widely used pesticides in the United States are classified as possible or probable carcinogens by the EPA based on studies of laboratory

animals. In (E), a study found carcinogens, neurotoxins, and pesticides that disrupt the endocrine or reproductive system in fruits and vegetables that had been washed, peeled, and prepared for consumption. The foods most likely to be contaminated were (in declining order) peaches, apples, celery, potatoes, grapes, and oranges.

46. **(C)** Emissions trading (also known as cap and trade) is an approach used to control pollution by providing economic incentives for achieving reductions in emissions of pollutants. There are active trading programs in several air pollutants. For greenhouse gases, the largest is the European Union Emission Trading Scheme. In the United States there is a national market to reduce acid rain, as well as several regional markets in nitrogen oxides.

47. **(C)** To get the estimate of the population size N, multiply the number marked in the first catch, M_1, by the total number caught in the second catch, C, and divide that by the number of marked recaptures in the second catch, M_2:

$$N = (M_1 \times C) / M_2 = (150 \times 100) / 20 = 750$$

48. **(B)** Population density is calculated as population size divided by area: 750 butterflies / 2 hectares = 375 butterflies per hectare.

49. **(B)** Clear-cutting removes *all* trees from an area in a single cutting. Seed-tree cutting occurs when loggers harvest nearly all of a stand's trees in one cutting but leave a few uniformly distributed seed-producing trees to regenerate the stand. Shelterwood cutting removes all mature trees in an area in two or three cuttings over a period of time. Selective cutting occurs when intermediate-aged or mature trees in an uneven-aged forest are cut singly or in small groups.

50. **(C)** Dams create a series of lakes, slowing the current and delaying downstream migration. The delay interferes with internal biological changes that enable the young salmon to survive in saltwater. In addition, the slow water exposes them to predators and disease. To increase salmon populations, one should release extra water from dams to wash juvenile salmon downstream.

51. **(A)** Today, 3 million fishing boats operate and are greatly depleting the supply of fish and other aquatic life-forms. Over the past 40 years, fishing quotas have more than tripled. In 1950, 20 million tons of fish and marine products were harvested. By 1990, this amount had increased to 100 million tons per year. The depletion of fish stocks has led to overfishing in all oceans and other bodies of water. Past catches were generally higher than those of today due to overfishing. Therefore, to base today's harvest limits on harvest numbers that were probably higher in the past would only accelerate depletion of current fishing stocks.

52. **(C)**

Element	Approximate % by weight
Oxygen	46.6
Silicon	27.7
Aluminum	8.1
Iron	5.0
Calcium	3.6
Sodium	2.8
Potassium	2.6
Magnesium	2.1
All others	1.5

53. **(A)** Smog is a type of air pollution derived from vehicular emission from internal combustion engines and industrial fumes that react in the atmosphere with sunlight to form secondary pollutants that also combine with the primary emissions to form photochemical smog. It is produced by the chemical reaction of sunlight, nitrogen oxides, and volatile organic compounds in the atmosphere. Air pollutants produced during the formation of photochemical smog can include nitrogen oxides (NO_x), peroxyacyl nitrates (PANs), ozone, and volatile organic compounds, all of which are usually highly reactive and oxidizing. It is present in all modern cities, but it is more common in cities with sunny, warm, dry climates and a large number of motor vehicles. Because it travels with the wind, it can affect sparsely populated areas as well. Since there is more sunlight during the summer, photochemical smog is more common during the summer months.

54. **(D)** Virulence is the harm that parasites and diseases cause to their host (e.g., parasite-induced host mortality or reduced fecundity). Parasite virulence is, in general, proportional to the degree that the parasite exploits the host. Parasite offspring are produced by exploiting the host; therefore, some virulence is inevitable. However, too-strong host exploitation leads to high virulence that jeopardizes survival of the host and the parasite itself. Thus, there should be an optimal level of host exploitation and virulence by the parasite.

55. **(D)** To find the doubling time of a population at any given annual rate of growth, divide 72 by the annual growth rate (in this case 1.2): $72 \div 1.2 = 60$ years. $1994 + 60 = 2054$.

56. **(D)** In the United States, roughly two-thirds of all SO_2 and one-fourth of all NO_x come from burning of fossil fuels, especially coal, in electric power plants. Ammonia (NH_3) plays a double role in acidification. First, it neutralizes in the atmosphere to a large extent the acid formed by oxidation of sulfur dioxide and nitrogen oxides, to form particulate ammonium (NH_4^+). Ammonia (NH_3) as such is not an acid but a weak base. However, it does react in the air with strong acids like sulfuric acid or nitric acid, forming slightly acidic ammonium salts, $(NH_4)_2SO_4$ and NH_4NO_3. They are less volatile, form particles, and sink in the end to the ground or rain out. When NH_4^+ is deposited and enters the soil, nitrification can occur. The hydrogen ions from the atmospheric acid that was neutralized by NH_3 in the atmosphere are released and additional acid is formed:

$$NH_4^+ + 2O_2 \rightarrow 2H^+ + NO_3^- + H_2O$$

As a result, NH_4^+ deposition and subsequent nitrification leads directly to soil acidification. Normal rain has a slightly acidic pH of about 5.2, because carbon dioxide and water in the air react together to form carbonic acid, a weak acid (pH 5.6 in distilled water): $H_2O_{(l)} + CO_{2(g)} \rightarrow H_2CO_{3(aq)}$. Carbonic acid then can ionize in water, forming low concentrations of hydronium and carbonate ions: $2H_2O_{(l)} + H_2CO_{3\ (aq)} \rightarrow CO_3^{2-}{}_{(aq)} + 2H_3O^+{}_{(aq)}$. Ozone ($O_3$) oxidizes nitrous oxide (NO) to nitrogen dioxide: $O_{3(g)} + NO_{(g)} \rightarrow NO_{2(g)} + O_{2(g)}$. Nitrogen dioxide can then react with oxygen gas to form nitric acid and nitrous acid, both contributors to acid rain: $2NO_{2(g)} + H_2O_{(g)} \rightarrow HNO_{3(l)} + HNO_{2(l)}$. Lead does not contribute to the formation of acid rain.

57. **(A)** Large-scale circulations develop in Earth's atmosphere due to uneven heating of its surface by the sun's rays. Daytime solar heating is greatest near Earth's equator, where incoming sunlight is nearly perpendicular to the ground. Near the poles, heat lost to space by radiation exceeds the heat gained from sunlight, so air near the poles is losing heat. Conversely, heat gained from sunlight near the equator exceeds heat losses, so air near the equator is gaining heat. The heated air near the equator expands and rises, while the cooled air near the poles contracts and sinks. Rising air creates low pressure at the equator. Air cools as it rises, causing water vapor to condense (rain) as the air cools with increasing altitude. As an air mass cools, it increases in density and descends back to the surface in the subtropics, creating high pressure.

58. **(C)** See the table in Chapter 6, "Populations" that shows differences between *r*-strategists and *K*-strategists.

59. **(B)**

$$\frac{(\text{births} + \text{immigrants}) - (\text{deaths} + \text{emigrants})}{\text{starting population}} \times 100\%$$

$$= \text{net annual percentage growth rate}$$

$$\frac{(1{,}000 + 500) - (600 + 100)}{100{,}000} \times 100\% = 0.8\%$$

60. **(D)** Climax communities are characterized by: stability, high species diversity, low competitive interaction, limited niche overlap, large body size, few offspring per year, one reproductive cycle per year, and *K*-selective species.

61. **(E)** Environmental mitigation is the process of addressing damage to the environment generally caused by public works projects. Environmental mitigation is used to describe projects or programs intended to offset known impacts to an existing natural resource, such as a forest, stream, wetland, or endangered species. Environmental mitigation typically involves a trading system: Debits are recorded when a natural resource has been destroyed or severely impaired, and credits are given when a natural resource has been deemed to be improved or preserved. Actions taken to avoid or minimize environmental damage are considered the most preferable method of mitigation. Potential environmental mitigation activities may include minimizing a proposed activity; restoring temporary impacts; precautionary and/or abatement measures to reduce construction impacts; and providing a suitable replacement or substitute environmental resource of equivalent or greater value.

62. **(C)** Carbon dioxide is the ultimate result of perfect combustion of any carbon-based fuel. The only ways to reduce carbon dioxide emissions are to make vehicles more fuel efficient and/or drive less, to use a noncarbon fuel such as hydrogen, or to use a green fuel such as ethanol that is produced from crops that absorb carbon dioxide as they grow.

63. **(B)** The median lethal dose, LD_{50}, of a toxic substance is the dose required to kill half the members of a tested population after a specified test duration and is frequently used as a general indicator of a substance's acute toxicity.

$$\frac{400 \, \text{mg}}{4 \, \text{kg}} \times \frac{1 \, \text{g}}{1{,}000 \, \text{mg}} = 0.1 \, \text{g/kg}$$

64. **(E)** The Kyoto Protocol is aimed at reducing global warming. The Protocol was initially adopted in 1997 in Kyoto, Japan, and entered into force in 2005. The United States did not agree to the Protocol, and in 2011, Canada withdrew. The Kyoto Protocol contains mandatory aims for the countries that have signed it. The Protocol commits Annex I countries (developed countries and countries in economic transition) to individual, legally-binding commitments on the reduction of their greenhouse gas emissions. The general target that the developed countries have to meet is to reduce their greenhouse gas emissions by approximately 5% below their 1990 levels between 2008–2012. The Kyoto Protocol allows for flexibility in terms of the methods countries could use to meet their gas reduction commitments. Such methods include: (1) compensating for emissions by increasing the number of a country's carbon sinks; (2) trading of emission allowances between countries; (3) promote environmentally-friendly foreign investments from industrialized countries into developing countries; and (4) allow developed countries to sponsor research to decrease emission levels in countries of economic transition. In exchange for the developed country's investment, the host country provides the investor with emission reduction units, also known as carbon credits. The developed economies can afterwards use their carbon credits towards meeting their emission-reduction requirements.

65. **(A)** Sediment is the largest contaminant of surface water by weight and volume in the United States, and is identified as the leading pollution problem in rivers and streams and the fourth leading problem in lakes. Sediment in surface water is largely a result of soil erosion, which is influenced by soil properties and agricultural practices. Sediment buildup reduces the useful life of reservoirs, can clog irrigation canals, block navigation channels, and increase dredging costs. By raising streambeds and burying streamside wetlands, sediment increases the probability and severity of floods. Suspended sediment can increase the cost of water treatment for municipal and industrial water uses. Sediment can also destroy or degrade aquatic wildlife habitat, reducing diversity and damaging commercial and recreational fisheries.

66. **(C)** Municipal Solid Waste (MSW)—more commonly known as trash or garbage—consists of everyday items we use and then throw away, such as product packaging, grass clippings, furniture, clothing, bottles, food scraps, newspapers, appliances, paint, and batteries. In 2010, Americans generated about 250 million tons of trash and recycled and composted over 85 million tons of this material, equivalent to a 34% recycling rate. On average, Americans recycled and composted 1.5 pounds from the 4.43 pounds of MSW generated per person per day.

67. **(A)** Control rods regulate the number of neutrons in the core of a nuclear reactor and so control the rate of the reaction. They are plates or tubes holding material, such as cadmium or boron, that absorb neutrons.

68. **(B)** A moderator is a medium that reduces the speed of fast neutrons, thereby turning them into thermal neutrons capable of sustaining a nuclear chain reaction. Commonly used moderators include regular (light) water (roughly 75% of the world's reactors), solid graphite (20% of reactors), and heavy water (5% of reactors).

69. **(C)** A moderator is a substance, such as water or graphite, that is used in a nuclear reactor to decrease the speed of fast neutrons and increase the likelihood of fission. The reactor control system requires the movement of neutron-absorbing rods (control rods) in the reactor under carefully controlled conditions. They must be arranged to increase reactivity (increase neutron population) slowly and under good control. They must be capable of reducing reactivity, both rapidly and slowly.

70. **(B)** Satellite imaging of the concentrations of PM_{10} air particulates (μg per m^3) from NASA show that cities in China have the highest air pollution concentrations in the world. The World Health Organization (WHO) upper safety level of PM_{10} is 20 μg per m^3. The city with the highest average PM_{10} concentration in the world is Ahvaz, Iran, with a concentration of 372 μg per m^3.

71. **(C)** $N_1 = N_0 + B - D + I - E$

    $$N_1 = 40,000 + 35(40) - 10(40) + 100 - 50$$

    $$N_1 = 41,050$$

72. **(D)** Biodiversity is the degree of variation of life forms with a given species, ecosystem, or biome and is a measure of the health of ecosystems. In terrestrial habitats, tropical and temperate regions typically support a large variety of different species whereas polar regions with climatic conditions not being optimal for producers, support fewer species. Biodiversity is not evenly distributed; rather it varies greatly across the globe as well as within regions. Among other factors, the diversity of all living things (biota) depends on temperature, precipitation, altitude, soils, geography, and the presence of other species. Generally, there is an increase in biodiversity from the poles to the tropics. Thus localities at lower latitudes have more species than localities at higher latitudes.

73. **(B)** Of more than 50,000 edible plant species in the world, only a few hundred contribute significantly to food supplies. Just fifteen crop plants provide 90% of the world's food energy intake, with three—rice, maize, and wheat—making up two-thirds of this. These three are the staples of over 4 billion people. Roots and tubers are important staples for over 1 billion people in the developing world. They account for roughly 40% of the food eaten by half the population of sub-Saharan Africa. They are high in carbohydrates, calcium, and vitamin C, but low in protein. One example, cassava has become one of the developing world's most important staples providing a basic diet for around 500 million people, and plantings are increasing faster than for any other crop. Cassava's starchy roots produce more food energy per unit of land than any other staple crop. Nutritionally, the cassava is comparable to potatoes, except that it has twice the fiber content and a higher level of potassium.

74. **(D)** Everyday, 3,000 Americans under the age of 18 become regular smokers, and a third of them will eventually die of nicotine-related causes. Two out of three 12 to 17 year olds who smoked nicotine in the last year show signs of addiction. Everyday, over 1,000 Americans die as a result of nicotine addition. In 1900, 4 billion cigarettes were produced in the United States. By the year 2000, the number had increased to 720 billion.

75. **(B)** Nitrification is a microbial process by which reduced nitrogen compounds (primarily ammonia) are sequentially oxidized to nitrite and nitrate.

The nitrification process is primarily accomplished by two groups of autotrophic nitrifying bacteria that can build organic molecules using energy obtained from inorganic sources, in this case ammonia or nitrite. In the first step of nitrification, ammonia-oxidizing bacteria (*Nitrosomonas*) oxidize ammonia to nitrite $NH_3 + O_2 \rightarrow NO_2^- + 3H^+ + 2e^-$. In the second step of the process, nitrite-oxidizing bacteria (*Nitrobacter*) oxidize nitrite to nitrate $NO_2^- + H_2O \rightarrow NO_3^- + 2H^+ + 2e^-$.

76. **(E)** The mantle is the thick layer of hot, solid rock between the crust and the molten iron core. The mantle makes up the bulk of the Earth, accounting for two-thirds of its mass. The mantle starts about 19 miles (30 km) down and is about 1,800 miles (2,900 km) thick.

77. **(B)** The lynx and hare populations have a predator-prey relationship. Disease, food supply, and other predators are variables in this complex relationship. The flux in this cyclic relationship is what allows the ecosystem dynamic to work. Without flux, vegetation wouldn't have a chance to recover from the hare population's continuous eating, and without vegetation, the hare population could no longer exist in its habitat, and therefore neither could the lynx population that depends upon the hare population for food. Every ten years or so, the hares' reproduction rate increases. As more hares are born, they eat more of their food supply. They eat so much food that they are forced to supplement their diet with less desirable and nutritious food. As the hare population grows, the lynx population begins to increase in response. Because there are so many hares, other predators opportunistically begin to hunt them along with the lynxes. The hares' less nutritious and varied diet begins to have an effect; the hares begin to die due to illness and disease. Fewer hares are born because there is less food. The hare population begins to go into a steep decline. Therefore, the lynx population also begins to decline. Some lynxes starve and others die due to disease. Both the lynx and hare populations have fewer offspring, and this decrease in population gives the vegetation a chance to recover. Once there is enough vegetation for the hares to begin to increase their population, the whole cycle begins again.

78. **(C)** The first law of thermodynamics states that energy can be changed from one form to another but it cannot be created or destroyed. The total amount of energy and matter in the universe remains constant, merely changing from one form to another. The second law of thermodynamics states that in all energy exchanges, if no energy enters or leaves the system, the potential energy of the state will always be less than that of the initial state. This is also commonly referred to as entropy.

79. **(C)** Less-developed countries typically have a large proportion of their population in the pre-reproductive age category.

80. **(C)** The term *territory* refers to any area that an animal of a particular species consistently defends against members of its own species (and occasionally against animals of different species). Animals that defend territories in this way are referred to as territorial. For a songbird to expend energy defending a territory larger than what is required to derive adequate resources would not be energy-efficient. To defend a territory smaller than what is required to derive adequate resources puts the songbird and the clutch in jeopardy.

81. **(C)** Although communities are variable and dynamic in both space and time, the interactions between species lead to patterns of community structure

that are characteristic of particular community types. For example, North American forests have vertical layers—canopy, midstory, and understory—that are composed of different life forms: trees, shrubs, herbs, and so on. Midwestern prairies have species that are active in the spring and fall (cool-season species) and others that are active in midsummer (warm-season species). In the northeast, many communities have some species that bloom in spring, some in summer, and some in fall. These other structures are thought to be a result of resource partitioning. This evolutionary strategy allows species that are potential competitors for a resource (space, light, pollinators) to coexist by specializing in different aspects of the resource.

82. **(C)** Selling pollution permits is essentially the same as charging each firm for every ton that they emit. Essentially, it is charging them in advance. If the pollution permit costs more than abatement, then they will abate. If it costs less, they will buy the permit and continue to pollute. In (A), the copper smelter would have to spend $450 and the electric utility would have to spend $250. In (B), the minimum each company would pay would be $251. In (C), each firm would be charged $151. In (D), halting production of an entire plant would be prohibitive.

83. **(C)** Of the organisms listed, the hawks occupy the highest trophic level as carnivorous consumers. Biomagnification, also known as bioamplification or biological magnification, is the increase in concentration of a substance that occurs in a food chain as a consequence of: (a) persistence (can't be broken down by environmental processes); (b) food chain energetics; and/or (c) low (or nonexistent) rate of internal degradation/excretion of the substance (often due to water insolubility). Lipid, (lipophilic) or fat soluble substances cannot be diluted, broken down, or excreted in urine, a water-based medium, and so accumulate in fatty tissues of an organism if the organism lacks enzymes to degrade them. When eaten by another organism, fats are absorbed in the gut, carrying the substance, which then accumulates in the fats of the predator. Since at each level of the food chain there is a lot of energy loss, a predator must consume much prey, including all of their lipophilic substances.

84. **(E)** Negative feedback causes the initial change to be reduced, slowing down change and helping to regulate the system and keep it in balance. If it is cold outside, the internal temperature of the house eventually drops as cold air seeps in through the walls. When the temperature drops below the point at which the thermostat is set, the thermostat turns on the furnace. As the temperature within the house rises, the thermostat again senses this change and turns off the furnace when the internal temperature reaches the pre-set point. Choices A, B, and D represent positive feedback loops.

85. **(D)** Refer to "Demographic Transition" in Chapter 6.

86. **(A)** Carbon dioxide and nitrous oxide levels are also linked to industry, automobile usage, and the use of fossil fuels to generate power at electrical plants. Nitrous oxide emissions can also be reduced by decreasing the amount of nitrogen-based fertilizers.

87. **(E)** Heap leaching is an industrial mining process that is used to extract copper, nickel, gold, and uranium, and other compounds from ore. The advantage of heap leaching is its relatively low operational costs and the ability to extract valuable mineral resources from what otherwise would be low grade ores. The

mined ore is usually crushed into small chunks and heaped on an impermeable plastic and/or clay lined leach pad where it can be irrigated with a leach solution to dissolve the valuable metals. The environmental consequences of the heap leaching processes are cyanide leaks and infiltration of cyanide and other toxins into the water supply. The production of one gold ring through the heap leaching method can generate up to 20 tons of waste material.

88. **(C)** Solids refer to matter suspended or dissolved in water or wastewater. Solids may affect water or effluent quality in a number of ways. Water with high dissolved solids is generally of inferior palatability and may cause health problems. Highly mineralized waters are unsuitable for many industrial applications. High suspended solids content can also be detrimental to aquatic plants and animals by limiting light and deteriorating habitat. Total dissolved solids are the amount of filterable solids in a water sample. To calculate total suspended solids:

Let A = weight of crucible + filter + residue after 24 hrs at 105°C (mg)
Let B = weight of crucible + filter (mg)

$$\text{Total suspended solids} = \frac{(A-B) \times 1{,}000 \text{ mL/L}}{\text{sample volume (mL)}} \times \frac{1{,}000 \text{ mg}}{\text{g}}$$

$$= \frac{(1{,}000 \text{ g} - 80.00 \text{ g}) \times 1{,}000 \text{ mL/L}}{100.00 \text{ mL}} \times \frac{1{,}000 \text{ mg}}{\text{g}}$$

89. **(E)** In a food chain, only about 10% of the energy is transferred from one level to the next. The other 90% is used for respiration, growth, digestion, escaping from predators, maintaining body heat, reproduction, etc. Plant (1,000) → mouse (100 available) → snake (10 available) → bird (1 available) → carnivore (0.1 available).

90. **(E)** A green manure is a type of cover crop grown primarily to add nutrients and organic matter to the soil. Typically, a green manure crop is grown for a specific period of time, and then plowed under and incorporated into the soil while green or shortly after flowering. Green manure crops are commonly associated with organic agriculture, and are considered essential for annual cropping systems that wish to be sustainable. In addition to the advantages covered in the question, green manures also function in weed suppression and prevention of soil erosion. Traditionally, the practice of using green manure can be traced back to the fallow cycle of crop rotation, which was used to allow soils to recover.

91. **(C)** Phosphates and nitrates are required for photosynthesis and vary in concentration due to biological activity. In surface waters, where plants are actively involved in the process of photosynthesis, nitrates and phosphates can be in short supply, limiting the amount of biological activity that can take place. The cycling of materials (in this case, nitrates and phosphates) is the function of organisms of decay. Microscopic bacteria in seawater can contain up to 1,000,000 bacterial cells per cubic centimeter.

92. **(D)** Ocean temperatures increase more slowly than land temperatures because of the larger effective heat capacity of the oceans and because the oceans lose

more heat by evaporation. Heat capacity tells you how much heat you can put in before the temperature has risen by 1 degree. If something has a large heat capacity, then more heat can be absorbed before it rises in temperature by 1 degree. The heat capacity of ocean water is about four times that of air. Land temperatures have increased about twice as fast as ocean temperatures (0.25°C per decade for land vs. 0.13°C per decade for oceans). The Northern Hemisphere warms faster than the Southern Hemisphere because there is more land in the Northern Hemisphere.

93. **(E)** Nitrous oxide is emitted by bacteria in soils and oceans and has 300 times more impact as a greenhouse gas than carbon dioxide. Nitrous oxide is the fourth largest contributor to greenhouse gases ranked behind water vapor, carbon dioxide, and methane.

    Radon (Rn) is a radioactive element and occurs naturally as the decay product of uranium and thorium. Radon sticks to dust particles in the air. If contaminated dust is inhaled, these particles can stick to the airways of the lung and increase the risk of developing lung cancer. Radon is responsible for the majority of the public exposure to radiation. Radon gas from natural sources can accumulate in buildings, especially in confined areas such as attics and basements. It can also be found in some spring waters and hot springs. According to the Environmental Protection Agency (EPA), radon is the second most frequent cause of lung cancer, after cigarette smoking, causing 21,000 lung cancer deaths per year in the United States.

    Sulfur dioxide (SO_2) is a poisonous gas with a pungent, irritating smell that is naturally released by volcanoes. Further oxidation of SO_2, usually in the presence of a catalyst such as NO_2, forms H_2SO_4, and thus acid rain. Sulfur dioxide emissions are also a precursor to particulates in the atmosphere.

    Formaldehyde (CH_2O) is found in plants, fruits, vegetables, animals (including humans), and seafood. The highest concentrations of formaldehyde are found in dried shiitake mushrooms (up to 400 mg formaldehyde per kg of mushroom). Formaldehyde is essential in human metabolism and is required for the synthesis of DNA and amino acids. Human blood contains ~2.5 µg of formaldehyde per ml of blood.

94. **(A)** Primary waves (P-waves) travel faster than other waves through the Earth and arrive at seismograph stations first hence the name "Primary." These waves can travel through any type of material, including fluids, and can travel at nearly twice the speed of S waves. Secondary waves (S-waves) arrive at seismograph stations after the faster moving P waves during an earthquake and displace the ground perpendicular to the direction of propagation. S waves can travel only through solids. S waves are about 60% slower than P waves. Surface waves are analogous to water waves and travel along the Earth's surface. Because of their low frequency, long duration, and large amplitude, they can be the most destructive type of seismic wave. They are called surface waves because they diminish as they get further from the surface.

95. **(D)** The more ozone in a given parcel of the stratosphere, the more heat it retains. Ozone generates heat in the stratosphere, both by absorbing the sun's ultraviolet radiation and by absorbing upwelling infrared radiation from the lower atmosphere (troposphere). Consequently, decreased ozone in the stratosphere results in lower stratospheric temperatures and increased UV radiation

reaching the surface of Earth. Observations show that over recent decades, the mid to upper stratosphere (from 30 to 50 km above Earth's surface) has cooled by 2°F to 11°F (1°C to 6°C). This stratospheric cooling has taken place at the same time that greenhouse gas amounts in the lower atmosphere (troposphere) have risen.

96. **(B)** In primary succession, which occurs in virtually a lifeless area (e.g., area below a retreating glacier), pioneer species like lichen, algae, and fungus, as well as other abiotic factors like wind and water, start to change or create conditions better suited for vascular plant growth. These changes include accumulation of organic matter in litter or humic layer, alteration of soil nutrients, and change in pH or water content of soil. These pioneer plants are then dominated and often replaced by plants better adapted to less austere conditions, such as grasses and shrubs that are able to live in thin, mineral-based soils. The structure of the plants themselves can also alter the community. For example, when larger species like trees mature, they produce shade on the developing forest floor, which tends to exclude light-requiring species. Shade-tolerant species will then invade the area.

97. **(D)** Most indoor air pollution experts agree that carpet should be avoided whenever feasible. The reason for this is that carpet is made up of some 120 different chemicals, many of which can cause health problems. Once installed, carpet can collect dust and even lead (tracked in from shoes) and can grow mold and dust mites. Methylene chloride is given off by paint thinners and paint strippers. Methylene chloride has a fairly long atmospheric half-life, 3 to 4 months, indicating the fairly long persistence typical of transported pollutants.

98. **(A)** Uniform distribution means that organisms of a population are generally equally spaced apart. In most situations, a natural uniform dispersion occurs when organisms that comprise a population compete for a common resource. In (A), red squirrels might be found anywhere within their geographic territory, defending it. In (B), cattails would be found clumped in only one area (near the water). In (C), dwarf mistletoe would only be found where a particular species of forest tree is growing. In (D), moths would have a higher density near lighted areas rather than in darker areas, as most moths are attracted by light. In (E), the lake trout would be found more in the deeper areas of the water.

99. **(B)** A density-independent factor is one where the effect of the factor on the size of the population is independent of and does not depend upon the original density or size of the population. The effect of weather is an example of a density-independent factor. A severe storm and flood coming through an area can just as easily wipe out a large population as a small one.

100. **(D)** High-quality energy is capable of performing a large amount of work, while low-quality energy is capable of performing less work. Energy always changes from high to low quality when work is performed. During the change, some energy is lost in the form of heat, which cannot do work (second law of thermodynamics). The amount of energy lost as heat is often as high as 90% of the total energy involved. The reason that (D) is the answer is because of the word "dispersed." Dispersed energy is not concentrated, therefore it is of low quality. An analogy would be ore deposits. A rich gold mine has a lot of gold in concentrated form. A stream might also have a lot of gold (maybe even more than the mine). However, if there are only a few small nuggets every few hundred feet, it is not as rich.

FREE-RESPONSE EXPLANATIONS

Question 1

(a) 3 points maximum

(i) 1 point

Restatement: Number of cubic feet of natural gas required to heat the house for the winter.

$$4{,}000 \; \cancel{\text{square feet}} \times \frac{100{,}000 \; \text{Btus}}{1 \; \cancel{\text{square foot}}} = \mathbf{400{,}000{,}000 \; Btus}$$

(ii) 1 point

Restatement: Cost of heating the home for one winter using natural gas.

$$400{,}000{,}000 \; \cancel{\text{Btus}} \times \frac{1 \; \cancel{\text{cubic foot natural gas}}}{1{,}000 \; \cancel{\text{Btus}}} \times \frac{\$5.00}{1{,}000 \; \cancel{\text{cubic feet}}} = \mathbf{\$2{,}000}$$

(iii) 1 point

Restatement: Cost of heating the home for one winter using electricity.

$$400{,}000{,}000 \; \cancel{\text{Btus}} \times \frac{1 \; \cancel{\text{kWh}}}{10{,}000 \; \cancel{\text{Btus}}} \times \frac{\$50}{500 \; \cancel{\text{kWh}}} = \mathbf{\$4{,}000}$$

NOTE:
1. If you do NOT show calculations, no points are awarded.
2. No penalty assessed if you do not show units. However, you risk setting up the problem incorrectly if you do not show units so that they cancel properly.
3. If your setup is correct but you make an arithmetic error, no penalty is assessed.

NOTE: For Parts (b) and (c), there are no points for just listing ideas. Each idea MUST be explained. The following are ideas that you could use to write your paragraph(s). Take these ideas, and create an outline of the order in which you wish to answer them. You do NOT need to use all ideas. Before you begin, decide on the format of how you wish to answer your question (pros versus cons, chart format with explanations within the chart, compare and contrast, etc.).

(b) 4 points maximum

(i) (1 point definition) In active solar heating applications, heat from the sun is collected, stored, and used primarily for domestic hot water and space heating. The reason the system is called active is because pumps and fans are used to transfer the captured heat to an area where it can be stored or used. The main components of an active solar system are the collectors, the collector controls, the storage tank, and the distribution system.

Active Solar System (1 point) Choose ideas from the list below:

* Active liquid solar systems heat water or an antifreeze solution in a collector. Liquid solar collectors are most appropriate for central heating. Flat-plate collectors are the most common. In the collector, a heat transfer fluid such as a water or antifreeze absorbs the solar heat. At the appropriate time, a controller operates a circulating pump to move the fluid through the collector. The heated liquid flows to either a storage tank or a heat exchanger for immediate use.
* Air collectors produce heat earlier and later in the day than liquid systems. Therefore, air systems may produce more usable energy over a heating season than a liquid system of the same size. Solar air collectors have a black metal plate for absorbing heat, which in turn heats air in the collector. A fan then blows this heated air into the house. Air collectors can be installed on a roof or on a south-facing exterior wall. These systems are easier and less expensive to install than a central heating system. They do not have a dedicated storage system or extensive ductwork. The floors, walls, and furniture will absorb some of the solar heat, which will help keep the room warm for a few hours after sunset. Masonry walls and tile floors will also provide more thermal mass and thus provide heat for longer periods. A well-insulated house will make a solar room air heater more efficient.

(ii) (1 point definition) In passive solar building design, windows, walls, and floors are made to collect, store, and distribute solar energy in the form of heat in the winter and reject solar heat in the summer. This is called passive solar design or climatic design because, unlike active solar heating systems, it doesn't involve the use of mechanical and electrical devices. The key to designing a passive solar building is to best take advantage of the local climate. Elements to be considered include window placement and glazing type, thermal insulation, thermal mass, and shading.

Passive Solar System (1 point definition) Choose ideas from the list below:

* There are two basic approaches to passive solar systems. First, direct gain—solar energy is transmitted through south-facing windows. Works best when the south window area is double-glazed and the building has considerable thermal mass in the form of concrete floors and masonry walls. Second, indirect gain in which a storage mass collects and stores heat directly from the sun and then transfers the heat to the living space(s).

(iii) Choose ONE idea from the list below: (1 point)

Advantages

- Reduces the use of fossil fuels and their emission of greenhouse gases.
- Solar energy is a renewable energy resource.
- The financial costs of solar systems are falling as technology develops.
- Fossil fuels, being a nonrenewable energy source, are running out. As they become more scarce, prices will rise, making solar energy more cost effective in the future.
- Suitable for remote areas that are not connected to energy grids or pipelines.
- Zero risk of indoor pollution such as carbon monoxide.

Disadvantages

- Higher initial cost of equipment than conventional heating systems.
- Clouds may reduce effectiveness.
- A solar energy installation requires a large area for the system to be efficient in providing a source of electricity. This may be a disadvantage in areas where space is short, or expensive.
- The location of solar panels can affect performance, due to possible obstructions from the surrounding buildings or landscape.

(c) 3 points maximum

(i) (1 point definition): A green building focuses on a whole-system approach, including energy conservation, resource-efficient building techniques and materials, indoor air quality, water conservation, and designs that minimize waste while utilizing recycled materials. Green buildings are a product of a good design that minimizes a building's energy needs while reducing construction and maintenance costs over the life cycle of a building.

(ii) (2 points) Choose any TWO ideas from the following:

- Solar collectors for space heating.
- Solar collectors for water heating.
- Photovoltaics to supply electrical energy.
- Hybrid systems that incorporate more than one power source such as wind and solar.
- Use of energy efficient (Energy Star) appliances.
- Products made from environmentally sustainable materials.
- Products that do not contain toxins (ex: PVC pipes, ozone-depleting chemicals, formaldehyde in plywood, paints and stains with VOCs).
- Products that reduce the environmental impact on building operations (ex: new energy saving thermostats, newer landscape irrigation systems that only activate based on soil moisture content).
- Products that remove or warn of indoor pollution hazards (ex: carbon monoxide detectors, air filtration systems, water purifying systems).
- Homes that meet or exceed insulation requirements.
- Green landscaping—landscaping that provides shade during the summer and allows sunlight to warm the house during winter and the use of plants that do not require large amounts of water.

Question 2

(a) Define "aquafarming" (1 point)

Aquafarming is the commercial farming of freshwater and saltwater fish, mollusks, crustaceans, and aquatic plants. Also known as aquaculture, it involves cultivating aquatic populations under controlled conditions.

(i) Describe TWO advantages of aquafarming over traditional wild harvesting. (2 points)

In aquaculture, the life cycle of organisms occurs under controlled conditions. Advantages of having control over the life cycle of organisms include: (1) more intensive farming is possible, which often results in higher yields and greater profits; (2) more uniformity in the product since environmental conditions are controlled and managed, which results in less waste and higher profits; (3) predator control; (4) the ability to accelerate growth and maturation by controlling the climate, especially on farms located in more temperate zones; and (5) most traditional expenses and variables inherent to wild harvesting (the costs of boats, crews, nets, weather conditions, searching for areas with enough of the species that can be captured to be profitable, etc.) are minimized or eliminated.

(ii) Describe TWO negative environmental impacts of aquafarming. (2 points)

Fish waste is organic and composed of nutrients necessary in all components of aquatic food webs. Aquaculture often produces much higher-than-normal fish waste concentrations, as the habitat is confined. The waste collects on the ocean bottom, damaging or eliminating bottom-dwelling life. Waste can also decrease dissolved oxygen levels in the water column, putting further pressure on other organisms. Waste products from aquafarming are often discharged untreated directly into the surrounding aquatic environment and frequently contain antibiotics and pesticides.

Growers often supply their animals with antibiotics to prevent disease, which can accelerate the evolution of bacterial resistance.

Other acceptable answers include:
- *Fish can escape from coastal pens, where they can interbreed with their wild counterparts, diluting wild genetic stocks.*
- *Escaped fish can become invasive, outcompeting native species.*
- *Aquaculture is becoming a significant threat to coastal systems; e.g., about 20% of mangrove forests worldwide have been destroyed since 1980, partly due to shrimp farming.*

(b) Describe TWO methods that can be employed in aquaculture to lessen the environmental impact. (2 points)

Onshore recirculating aquaculture facilities using polyculture techniques and properly sited facilities (e.g., offshore areas with strong currents) can minimize the negative environmental effects.

Other acceptable methods include:

- *Formulate coastal aquaculture development and management plans.*
- *Formulate integrated coastal zone management plans.*
- *Assess the capacity of the local ecosystem to sustain aquaculture development with minimal ecological change, and establish a permit system based on the local ecosystem's capacity for aquafarming.*
- *Establish guidelines governing the use of wetlands and mangrove forests.*
- *Establish guidelines for the use of bioactive compounds (pesticides, hormones, pharmaceuticals, herbicides, etc.)*
- *Assess and evaluate the true consequences of transfers and introduction of exotic organisms.*
- *Regulate discharges from land-based aquaculture through the enforcement of effluent standards.*
- *Establish quality control measures for aquaculture products.*
- *Apply incentives and deterrents to reduce environmental degradation from aquaculture activities.*
- *Monitor water quality using established protocols for any signs of ecological change.*

(c) How is raising and harvesting kelp and bivalve mollusks a more sustainable method of aquafarming than raising and harvesting higher trophic order fish such as salmon? (1 point)

Raising seaweed and filter-feeding bivalve mollusks, such as oysters, clams, mussels, and scallops, has a very low impact on the environment and may even be environmentally restorative, as these filter-feeders filter pollutants and excessive nutrients from the water, often improving water quality. Kelp extract nutrients such as inorganic nitrogen and phosphorus directly from the water, while filter-feeding mollusks can extract nutrients as they feed on particulates such as phytoplankton and detritus.

(d) The newspaper announcement claims that aquafarming is the answer to world hunger. Give TWO examples of how this statement may be contrary to environmental sustainability. (2 points)

Salmon farming currently involves a high demand for wild forage fish for feed, as well as their by-products—fish meal and fish oil. As carnivores high on the food chain, salmon require a lot of protein, and farmed salmon consume more protein than they produce; e.g., each pound of farmed salmon requires up to 6 pounds (2 kg) of wild fish. About 75% of the world's monitored fisheries are already near to or have exceeded their maximum sustainable yield. The industrial-scale extraction of wild forage fish for salmon farming also impacts the survivability of the wild predator fish that rely on them for food.

Question 3

(a) Describe the role and importance of stratospheric ozone to life on Earth. (1 point)

The ozone (O_3) layer, located in the lower portion of the stratosphere, absorbs up to 99% of the sun's high-frequency ultraviolet radiation, which is damaging to life on Earth. About 90% of the atmospheric ozone resides in a layer approximately 6–30 miles (10–50 km) above the Earth's surface, in the region of the atmosphere known as the stratosphere. This stratospheric ozone is commonly referred to as the "ozone layer." The remaining ozone is in the lower region of the atmosphere, the troposphere, which extends from the Earth's surface up to about 6 miles (10 km). Stratospheric ozone plays a beneficial role by absorbing most of the biologically damaging ultraviolet sunlight (UVB and UC) before it reaches the Earth's surface. The absorption of ultraviolet radiation by ozone in this zone creates a source of heat in the stratosphere that increases with altitude. Ozone thus plays a key role in the temperature structure of the Earth's atmosphere. Without the filtering action of the ozone layer, more of the sun's UVB and UVC radiation would penetrate the atmosphere and reach the Earth's surface.

(b) Suggest ONE possible reason for the appearance of the hole in the ozone layer, which was first observed in the 1980s. (1 point)

Scientific evidence has shown that human-produced chemicals are most likely responsible for the observed depletion of the ozone layer. CFCs (chlorofluorocarbons) were the first mass-produced chemical compounds shown to destroy ozone. CFCs were primarily used in refrigeration and air-conditioning (e.g., Freon) and in the dry cleaning industry (e.g., carbon tetrachloride). Later, halocarbons, chemical compounds similar to CFCs but which contain bromine, were also shown to destroy ozone molecules. Halocarbons were used in foam blowing, soil fumigants and pesticides (e.g., methyl bromide), fire retardants and extinguishers, and as a solvent in producing circuit boards (e.g., methyl chloroform).

(c) Ultraviolet (UV) radiation occurs in three forms, known as UVA, UVB, and UVC.

(i) Describe at least ONE characteristic of EACH form of ultraviolet radiation. (3 points)

The three types of UV radiation are classified according to their wavelength. They differ in their biological activity and the extent to which they can penetrate the skin; i.e., the shorter the wavelength, the more harmful the UV radiation. However, shorter wavelength UV radiation is less able to penetrate the skin. Short-wavelength UVC is the most damaging type of UV radiation. However, it is completely filtered by the atmosphere and does not reach the Earth's surface. Medium-wavelength UVB is very biologically active but cannot penetrate beyond the superficial skin layers. Most solar UVB is filtered by the atmosphere. The relatively long-wavelength UVA accounts for approximately 95% of the UV radiation reaching the Earth's surface.

(ii) Describe the effects that increased UV radiation would have on:

(a) Human health (1 point)

Changes in the skin's DNA due to UVA and UVB rays can cause serious, long-term skin damage. UVA and UVB radiation can penetrate into the deeper layers of the skin and is responsible for the immediate tanning effect. Furthermore, it also contributes to skin aging and wrinkling, and recent studies suggest that it may also enhance the development of skin cancers. One effect of UV exposure is that it advances the signs of aging. Premature wrinkling is common in people who have been exposed to the sun over long periods of time, as are age spots and uneven complexions. UV radiation can also cause cataracts and weaken the immune system.

(b) Terrestrial ecosystems (1 point)

Excessive exposure to UV radiation can cause cancers in animals and damage their eyesight. Increase UV exposure can harm DNA and proteins, and affect organisms in their developmental stages. These direct effects may lead to indirect effects, such as decreased primary productivity, changes in biodiversity, decreased nitrogen uptake by microorganisms, and reduced capacity for oceans to fix carbon dioxide. Research on the effects of increased UVB exposure in higher trophic levels has shown reduced reproductive capacity, growth, and survival rates. Experiments on certain food crops (e.g., rice and soybeans) have shown lower crop yields when exposed to higher levels of UVB radiation. The plants minimize their exposure to increased UV radiation by limiting the surface area of their foliage, which in turn impairs growth.

(c) Marine ecosystems (1 point)

Phytoplankton, which are located at the base of the aquatic food pyramid, account for approximately 30% of the world's intake of animal protein. Phytoplankton productivity is restricted to the upper layer of the water, where sufficient light is available. A small increase in UVB exposure brought about by less stratospheric ozone could significantly reduce the size of plankton populations, affecting the environment in several ways: (1) with less organic matter in the upper layers of the water, UV radiation can penetrate deeper into the water and affect more complex plants and animals living there; (2) UV radiation directly damages animals during their early development; (3) pollution of the water by toxic substances may synergistically magnify the adverse effects of UV radiation, working its way up the food chain; (4) less plankton means less food for the animals that prey on them and a reduction in fish stocks already depleted by overfishing; (5) decreased bacterial and plankton activity may lead to an increase in dissolved organic matter in ocean waters, as assimilation of dissolved organic matter is reduced. Cyanobacteria, organisms important in nitrogen fixation, are also at risk. Cyanobacteria transform dissolved nitrogen in ocean water to nitrates and other forms that are accessible by higher plants; (6) a decrease in phytoplankton growth reduces the uptake of carbon dioxide by the oceans, thus leaving more CO_2 in the atmospheric reservoir. Increased atmospheric CO_2 has implications for global warming; and (7) decreased amounts of plankton causes a decrease in the amount of dimethyl sulfide (DMS) they release, an important source of cloud condensation nuclei.

(d) Identify and describe ONE alternative to ozone-destroying chemical compounds. (1 point)

Compounds containing C-H bonds (HCFCs) have been designed to replace chlorofluorocarbons. These compounds are more reactive and less likely to survive long enough in the atmosphere to reach the stratosphere, where they would affect the ozone layer. While less damaging than CFCs, HCFCs also have a significant negative impact on the ozone layer. HCFCs are also being phased out.

(e) Identify and briefly describe ONE international treaty that addresses the release of chemical compounds that destroy stratospheric ozone. (1 point)

The 1987 Montreal Protocol on Substances that Deplete the Ozone Layer was an international treaty designed to protect the ozone layer by phasing out the production of a number of substances, especially chlorofluorocarbons (CFCs), believed to be responsible for ozone depletion. Stratospheric ozone depletion can be expected to continue (but at a slower pace) because of CFCs used by nations that have not banned them, and because of gases that are already in the stratosphere. CFCs have a very long atmospheric lifetime, ranging from 50 to more than 100 years, so the complete recovery of the ozone layer is expected to require several lifetimes.

Question 4

(a) (i) 2 points maximum (1 point for each correct event)

Restatement: Two possible events that have occurred within the last 100 years that can account for the increase in human life span.

Since 1900, the average life span of persons living in the United States has increased by more than 30 years, with most of this gain attributable to advances in public health. The three advances in public health that have had the greatest impact on world health and human life span have been the availability of clean water, improvements in sanitation, and the development of vaccines.

(ii) 1 point maximum (1 point for a correct and accurate description of one event listed above)

Restatement: Describe how public sanitation increased average human life span.

At the turn of the 20th century, it was a common practice to dispose of garbage, industrial wastes, and raw sewage by dumping them into waterways. Few municipalities treated wastewater because it was widely believed that running water purified itself. Typhoid, dysentery, and diarrhea were the most common waterborne diseases and were also the third leading cause of death in the United States at that time. As a result of disinfection of public drinking water and improved sanitation methods, the major waterborne diseases have all but ceased to exist in the United States. Worldwide, the median reduction in deaths from water-related diseases is approximately 70% among people with access to potable water and proper sanitation. Yet waterborne diseases continue to be major killers in less-developed countries, causing half of all deaths of children in poor countries.

(b) (i) 1 point maximum (1 point for a correct explanation of how malaria is transmitted)

Restatement: How malaria is transmitted through the human population.

Malaria is naturally transmitted by the bite of a female *Anopheles* mosquito. When this species of mosquito bites an infected person, a small amount of blood is taken, which contains malaria parasites (*Plasmodium*). These parasites develop within the mosquito, and about one week later, when the mosquito takes its next blood meal, the parasites are injected with the mosquito's saliva into the person being bitten. After a period of between two weeks and several months (occasionally years) in the human liver, the malaria parasites start to multiply within red blood cells, causing symptoms that include fever and headache. In severe cases, the disease worsens, leading to coma and death.

(ii) 4 points maximum (1 point for each correct environmental factor, maximum of 2 points; 1 point for each correct explanation of how each factor influenced the increase of the disease, maximum of 2 points)

Restatement: Two environmental factors that contribute to the emergence of malaria and how those factors influence the increase in the incidence of the disease.

Environmental factors that can contribute to the emergence of malaria include:

1. A decrease in the population of mosquito predators, which can increase the mosquito population and the rate of transmission.
2. Genetic resistance to pesticides, which can also increase the mosquito population and the rate of transmission.
3. Changes in the climate (e.g., global warming), which can increase the number of suitable habitats, resulting in increased rates of transmission.
4. Changes in the current habitats (e.g., agriculture, logging), which can increase the number of breeding sites.
5. An increase in the population density, which allows for a greater opportunity for transmission.
6. The emergence of *Plasmodia* strains that are resistant to anti-malarial drugs, which can then lead to an increase in the incidence and transmission of malaria.

(iii) 2 points maximum (1 point for a correct description of biological control and 1 point for a correct method for controlling the spread of malaria)

Restatement: Description of biological control and an example of one biological control method used for controlling the spread of the disease.

Biological control is the reduction of pest populations by natural enemies and typically involves an active human role. It is a method of controlling pests that relies on predation, parasitism, or other natural mechanisms and can be an important component of integrated pest management programs. Effective biocontrol agents that can be used to decrease the incidence of malaria include predatory fish that feed on mosquito larvae; other insects, such as dragonfly nymphs, that consume mosquito larvae in the breeding waters; adult dragonflies that eat adult mosquitoes; other mosquito species that prey on the *Anopheles* mosquito; predatory crustaceans; birds; bats; lizards; and frogs. Microbial pathogens that can be used to target and control *Anopheles* mosquitoes include viruses, bacteria, fungi, protozoa, nematodes, and microsporidia. Dead spores of varieties of the natural soil bacterium Bt (*Bacillus thuringiensis*) can be used to interfere with the digestive systems of *Anopheles* larvae and can be dispersed by hand or dropped by air over large areas.

Index

How to Use the CD-ROM

The software is not installed on your computer; it runs directly from the CD-ROM. Barron's CD-ROM includes an "autorun" feature that automatically launches the application when the CD is inserted into the CD-ROM drive. In the unlikely event that the autorun feature is disabled, follow the manual launching instructions below.

Windows®

1. Click on the Start button and choose "My Computer."
2. Double-click on the CD-ROM drive, which is named **AP_Environmental_Science.exe.**
3. Double-click **AP_Environmental_Science.exe** to launch the program.

MAC®

1. Double-click the CD-ROM icon.
2. Double-click the **AP_Environmental_Science** icon to start the program.

SYSTEM REQUIREMENTS

Microsoft® Windows®
Intel Pentium II 450 MHz,
AMD Athlon 600 MHz or faster processor
(or equivalent).
Memory: 128MB of RAM.
Graphics Memory: 128MB.
Color Display: 1024 × 768
Platforms:
Windows 7, Windows Vista®, Windows XP

MAC®
Intel Core™ Duo
1.33GHz or faster processor.
Memory: 256MB of RAM.
Graphics Memory: 128MB.
Color Display: 1024 × 768
Platforms:
Mac OS 10.4 or higher.